SMART MATERIAL
STRUCTURES
MODELING, ESTIMATION AND CONTROL

RAM Research in Applied Mathematics
Series Editors : P.G. CIARLET and J.-L. LIONS

SMART MATERIAL STRUCTURES

MODELING, ESTIMATION AND CONTROL

H.T. BANKS

University Professor and Drexel Professor of Mathematics
North Carolina State University
Raleigh, North Carolina

R.C. SMITH

Assistant Professor of Mathematics
Iowa State University
Ames, Iowa

Y. WANG

Mathematician USAF
Armstrong Laboratory
Brooks AFB, Texas

JOHN WILEY & SONS
Chichester ● New York ● Brisbane ● Toronto ● Singapore

MASSON
Paris ● Milan ● Barcelona

La collection **Recherches en Mathématiques Appliquées** a pour objectif de publier dans un délai très rapide des textes de haut niveau en Mathématiques Appliquées, notamment :

— des cours de troisième cycle,
— des séries de conférences sur un sujet donné,
— des comptes rendus de séminaires, congrès,
— des versions préliminaires d'ouvrages plus élaborés,
— des thèses, en partie ou en totalité.

Les manuscrits, qui doivent comprendre de 120 à 250 pages, seront reproduits directement par un procédé photographique. Ils devront être réalisés avec le plus grand soin, en observant les normes de présentation précisées par l'Éditeur.

Les manuscrits seront rédigés en français ou en anglais. Dans tous les cas, ils seront examinés par au moins un rapporteur. Ils seront soumis directement soit au

The aim of the **Recherches en Mathématiques Appliquées** series (Research in Applied Mathematics) is to publish high level texts in Applied Mathematics very rapidly :

— Post-graduate courses
— Lectures on particular topics
— Proceedings of congresses
— Preliminary versions of more complete works
— Theses (partially or as a whole)

Manuscripts which should contain between 120 or 250 pages will be printed directly by a photographic process. They have to be prepared carefully according to standards defined by the publisher.

Manuscripts may be written in English or in French and will be examined by at least one referee.

All manuscripts should be submitted to

Professeur P.G. Ciarlet, Analyse numérique, T. 55,
Université Pierre et Marie Curie, 4, place Jussieu, 75005 Paris
soit au / or to
Professeur J.-L. Lions, Collège de France,
11, place Marcelin-Berthelot, 75005 Paris

© Masson, Paris, 1996

ISBN Masson : 2-225-85214-6   **ISBN Wiley** : 0-471-97024-7

ISSN : 0298-3168

MASSON S.A.
JOHN WILEY AND SONS Ltd

120, bd Saint-Germain, 75280 Paris Cedex 06
Baffins Lane, Chichester, West Sussex PO 19 1UD, England

Contents

Preface

The field of *smart material* systems (also referred to as *controllable material* systems, *adaptive material* systems, *compliant* systems, *intelligent* systems by many contributors to the field) is an emerging science. It is in its infancy with regard to the fundamental science and mathematics necessary to insure reliable implementation in routine applications. Moreover, there is much exciting and challenging work for control theorists with an interest in science and engineering. As we hope the reader will discern from this monograph, there is an incredible opportunity for new modeling, analysis, estimation and feedback control ideas to play a fundamental role in this scientific endeavor. It is this view that prompted our monograph at this time, in hopes to provide a stimulus for other applied mathematicians to consider some of the plethora of problems and challenges in the development of smart material technology.

When we first started writing sections of this monograph in November, 1994, a number of our own initial efforts reported here were not yet completed. As the reader will see, they involve mathematically based investigations related to computation and implementation that utilize parts of the theoretical developments on estimation and control of distributed parameter systems that have been developed by theoreticians over the past two decades. We believe that our own findings substantiate the importance and the relevance of those theoretical efforts. In particular, we cite the account of PDE model-based damage detection in Chapter 6 and the (first to our knowledge) successful physical implementation of a PDE-based feedback control and compensator for attenuating periodic and transient structural vibrations. These latter results are summarized in Chapter 8.

In Chapter 1 we give brief summaries of a number of "smart materials" that are of current interest in the research and applied engineering community. The term "smart materials" which we, following convention in the literature, will often use in our discussions is something of a misnomer for our (and others') actual interest. While some materials (e.g., piezoceramics, shape memory alloys) can truly sense and inherently respond or actuate, it is *controlling* these phenomena in some systematic manner that is really the focus of much of the attention of the "smart material" community. Thus the development of smart material structures or systems often involves multiple components of "smart materials" and actually is about *control* (and associated subjects such as observation, sensor or measurement phenomena, form of input controls, estimation and parameter identification) of this new type of structures.

After the summary in Chapter 1, we focus on piezoceramic-based smart structures in the remainder of the monograph. Chapters 2 and 3 are devoted to a summary of structural modeling (beams, plates, shells, and sections thereof) and details for inclusion of both inactive and active dynamic effects of surface-mounted piezoceramics on these structures. In Chapter 4 we give a theoretical framework (variational or weak formulations and semigroups) that provides a basis for the parameter estimation techniques presented in Chapter 5 as well as the feedback control theoretic discussions of Chapters 8 and 9. The parameter estimation techniques, which must be used as precursors to the control and compensator efforts of Chapters 8 and 9, can also be used in PDE model-based damage detection algorithms and these are discussed in Chapter 6. A substantial validation for these "theories" is provided by examples of use with experimental data presented in Chapters 5, 6, and 8.

While much of the control and estimation theory needed in our discussions has already been developed, there are (in spite of a large literature as surveyed in Chapter 7) theoretical gaps in the rigorous mathematical development. Some of these are pointed out in Chapters 7 and 9 and are the focus of current investigations.

There is a serious omission in our presentation regarding hysteresis and the identification and control of materials and structures involving hysteresis-related phenomena. As the reader will easily learn from reading Chapter 1, this is a critical topic for many smart material structures. While there is a substantial mathematical research literature on hysteresis (e.g., see [119, 179] and the extensive bibliographies therein), we are only beginning to utilize these ideas for practical estimation and control in the context of smart structures. Our own initial efforts [27] are a meager beginning on what is really needed in this area and we have not included a discussion in the present version of this monograph.

We owe gratitude to many: our numerous students, postdocs and professional colleagues who have read and commented on parts of this manuscript and who, in some cases, have collaborated substantially in either modeling, theoretical, computational or implementational components of the research reported here. Some of these individuals are acknowledged by name in the publications cited throughout the monograph.

We have enjoyed the generous support of our home institutions and a number of government funding agencies and labs. These include the Air Force Office of Scientific Research (Mathematical and Geosciences Directorate), Air Force Armstrong Laboratories (Brooks AFB), National Aeronautics and Space Agency (NASA Langley Research Center), and the National Science Foundation.

H.T. Banks
Ralph C. Smith
Yun Wang
March, 1996

Chapter 1

Smart Materials Technology and Control Applications

Recent technological advances in materials science in combination with stringent demands on controller design have led to the development of potentially successful controllers employing advanced sensors and actuators. By utilizing physical properties of the sensors and actuators, significant control of system dynamics can often be attained with a fraction of the hardware and time delays associated with traditional servomechanisms. These "smart material structures" (structures possessing the capability to sense and actuate in a controlled manner in response to variable ambient stimuli) involve combinations of advanced sensors, actuators and microprocessors. They have already revolutionized the design of many feedback control systems and promise to have an even more profound effect on design and control in the future.

In this monograph, mathematical issues concerning modeling, parameter estimation and control in smart material systems are discussed. As noted in the survey given in this chapter, the field of smart material systems is vast, and attention in specific applications of remaining chapters is focussed on a few representative materials and systems. Specifically, we will focus on systems in which structural vibrations or structural interactions with adjacent fields are controlled using surface-mounted actuators and sensors. The mathematical framework is general, however, and can be used when theoretically designing controllers for many general systems having a variety of actuators and sensors in various configurations. In this perspective, the modeling, parameter estimation and control techniques detailed here are intended to benefit both scientists who are implementing smart material controllers and those interested in a rigorous abstract theory concerning such feedback control systems.

1.1 Control Requirements

The prominence of smart material control systems can be motivated by a consideration of criteria facing the control engineer in typical applications. At a basic level, the smart material system is comprised of actuators, sensors and control laws. The sensors may be bonded to the surface or embedded in the structure and ideally should be chosen so as to have negligible effect on the passive system dynamics. Their purpose is to monitor the system response in a manner which is compatible with the control law; that is, they must yield measurements which are adequate for state reconstruction and determination of a control gain. Because of the distributed nature of structural, structural acoustic and fluid/structure systems, sensors providing distributed state measurements are advantageous in many applications.

Like the sensors, ideal actuators should be lightweight, relatively inexpensive, and have minimal effect on the passive system dynamics. They too can be either surface-mounted or embedded in the underlying structure and may be designed to assume some of the mechanical strength and load-bearing requirements for the system. The actuators ultimately respond to a control gain acting on sensed data, and hence their configuration and design must be in accordance with the other two components of the control system. For example, the control law may rely upon collocated sensors and actuators, or actuators providing distributed input. Moreover, the mechanisms through which the actuators provide energy dissipation must be adequate for the application. For example, lightweight actuators providing a broadband response are required in many aerospace and automotive applications whereas actuators providing large strains are necessary when designing smart material engine mounts or controllers for stabilizing industrial machinery. Finally, the actuators must be rapid in the sense that they respond with minimal destabilizing delays.

Both the sensors and actuators must be able to operate in environments which are quite often extreme. Components in aerospace applications must be robust with regard to large thermal gradients and radiation whereas actuators in applications involving chemical deposition chambers for crystal growth must be able to withstand chemical, pressure and thermal extremes. These components must also be flexible in the sense that they are able to perform a variety of agendas. For example, actuators and sensors in aerospace applications may be required to provide vibration, structural acoustic, boundary layer and fatigue attenuation as well as perform damage detection diagnostics simply through changes in the applied software.

To provide optimal efficiency, the control law must be designed in accordance with the sensors and actuators under consideration. For example, many currently employed sensors and actuators yield discontinuous (unbounded) output and input operators in the mathematical problem formulation; such discontinuities must be reconciled when formulating and implementing the controller. Similarly, the control law must account for potential nonlinearities, hysteresis, phase shifts, and any other properties of the smart material sensors and actuators in order to obtain the desired performance.

1.2 Smart Materials

A list of materials employed in smart material systems includes piezoelectric, electrostrictive and magnetostrictive elements, electrorheological fluids and solids, shape memory alloys, fiber optics et cetera. In some cases, the underlying material properties have long been known but only recently has the material been refined for smart material applications. For example, the piezoelectric effect was documented by Pierre and Jacques Curie in 1880 while piezoelectric crystals were employed in radios in the 1940's. Other materials such as fiber optic cables and electrorheological fluids were only recently developed and a significant number of their material properties are still being characterized. In both cases, however, the incorporation of the material as a component, i.e., an actuator and/or sensor, in a smart material system is new, and issues regarding such are currently under intense investigation from both mathematical and implementational perspectives.

We discuss in the next several sections some of the pertinent attributes of the various materials. This will illustrate the capabilities of the materials for satisfying the previously defined control criteria and indicate issues to be considered when modeling their interactions as components in smart material systems.

1.2.1 Piezoelectric Elements

Piezoelectric materials belong to a class of dielectrics which exhibit significant material deformations in response to an applied electric field as well as produce dielectric polarization in response to mechanical strains. In current technology, piezoelectric sensors and actuators can be created by poling an appropriate substrate through the application of a large electric field at high temperatures. Substrates for the process are chosen to have a crystalline, ceramic or polymeric lattice structure in which the atomic structure along at least one axis differs from that in the remaining coordinates; hence the material is anisotropic and typically orthotropic. Poling has the effect of partially aligning the polar axes of the domains to yield a macroscopic polarization which facilitates the electromechanical coupling. As a result of this coupling, the piezoelectric material will deform in response to an applied electric field; this gives the material its actuating properties. The sensing capabilities come from the converse effect in which the mechanical stresses in the material cause rotations of the partially aligned dipoles to generate an electric field.

As depicted in Figure 1.1 and illustrated experimentally in [89], these field-strain relations are nearly linear for small electric fields which proves advantageous when employing the materials in control systems. At higher fields, however, the polarization saturates and domains expand and switch. This leads to significant hysteresis and strain-based nonlinearities (see again, Figure 1.1) which can be detrimental when employing piezoelectric elements in control regimes requiring high electric fields.

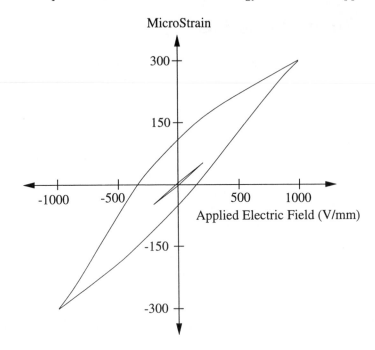

Figure 1.1: Strain distribution of G-1195 for moderate and large electric fields.

Phenomenological relations modeling the direct and converse piezoelectric effects can be obtained from the thermodynamic formalism proposed by Devonshire [78] which expresses the constitutive relations as derivatives of the total system energy. While piezoelectric elements exhibit nonlinear hysteresis at high excitation levels, the response required in current typical structural applications is very nearly linear. We therefore restrict our discussion to the linear constitutive laws formulated by Voigt in 1910.

As detailed in [61, 89], the linear direct and converse constitutive relationships for piezoelectric elements are

$$
\begin{aligned}
D_m &= \varepsilon_{mn}^T E_n + d_{mk\ell}\,\sigma_{k\ell} \qquad &\text{(Direct Effect)} \\
e_{ij} &= s_{ijk\ell}\,\sigma_{k\ell} + d_{mij}\,E_m \qquad &\text{(Converse Effect)}
\end{aligned}
\tag{1.1}
$$

where D and E denote the electric displacement and field, respectively, while ε and d denote the material dielectric tensor and piezoelectric strain tensor, respectively. The material strain e is related to the stress σ through the elastic compliance s. Following usual tensor convention, the repeated subscripts indicate summation. We point out that the direct relationship is employed when modeling the sensing capabilities of the piezoelectric element whereas actuator interactions are modeled using the converse relationship.

One advantage of piezoelectric elements in control and damage detection applications results from their relative insensitivity to temperature variations when employed below a transition temperature termed the Curie temperature.

As experimentally illustrated in [89], the variation of d for a standard piezoceramic material under constant loads is less than 12% over a temperature range of $120°\,C$. Because d relates the generated strain to the input electric field, such temperature stability is crucial in control applications since uncertainties in d are equivalent to uncertainties in the magnitude of the control gain. This temperature insensitivity gives piezoelectrics an advantage over electrostrictive elements in many applications since the coupling coefficients in the latter are typically highly temperature-dependent.

A second advantage of piezoelectric elements results from their flexibility in a variety of applications. This is partially due to the range of materials which can be poled to have piezoelectric properties. Two widely used piezoelectric materials are piezoceramics and piezopolymers of which lead zirconate titanates (PZT's) and polyvinylidene fluorides (PVDF's) are respective examples. A piezoceramic material to which we will be referring throughout this monograph is G-1195 since it is employed in many current laboratory applications. The reader is referred to Chapter 5 of [93] for a summary of material properties of various piezoelectric materials.

Due to their ceramic nature, PZT patches have rigidities that are comparable to and often exceed those of the underlying structures. One result of this is efficient conversion of electrical to mechanical energy; this makes the elements very efficient actuators in a variety of applications. While the generated strains are less than those of magnetostrictives, piezoceramic patches are often quite sufficient actuators for thin structures without the mass-loading associated with the bulky magnetostrictives. Moreover, the PZT's are effective over a large frequency range and this adds to their attractiveness as control elements. Finally, the dual piezoelectric properties exhibited by the patches makes them amenable to use as self-sensing actuators [80] or elements in controllers requiring collocated sensors and actuators [1].

While advantageous in many applications, piezoceramics suffer from their inherent brittleness and the expense associated with molding them to complex configurations. Alternatively, PVDF films have the consistency of a thin plastic wrap and can be bonded to virtually any geometry; hence they are effective sensors in a wide range of applications. As sensors, the PVDF films provide flexibility and a range of sensitivities that can accurately measure excitations ranging from impacts of micron-sized particles to armor-piercing missiles in a variety of applications [65, 66, 144]. Moreover, they can be utilized as modal sensors due to the ease with which they can be cut and bonded to various surfaces [67, 132]. Unlike electrostrictive sensors, however, they cannot be interactively tuned to the modes; hence they are used for modal sensing in structures whose modes have been previously determined through other means.

The direct application of PVDF films as actuators has been limited by the fact that they are much more compliant than most underlying substrates and exhibit significantly weaker electromechanical coupling coefficients than the PZT's (PZT: $d_{31} = 110 \times 10^{-12}\ m/V$, PVDF: $d_{31} = 23 \times 10^{-12}\ m/V$). On the other hand, the dielectric strength of PVDF's is larger than that of PZT's

(40 $V/\mu m$ for PVDF's versus 2 $V/\mu m$ for PZT's); hence they can be exposed to much higher electric fields.

Because of their approximately linear behavior in low electric fields, relative temperature insensitivities and flexibility as sensors and actuators, piezoelectric elements satisfy many of the control criteria defined in the previous section. Their utility in aerospace and automotive applications is augmented by the fact that they are lightweight and can be bonded to a variety of structures (while their effect on the passive structural dynamics is minimal, appropriate physical parameters for the elements must still be determined through data fitting techniques to ensure an accurate model fit). They also provide a broadband response and, due to their electrical nature, have minimal response delays as compared to traditional servomechanisms.

In typical structural, structural acoustic and fluid structure applications, the piezoelectric elements are either surface mounted or embedded within the structure. As depicted in Figure 1.2, bending moments, in-plane forces, or a combination of the two can then be generated through input voltages to the elements. The goal when modeling the system and designing a control law is to accurately determine an input voltage and the structure/piezoelectric element interactions in response to the voltages so as to obtain adequate control authority.

The interactions between the elements and structure are time-dependent with possible transient effects; hence dynamic models describing the phenomena are necessary for many applications. Distributed models are also required since the underlying structural, structure/field and actuator/sensor dynamics have spatial variability. Moreover, the modal characteristics for the systems differ from those of the isolated structure due to coupling between the structure and actuators/sensors as well as coupling with any adjacent fields. This makes PDE-based models, which incorporate the coupling, preferable for a variety of applications.

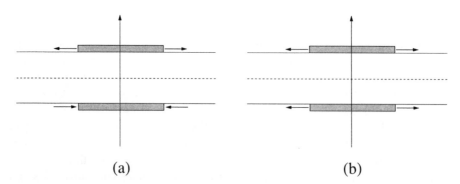

(a)   (b)

Figure 1.2: Identical surface-mounted piezoceramic actuators; (a) Out-of-phase voltages produce pure bending moments, (b) In-phase voltages produce purely in-plane forces.

Appropriate model-based control methods must be designed in the realm of unbounded (discontinuous) control operators since the piezoelectric actuators provide control input only in discrete regions of the structure. Similar accommodations with the output operators must be made when employing piezoceramic patches as sensors. Alternatively, PVDF sensors can be used to obtain more extensive state measurements and some current efforts are being made to utilize them as analog state sensors.

A survey of piezoceramic patch usage in structural problems [3, 68, 70, 71, 72, 91, 112, 161, 176, 177], structural acoustic applications [79, 92] and fluid/structure interactions [18] illustrates some of their current applications. Additional examples illustrating their usage are also given later in this chapter as well as in subsequent chapters. Modeling, parameter estimation and control issues concerning the piezoelectric elements are detailed in remaining chapters.

Advantages:

- Relative temperature insensitivity and linear response at low excitation levels;
- Lightweight construction and flexibility as sensors and actuators in a large variety of applications;
- Broadband frequency response;

Disadvantages:

- Significant hysteresis at large electric field levels;
- Brittleness and small tensile strength of PZT's;
- Weak electromechanical coupling coefficients for PVDF's;
- Piezoelectric effect generated through poling can decay, thus leading to aging effects and performance degradation;

Current Control Applications: Sensors and actuators in structural, structural acoustic and fluid/structure systems.

1.2.2 Electrostrictive Elements

Electrostriction is characterized generally as the mechanical deformation which occurs in a dielectric material when an electric field is applied. While characteristic of all dielectrics, the strains generated in most dielectrics are typically second-order and too small to yield satisfactory sensing and actuating capabilities. Materials such as relaxor ferroelectrics, however, have sufficiently large dielectric permittivities so as to generate polarizations, and hence strains, that can be utilized in applications. The current relaxor ferroelectrics which are used in this capacity are based upon lead magnesium niobate (PMN) compounds or lead-titanate enriched lead magnesium niobates (PMN-PT) (see [108] for the specific chemical composition of such compounds).

Unlike the piezoelectric materials, uncharged electrostrictives are isotropic and are not poled so the material exhibits no net polarization. Moreover, the electromechanical coupling is nonlinear in the sense that the generated strain is approximately proportional to the square of the induced polarization for low electric fields with saturation occurring as the field increases (see Figure 1.3). Due to the symmetry of the electrostrictive crystals and quadratic nature of the strain-field relations, field reversal does not lead to the generation of negative strains as is the case with piezoelectric elements. To attain bidirectional actuation, opposing elements must be employed or the field shifted through an applied DC bias field (this latter technique is used with magnetostrictives).

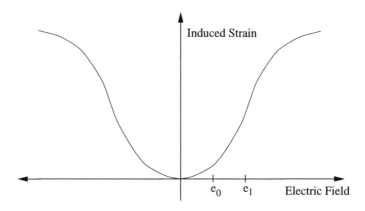

Figure 1.3: Strain-electric field distribution for an electrostrictive element near its Curie temperature.

The response of the electrostrictive materials also differs from that exhibited by piezoelectrics in that the spontaneous polarization for the former decays quite slowly as the temperature is increased through the transition Curie point. As a result, the electrostrictives can be operated in a range in which hysteresis is negligible but strong polarization remains. While this contributes to the highly temperature-dependent nature of the electrostrictives, it virtually eliminates the hysteresis effects which plague piezoelectrics in high electric fields (electrostrictives do exhibit significant hysteresis if operated much below the Curie temperature).

Constitutive relations modeling the electrostrictive effect can be determined through consideration of the free energy term in the phenomenological thermodynamic theory proposed by Devonshire in 1954 [78]. When formulating these relations, either the polarization or electric field can be chosen as independent variable with advantages and disadvantages to either choice. To remain consistent with notation for the corresponding piezoelectric relations, we summarize here a constitutive relation in terms of the applied electric field; the reader is referred to [89, 90] for further details regarding this formulation and [106, 107, 108] for derivations in terms of the polarization.

With e, σ, D, E and s again denoting the strain, stress, electric displacement, electric field and material compliance, respectively, a constitutive law for electrostriction with low electric fields is

$$
\begin{aligned}
D_j &= \varepsilon_{jn}^T E_n + 2m_{jnk\ell} E_n \sigma_{k\ell} \qquad \text{(Direct Effect)} \\
e_{ij} &= s_{ijk\ell} \sigma_{k\ell} + m_{pqij} E_p E_q \qquad \text{(Converse Effect)} .
\end{aligned}
\tag{1.2}
$$

A comparison of the piezoelectric relation (1.1) and electrostrictive law (1.2) reveals that while many similarities exist, the nonlinear behavior of the electrostrictive element is manifested through the squared relationship between the electric field and resulting strain. Moreover, while the electrostrictive coupling parameter m plays a role analogous to that of the piezoelectric strain constant d, it is highly temperature-dependent as compared with the relatively temperature insensitive d (see [89] for experimental data demonstrating the behavior of both parameters).

As noted in Figure 1.3, the quadratic behavior of the strain-field relationship is manifested only for low electric fields and strain gradients decrease at higher fields due to saturation. To model the constitutive behavior over a wider voltage range, various researchers have replaced the quadratic electric field dependence by a hyperbolic tangent law (see for example, [89]). While providing a more accurate model, this latter formulation introduces additional *temperature-dependent* parameters which must be estimated through least squares techniques with experimental data and updated as the temperatures change. In addition to temperature dependence, some parameters such as stiffness exhibit a dependence upon the applied field. The highly nonlinear and temperature-dependent behavior of the elements (yielding nonlinear strain-field laws and temperature and field dependent parameters) are the two primary challenges which must be addressed when incorporating electrostrictive elements in control systems.

In initial actuator development, the pernicious effects of the nonlinear and temperature-dependent response have been addressed through output linearization and temperature gain scheduling [89]. The success of output linearization results from the fact that while the strain-field behavior is nonlinear, it is relatively well-defined (modeled) by either the quadratic or hyperbolic tangent relations. By inverting these relations, a transform yielding linear output in response to linear input is obtained. In this manner, the output from the electrostrictive elements can be treated as essentially linear and can then be incorporated in linear control theories. The temperature gain scheduling is accomplished through systematic adjustment of control gains to obtain consistent performance at various temperatures.

As documented in [89] for control of beam vibrations, the implementation of electrostrictive actuators in this manner yielded performance levels comparable to those obtained with piezoceramic actuators. With regards to actual control output, the electrostrictive actuators are capable of generating significant strains since the large permittivity of the electrostrictives leads to large induced polarization levels. Moreover, strains adequate for structural control can be generated over a wide frequency range; this facilitates their use in many applications. In terms of consistency and longevity, the electrostrictives can be advantageous over piezoelectrics since they are not poled and hence avoid the decay in performance which can plague piezoelectric actuators. Thus while the incorporation of electrostrictive actuators in control systems lags behind that of piezoelectrics, initial research has indicated that in certain applications, the electrostrictives may be preferable.

Due to the unpoled and isotropic nature of the electrostrictive materials, polarization and hence sensing *cannot* be accomplished without an applied DC bias field. As illustrated in [148], the quadratic nature of the electromechanical coupling can be utilized to determine DC bias levels which facilitate tuning of the electric field-to-output strain ratios (the strain gradient at e_0 in Figure 1.3 differs from that at e_1). This can provide transduction sensitivities and sensing capabilities not attainable with piezoelectric sensors. For example, modal configurations must be determined *a priori* before piezoelectric modal sensors are cut, and new sensors must be fashioned if the modal configuration changes. Alternatively, electrostrictive sensors can be mounted in a fixed array and then tuned to the modal configuration by using optimal control theory to determine appropriate DC biases. The capacity for tuning the sensitivity also provides the electrostrictive sensors with a capability for filtering noise.

As noted previously, the use of electrostrictive sensors and actuators has been more limited than that of piezoelectrics due to the inherently nonlinear and temperature-dependent nature of the electrostrictive materials. In applications with controlled temperature environments, however (e.g, underwater or *in vivo*), their utilization has already been widespread with uses ranging from ultrasonic motors to medical ultrasonic probes. Moreover, their small thermal expansion, small hysteresis and good set point accuracy make them advantageous in applications such as non-hysteretic micropositioners, deformable mirrors, precision laser systems and tools, and impact dot-matrix printers.

The popularity of electrostrictives as control elements will likely increase as the design of tunable sensors is clarified and implementational issues regarding their inherent nonlinearity and temperature-dependence are resolved. While output linearization has been demonstrated as a viable option, it requires an accurate model of the nonlinearities in a form that can be inverted or transformed. Future research may provide alternative techniques for utilizing the inherent nonlinearities.

In addition to addressing the temperature dependence and nonlinear response of the electrostrictives, model-based control methods incorporating these elements must address many of the issues associated with piezoelectric elements. Specifically, dynamic interactions between the element and underlying structure must be modeled and control laws utilizing the properties of electrostrictives in the framework of unbounded operators must be constructed.

Advantages:

- Small hysteresis levels above Curie temperature; small thermal expansion;
- Lightweight construction permits control without severely effecting passive dynamics;
- Nonlinear constitutive relations permits tuning of field-strain ratios;
- Unpoled nature of materials can lead to extended performance over piezo-electric elements;
- Broadband frequency response;

Disadvantages:

- Nonlinear strain-field relations and field-dependent parameters;
- Temperature dependency of elements manifested in temperature dependent parameters;

Current Control Applications: Sensors and actuators in structural systems.

1.2.3 Magnetostrictive Transducers

The concept of magnetostriction is similar to that of electrostriction in that it refers to the tendency of certain materials to generate strains in response to an applied field, in this case, an applied magnetic field. Although significantly larger, the strain distributions generated by magnetostrictive elements are qualitatively similar to those generated by electrostrictive elements. To illustrate the nature of a typical magnetostrictor, we consider the transducer depicted in Figure 1.4 (the modeling and performance of this transducer are detailed in [100]).

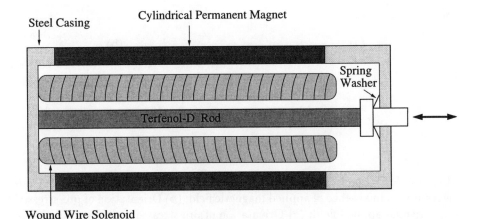

Figure 1.4: Cross section of the Terfenol-D magnetostrictive transducer.

In current transducers, the magnetostrictive material is typically composed of terbium and dysprosium alloyed with iron. A commonly employed material is Terfenol-D (Ter: terbium, fe: iron, nol: Naval Ordinance Laboratory, D: dysprosium) which is constructed as a cylindrical rod and placed in the center of the transducer. As depicted in Figure 1.5a, the material in such form contains oblong magnetic domains which are primarily oriented perpendicular to the rod's longitudinal axis when no field is applied. The number of domains perpendicular to this axis can be increased by pre-stressing the material (see Figure 1.5b), and this is accomplished by the spring-loaded washer at the end of the rod. In the presence of a magnetic field, the domains rotate so as to align with the field. Consequently, if the field is applied along the longitudinal rod axis, the domains rotate in the sense depicted in Figure 1.5c and significant strains are generated. This is termed the Joule effect and provides the actuator capabilities for the transducer. Sensing is accomplished through measurement of the magnetic fields which results when mechanical stresses cause rotation of the domains (Villari effect) (see [175] for details regarding these effects). While both capabilities exist for the materials, the primary use of magnetostrictives to date as been as actuators in applications requiring large strains. In recent analysis reported in [155], however, issues regarding a self-sensing magnetostrictive actuator analogous to the piezoceramic design of [80] were detailed, and sensing issues will likely receive further attention in the future.

The approximate relationship between the applied magnetic flux density B and resulting strains e is depicted in Figure 1.6 (see [113] for details regarding the exact relationships). As with electrostrictive elements, the relationship is highly nonlinear with saturation occurring at large field strengths. To facilitate the generation of bidirectional strains, a cylindrical permanent magnet and/ or a DC current in a solenoid surrounding the rod is used to generate a biasing

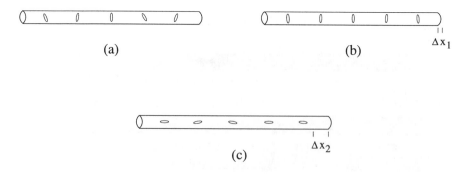

Figure 1.5: Magnetic domains in the Terfenol-D rod; (a) Orientation of unstressed rod in absence of applied magnetic field; (b) Orientation of pre-stressed rod with no applied field; (c) Orientation of pre-stressed rod when field is applied in direction of longitudinal rod axis.

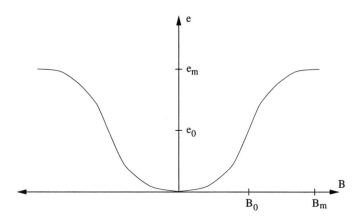

Figure 1.6: Strain distribution e generated by an applied magnetic flux density B.

magnetic flux density B_0 (this is similar to the use of biasing DC voltages in electrostrictives). The solenoid is then used to vary the flux density in the rod between 0 and B_m. This provides the capability of generating both positive and negative strains.

The quadratic strain-field relationships used with electrostrictive elements are inadequate for magnetostrictives. In initial investigations [100], an affine combination of linear and sinusoidal components was used to model the constitutive relationship between the applied bias B_0, the magnetic flux density B generated by the solenoid, and the resulting strains e. This model ignores the slight hysteresis that exists between B and e (this hysteresis is also omitted in Figure 1.6) but proved adequate for predicting harmonics and trends for the Terfenol-D actuator.

A second component of the modeling concerns the derivation of relationships between the current applied to the solenoid, the resulting magnetic field H, and the associated magnetic flux density B. As depicted in Figure 1.7a, the relationship between B and H is highly nonlinear and displays significant hysteresis. Such hysteresis then manifests itself in the strain-field relations as shown in Figure 1.7b.

Several models for these relations are discussed in [100]. For small current amplitudes, one model is

$$B = \mu H_0 + \mu n I \left[\cos(\omega t - \phi) + \frac{k}{\pi} \sin(\pi \cos(\omega t - \phi)) \right]$$

where μ denotes the magnetic permeability of Terfenol-D and $H_0 = (H_m + H_\ell)/2$ as depicted in Figure 1.7a (it should be noted that values of μ are highly dependent upon transducer operating conditions such as frequency, load and temperature). Hence the term μH_0 models the magnetic flux density due to the permanent magnet. The contributions from the solenoid are modeled by the

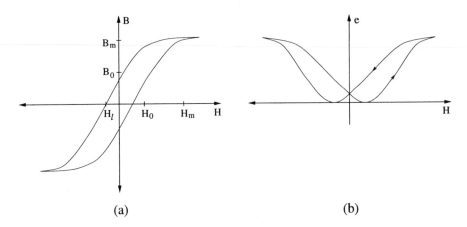

Figure 1.7: (a) Relationship between the magnetic field strength H and the magnetic flux density B; (b) Applied magnetic field H and resulting strain distribution e.

remaining periodic component where n indicates the number of wire turns per unit length in the solenoid, I is the amplitude of the applied current, ω denotes the circular frequency of the current and ϕ is the phase angle between B and H. This latter parameter incorporates hysteresis in the model and gives the width of the loop in Figure 1.7a. The parameter k determines the slope at either end of the hysteresis loop ($k = 0$ would yield elliptic loops). Experimental results in [100] demonstrate that with low current amplitudes, this model adequately predicts trends and the existence of harmonics in displacements generated by the magnetostrictive transducers. The model does not capture the sharp corners in the experimental hysteresis loops, however, and this may lead to loss of resolution when predicting the magnitude of certain harmonics.

We note that in some literature, the constitutive relations between H and e are described as linear with a proportionality "constant" d used to describe the slope between the applied magnetic field H and the resulting strain e. Care must be taken when employing such descriptions since d is *not* constant and in fact is highly dependent upon the applied current I and hence H (see Figure 6 of [100]). Hence for any significant range in the values of H, the constitutive relations are nonlinear and must be modeled as such when employing magnetostrictive transducers in applications.

A fact which must be considered when determining the utility of magnetostrictive transducers for a system concerns their size and mass. While significant variations exist, a typical transducer can have a mass on the order of 0.86 kg with lengths on the order of 110 mm (see [101]). Hence the magnetostrictive actuators will contribute more mass loading and alteration of passive dynamics than standard piezoelectric or electrostrictive elements. When compared with standard magnetic shakers, however, it is noted that magnetostrictive actuators can produce comparable forces while contributing roughly 1/10 the mass of the traditional shaker (again see [101]).

While the size, inherent nonlinearities and hysteresis provide significant challenges to be addressed when modeling and incorporating magnetostrictive transducers in control systems, significant advantages can be realized in certain applications due to the giant strains and forces generated by the transducers. For example, strains on the order of $1000 - 2000$ microstrain (parts per million) can be obtained with magnetic field amplitudes less than 1000 Oersted, which is readily obtainable in applications (see [84, 101, 155]). The value of 2000 microstrain yields a 0.2% strain which can be compared with the 1-sided strain of 0.03% for electrostrictives and 0.08% peak-to-peak G-1195 strain reported in Figure 1 of [89]. While such comparisons are rough due to temperature and frequency dependencies, they indicate relative magnitudes attainable with the three elements. Moreover, as detailed in [101], the output forces from a Terfenol-D transducer can be on the order of forces generated by significantly larger and more massive magnetic shakers.

Due to their capacity for generating large strains and forces, magnetostrictive actuators have been employed as ultrasonic transducers, positioners, sonar projectors (500-2000 Hz) and isolators (5-60 Hz). They have also been employed as actuators for reducing structural vibrations. Further development of magnetostrictive actuators and sensors will require additional modeling of the field-strain relations, dynamic models describing actuator/sensor/structure interactions, and the development of appropriate control laws.

Advantages:

- Actuators generate very large strains in comparison with those created by piezoelectric and electrostrictive actuators;

Disadvantages:

- The constitutive relationship between applied magnetic field strength and generated strains is highly nonlinear and exhibits significant hysteresis;

- Current magnetostrictive transducers are large and significantly mass load a structure;

Current Control Applications: Sensors and actuators in structural systems requiring the generation of large strains. Vibration suppressors and positioners for industrial machinery.

1.2.4   Electrorheological Fluids

Electrorheological or ER fluids are comprised of a diverse class of fluid dispersions which exhibit the "ER response" (sometimes referred to as the "Winslow effect" after the investigator [184] who first carried out careful experiments on the phenomenon). This response consists of significant but mainly reversible changes in rheological behavior of the fluid when it is subjected to external electric fields. In the extreme, these changes convert a low viscosity fluid to a solid like substance. (A magnetic analogue of ferromagnetic dispersions - the ferrofluids - exhibit similar properties.) ER fluids are colloidal dispersions, that is, dispersions of solid particulates in nonconducting or insulating oils where the diameters of the solid particulates (typically in the ranges of less than one μm to several hundred μm) are much greater than those of the solvent particles. These dispersions are constructed with a large mismatch between the dielectric properties of the solute material and the solvent. The ER phenomenon resides in the physics and chemistry of the particulates which are polarizable or conducting materials. The ER response is due to some form of induced polarization and subsequent interaction among the particulates.

While a complete and accepted theory is yet to be agreed upon by investigators, the phenomenological response is well documented. An applied field E enhances the shear stress τ and induces a Bingham fluid-like behavior in a dispersion that is essentially a Newtonian fluid in the absence of the field.

In a Newtonian fluid [86] there is a linear relationship $\tau = \mu\dot{\gamma}$ between the shear stress and the shear strain rate $\dot{\gamma}$ with the slope defined by the absolute viscosity μ. For a Bingham plastic or fluid, there is a static yield stress τ_s as well as the plastic viscosity μ_p which characterize the fluid in an affine relationship $\tau = \tau_s + \mu_p\dot{\gamma}$. For more general non-Newtonian fluids one may find that μ_p must be replaced by the apparent viscosity $\mu_a = \mu_a(\dot{\gamma})$ which depends on the shear rate. Repeated experiments demonstrate that the behavior of ER fluids (see [55, 94, 93] and the substantial number of references given in these reviews) involve a number of the general features of non-Newtonian fluids (see Figure 1.8). Typically ER fluids are Bingham like in that a substantial static yield stress τ_s ($> 10kPa$) is present in response to E fields in the range of kV/mm. However, once the fluid is in motion, there may be a somewhat smaller yield stress τ_d ($> 5kPa$) and the viscosity μ_a may become dependent on the shear rate $\dot{\gamma}$. These responses can be produced at a particulate volume fraction in the range 0.1 to 0.4. Moreover, one can expect a rather strong dependence of τ_s, τ_d and μ_a on the strength of the E field (perhaps in a nonlinear manner). The value of an ER fluid in a smart material application depends on the ability to regulate or control the shear stress and viscosity of the fluid as a function of the applied electric field E.

The above description is only a rather general outline of ER fluid behavior and individual behavior varies over a wide diversity of different dispersions. Indeed there are a large number of solute/solvent combinations which may be classified as an ER fluid; these include inorganic particles of silica, titania and other metal oxides or organic particles of flour, metal soaps, ion-exchange resins, microcrystalline cellulose in oils (silicone, cooking, or mineral), kerosene or halogenated hydrocarbons. Whatever the combination, the dispersions usually consist of a solvent that is a nonconducting low viscosity liquid and solid particulates or solute which are nonabrasive, nonconducting or semiconducting

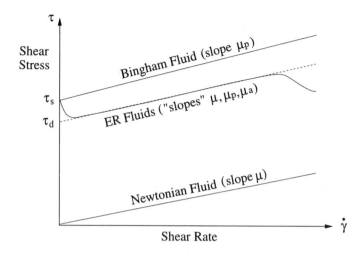

Figure 1.8: Typical shear stress/shear rate curves for various fluids.

but polarizable. Frequently the particulate surfaces are coated with organic activator compounds (water or other polar fluids such as ethanol or diathylamine) to enhance polarization. Moreover, small amounts of surfactants may be added. These are thought to contribute to stabilization of the particles, enhance E field induced attraction of the particulates and prevent sedimentation in the dispersed phase of the liquid.

The exact mechanisms by which the factors produce the ER response are not well understood. While numerous hypotheses exist on the origin and the etiology of the response, there are a paucity of predictive models and little agreement on the exact mechanisms. There is a generally agreed upon initial step in the response; the external E field induces electric dipoles in the particulates which then interact to form columnar fibers or fibrils (i.e., column like chains) parallel to the direction of the field. (These columns have been observed repeatedly in static experiments.) This particle fibrillation (also referred to as fibrination or fibrulation) alone, however, is not adequate to explain the excess of shear stress produced in flowing dispersions (e.g., the response is too slow for the rapid response time observed experimentally in ER fluids). Among suggested possibilities are coulombic interactions between polarized particulates to produce *clusters* (or chain like aggregates) that act to impede flow but extend throughout the bulk of the fluid in columnar form only in static fluids.

Whatever the situation, this lack of basic understanding has greatly inhibited the use of ER fluids as smart material actuators. A careful modeling is difficult since it requires an interdisciplinary effort involving rheology and dielectrics. The complex ER response depends on applied field strength, suspension volume fraction, particulate size (and perhaps shape) and dielectric properties, and temperature as well as the combined physical and chemical properties of the solvent, solute and solid/liquid interfaces. This complexity has resulted in significant difficulties in the design of practical ER fluids. The earliest careful efforts in this regard were due to W.M. Winslow [184] in the period 1939-1949. There were sporadic efforts in the period 1949-1980, with dramatically increased research activity in the 1980's (see the extensive bibliography in [93]). However, many design related questions remain including electrical conductivity of the fluid and power consumption, heat dissipation, and level of shear stress that the fluid can sustain as a function of the applied field. The search for an efficient practical ER fluid that exhibits low viscosity in absence of the E field and high viscosity in the presence of the field is still underway.

The potential applications, as recognized by Winslow in his pioneering research and patents, are enormous. They include electrically triggered clutches, pumps and hydraulic valves, miniature robotic joints and robotic control systems, vibration isolation devices such as shock absorbers and engine mounts, and other active feedback control damper systems.

Advantages:

- Relatively large forces with comparatively small electric fields;
- Flexible design; wide variety of dispersions possible;

Disadvantages:

- Temperature dependence, lack of long term stability of ER response;
- Complexity of system may impede practical actuator/control law design;

Current Control Applications: Actuators in large load-bearing structures.

1.2.5   Shape Memory Alloys (SMA)

Shape memory alloys are a class of metal compounds which possess the capability to sustain and recover relatively large strains ($\approx 10\%$) without undergoing plastic deformation. These unique material characteristics are due in large part to the materials' ability to undergo internal crystalline transformations in the presence of external applied stress and/or changes in temperature. These transformations from the parent phase, *austenite*, at stress-free, high temperature conditions, to several variants of the low temperature *martensite* phase, are also a function of the history of the material. While these materials are very much in the category of emerging technologies, several of them are currently commercially available and have been used in engineering applications. Among the most popular are the nickel-titanium alloy known as Nitinol (Ni: nickel, ti: titanium, nol: Naval Ordinance Laboratory) and copper zinc aluminum (CuZnAl) alloys. NiTi can be used in high performance devices with recoverable strain in the 6% range while CuZnAl performs well in low cyclic load situations with recoverable strains of approximately 2%. These values can be placed in perspective through comparison with strains generated by the piezoceramic material G-1195 which are on the order of 0.1%.

The phenomenological responses in SMA are reasonably well understood. If one takes a stress-free SMA at high temperature in the austenite or parent phase and cools it, a gradual transformation to the low temperature or martensite phase is achieved. Several variants of the martensite, including multiple twins (electron microscopic observable complementary dislocations in the lattice crystal structure), are obtained in this cooling process. Since for a given SMA, the transformation from austenite to martensite (and the reverse transformation due to heating) is gradual, an important state variable is the martensite fraction ξ with $0 \leq \xi \leq 1$ where $\xi = 1$ and $\xi = 0$ correspond to an all martensite phase and all austenite phase, respectively. Associated with this state variable ξ are four temperatures denoted by M_f, M_s, A_s, A_f and referred to as martensite finish, martensite start, austenite start, and austenite finish temperatures, respectively. When an SMA in austenite phase ($\xi = 0$) is cooled, M_s is the temperature at which the transition to martensite begins and M_f is that temperature at which the transition is completed, i.e., $\xi = 1$ so that the material is composed completely of the martensitic variants. An analogous description defines the temperatures A_s and A_f. For most currently available SMA, one has the order relationship $M_f < M_s < A_s < A_f$ although it is possible to have $M_s > A_s$. A schematic of the cooling/heating phase transition phenomena is given in Figure 1.9 for a stress-free SMA.

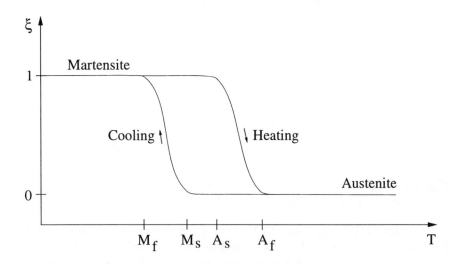

Figure 1.9: Temperature driven phase transitions in stress for SMA.

The thermally induced phase transitions described above are the basis for an explanation of the strain recovery features of SMA. If the SMA is in the martensite phase ($\xi = 1$) and a unidirectional stress σ is applied, at a temperature dependent critical stress $\sigma_{\mathrm{crit}} = \sigma_{\mathrm{crit}}(T)$, detwinning of the martensite variants begins.   This eventually results in a single variant of detwinned martensite aligned with the axis of the stress loading. A similar state is achieved under the loading if one starts with an SMA in the austenite phase. During these phase transformations, the internal stress in the SMA changes only slightly and a significant apparently plastic strain is achieved. If $T > A_f$ during this stress loading, the martensite is unstable at this temperature and the full strain can be recovered upon unloading. The loading/unloading stress-strain curves in this case are characteristic hysteresis loops as depicted in Figure 1.10(a) and the behavior is classified as "pseudoelastic". However, if $T < A_s$ during the stress-induced martensite phase transformation, a large residual strain ε_r remains after unloading as shown schematically in Figure 1.10(b). This strain can be recovered by heating the SMA to temperatures $T > A_f$; this is termed the "shape memory effect" or SME. This recovery can be free, fully restrained, or controlled. In the first case, no work is done while in the second case, the SMA is restrained from recovery of its original dimension and geometry thereby producing large internal stresses. In the final case, the SMA is constrained so partial recovery of the residual strain is achieved but some stress is present to prevent full recovery.

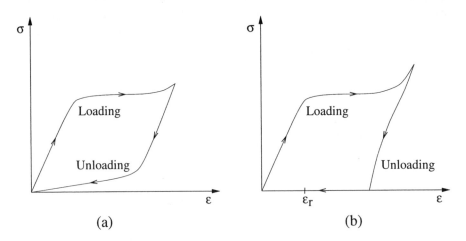

(a)   (b)

Figure 1.10: (a) Pseudoelasticity: stress-strain hysteresis loop $(T > A_f)$; (b) Shape memory effect: residual strain ε_r $(T < A_s)$.

It is this SME feature of SMA that can be exploited to develop distributed force actuators to be used as controllers in composite materials. In a typical example, one employs SMA fibers reinforcing a non-SMA composite structure, e.g., SMA fibers in an elastic matrix. The SMA fibers or "wires" are stress loaded and deformed at low temperatures $(T < A_s)$ in the martensitic phase. They are then unloaded to generate some martensitic residual strain. Once embedded in the elastic matrix they can be heated (using an electrical or thermal input) to achieve residual strain recovery. In Nitinol-embedded composite structures one can produce up to 10^6 psi in constrained recovery.

Potential applications for SMA based actuators range from space deployment of large antennas (the antennas are shaped while the embedded SMA fibers are in low temperature martensite phase and stress loaded; the antenna is then collapsed during unloading, is launched into space and then heated to "recover" its deployment configuration with the SMA in the austenite phase) to simple shape memory fasteners and pipe couplers. The latter examples are established engineering uses of SMA. In commercially available pipe couplers, a TiNiFe alloy tube coupler is fabricated as a ring or sleeve with inner diameter smaller than that of the pipes to be connected. At very low temperature the pipes are forced into the sleeve which has been stretched (in a stress induced martensitic phase transformation). Once the coupler joint is heated to $T > A_f$ (austenite phase), the residual strain in the SMA sleeve is recovered, producing a very effective seal at the pipe coupling. Other potential applications of SMA actuators (many in a discrete or switching format [81]) are in the automotive industry (thermal and electrical devices that combine sensing and actuation in response to changes in temperature), fire detector and alarm systems and fire protection gas valves (to cut off gas flow in a system in response to elevated temperatures). They can also be employed as sensor/actuator systems in air conditioners, as circuit breakers in electrical appliances, in robotics (SMA based mechanical force devices, SMA springs) and as proportional controllers in many mechanical devices.

A substantial impediment to the development of reliable control devices employing SMA resides in the fact that good dynamical models for SMA are not yet available. Modeling is quite difficult since the main features involve large strains and nonlinear mechanisms. Thus the usual infinitesimal theory of elasticity is not applicable and one must turn to the nonlinear theory of large deformations involving finite strains and finite stresses. A first principles approach entails writing basic momentum balance equations in terms of the Piola-Kirchhoff stresses. One must then develop constitutive relationships between these stresses and the Green-St. Venant strains [64, 97, 98, 142]. While large deformation theories have been developed in detail for certain specific applications involving rubber and polymers [150, 181], the general theory is quite tedious and can involve almost overwhelming complexity for general dynamics in the case of specific SMA.

Some progress in developing SMA constitutive laws has been noted in recent years. Efforts have been made relating experimental findings in a thermomechanical approach. Other approaches entail a macroscopic phenomenological formulation in terms of a Helmholtz free energy. Whatever the approach, one needs SMA constitutive laws of the form

$$d\bar{\sigma} = D(\bar{\varepsilon}, T; \xi)d\bar{\varepsilon} + \Theta(\bar{\varepsilon}, T; \xi)dT + \Omega(\bar{\varepsilon}, T; \xi)d\xi$$

where $\bar{\sigma}$ is the second Piola-Kirchhoff stress tensor, $\bar{\varepsilon}$ is the Green strain, D is the SMA (Young's) modulus, Θ is the thermoelastic tensor (or thermal expansion coefficient) and Ω is the transformation tensor (involving metallurgical parameters). Various assumptions can be made to derive constitutive laws from this differential or rate form and some significant efforts on empirically based formulations have been reported [56, 57, 136]. An example of the phenomenological approach is reported in [105] (see also [59]) where one assumes the Landou-Devonshire model for the free energy $\Psi = \Psi(\bar{\varepsilon}, T)$ and attempts to model the constitutive law in the form $\bar{\sigma} = \rho \frac{\partial \Psi}{\partial \bar{\varepsilon}}$. However, dynamical modeling of SMA is in its infancy and much remains to be done if one is to develop reliable distributed control devices based on SMA technology.

Advantages:
- Small size, internal material-based mechanisms for single unit actuators;
- Excellent cycling performance in repeated loading;
- Flexible actuation signal: heat via electrical signals, hot water, hot air, laser or infrared rays.

Disadvantages:
- Slow response time as compared to other smart material components (e.g., piezoceramics);
- Nonlinear hysteresis and thermomechanical response characteristics are an essential feature: this results in difficult modeling and design of actuators for consistent repeatable operation;
- Energy requirements for actuating shape memory effect are relatively high;
- SME is very much limited to a certain thermal range, requiring very precise actuation.

Current Control Applications: Switches in a discrete electrical or mechanical setting; numerous shape memory fasteners.

1.2.6   Fiber Optic Sensors

Essential components of any smart material system are the sensors. During the last decade, a class of sensors employing optical fibers and the refractive properties of light have been developed in the context of smart material technology [93]. While these sensors can also sense acoustical and thermal perturbations, developments have mainly focused on embedded fiber optics to measure mechanical strain during the dynamic responses of structures. In this technology, one replaces traditional strain devices involving copper wires and electrical signals by optical fibers and light beams or lasers. Light, which is of course a form of electromagnetic energy, is transmitted through gladded glass fibers at very high frequency, short wave lengths ($\lambda \approx 1~\mu m$). If these isotropic glass fibers are unstrained, an initially unpolarized light signal remains unpolarized. However, if the glass is strained, it becomes temporarily birefringent and the signal is polarized and its propagation is altered. Moreover, if the glass is strained, the path length that any light ray must travel (and possibly be reflected if mirrored fibers are used) is altered. Fiber optic sensing technology is based on observing and analyzing such modifications in amplitude, wave length, phase, polarization characteristics and/or modal distribution of the signals in structures (and the embedded fibers) that are being deformed.

There are several general classes of fiber optic strain sensors currently being advocated for smart material uses. Among these are *interferometric* strain sensors [60, 111, 169, 178] which typically involve a "sensor arm" fiber and a "reference arm" fiber. In a Mach-Zehnder interferometric sensor, a single beam input is split by two single mode fibers which act as sensor arm and reference arm. The reference arm is isolated from the strain deformation while the sensor undergoes the same deformation as the structure. Differences in the path length resulting from deformation of the sensor arm produce a relative phase shift. The output signals are collected and the signal modifications are calibrated to mechanical deformations. The nature of these sensors make them inherently difficult to simultaneously "localize" (measure a "point" strain) and make directionally sensitive. To the contrary, a Michelson interferometric sensor allows for excellent "point" sensing. In this sensor the signal is also split into two arms (sensor and reference) but the fibers ends are mirrored so that the device observes reflected signals. The arms are of different length ("gauge length") which allows for millimeter range localization of strain measurements. A Fabry-Perot interferometer sensor is similar to the Michelson device except that a single mirrored end fiber is used along with an embedded mirror in the sensor area.

Polarimetric strain sensors also employ a single mirrored end fiber and depend on analysis of reflected signals. However this class of sensors requires a polarized He-Ne laser input and a single mode high birefringence (HiBi) fiber. Mechanical deformation produces frequency dependent phase shifts in the signal which are analyzed for strain integrated over the path of the sensor. These sensors are especially useful for measurements of large strains and applications requiring lengthy spatial sensing.

Fiber optic strain sensors are much more versatile then traditional resistive strain gauges. They can be placed at various levels in a 3-dimensional structure to obtain information about principal strains as well as the full strain tensor. In one popular adaptation, they are constructed in rosette style: three 1-dimensional strain gauges in three distinct directions are used to obtain 2-dimensional or "in-plane" strains.

While fiber optic technology is still under development and offers no actuation capabilities, it may yet play an important role in emerging smart material technology.

Advantages:

- Lightweight, much smaller than copper wire cable;
- Readily embedded in structures; sufficiently small that they don't affect integrity of structure;

- Do not generate heat or electromagnetic interference;
- Do not require electrical isolation from structure, do not provide conductive pathway through the structure;
- Mechanically flexible, diverse geometry possible;
- Low attenuation of signals;
- Low maintenance, high reliability.

Disadvantages:
- No actuation capabilities;
- Fragile in some configurations; damage difficult to repair if embedded;
- Heat sensitive; analysis must compensate for possible heat fluctuation during strain measurements;
- Easily damaged by heat and pressures during composite material construction.

Current Control Applications: None; still a developing technology as a sensor component.

1.3   Current and Projected Smart Materials Applications

Initial applications of smart materials technology to structural, structural acoustic and fluid/structure systems involved a very small number of actuators and sensors. For example, initial experiments concerning the control of noise transmission through a shell using piezoelectric elements involved only two pairs of piezoceramic actuators with interior sound pressure levels measured by microphones [92]. As the complexity of experiments increased to model more realistic environments, more actuators and sensors were employed (8 pairs of piezoceramic patches in the experiments of [168]) with the number often limited by the input capabilities of the data acquisition hardware.

Challenging generalizations of these trends are quite natural. For example, piezoceramic-based smart "panels" or "walls" such as that depicted in Figure 1.11 are envisioned for near-future applications. These panels would have multiple sensor/actuator groups with computational chips (containing control law algorithms) to permit multiple locally responsive sensor/control/actuator clusters in a truly distributed smart material structure paradigm.

Another direction in the utilization of the previously-described smart material components involves the incorporation of multiple materials in the system to take advantage of their respective properties. For example, PVDF sensors and piezoceramic actuators were employed in the structural acoustic systems described in [66, 67] to reduce structure-borne noise. This tendency toward the combination of smart materials to create multiple-component controllers

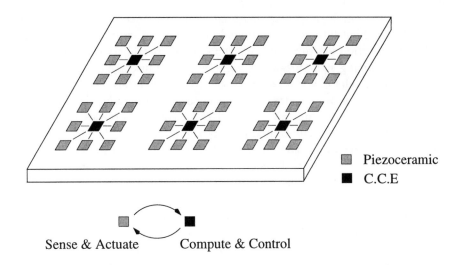

Figure 1.11: Smart "panel" employing computational control elements (C.C.E) and piezoceramic patches as sensors and actuators.

will likely continue as smart materials are refined and multiple input/output data acquisition hardware is improved.

To illustrate this philosophy, we consider possibilities for helicopter rotor design using smart material systems (details regarding this and other potential multiple-component smart material systems can be found in [93]). Rotor specifications in current aircraft are typically designed for worst-case conditions which promotes robustness but diminishes performance in standard operating environments. An alternative is to replace over-designed passive components by active components constructed from smart materials. The potential for extending the performance, robustness and integrity of rotors through smart material design can be illustrated through consideration of a rotor with fiber optic sensors and shape memory alloy, electrorheological fluid, magnetostrictive and piezoelectric actuators. The fiber optic sensors would provide a capability for measuring thermal gradients and vibration levels in the rotor. The electrorheological fluid and magnetostrictive transducers provide a capacity for vibration reduction through dynamic alteration of mass, stiffness and damping characteristics of the rotor. Large changes in the camber and geometry of the airfoil could be attained with the shape memory alloys while piezoceramic actuators would provide the capability for altering the surface geometry of the airfoil. In combination, the two could be used to significantly adapt the aerodynamical properties to changing flight conditions. Finally, hybrid control strategies based on models coupling the mechanical and aerodynamical rotor properties would be combined to address flutter, buffeting, gust fields and dynamic stall.

Associated with the smart material rotor design to enhance performance is the complementary problem of fatigue reduction. The consideration of this phenomenon in parallel with the vibration suppression is of paramount importance when employing smart material actuators since initial research has demonstrated high fatigue levels around the edges of actuators programmed solely for vibration attenuation. Smart material controllers can also be employed for reducing structure-borne noise inside the fuselage or cockpit, compensation for modal characteristics which change according to fuel consumption, and active control of surface properties to minimize radar signatures. Finally, actuators and sensors, already in place for control purposes, can be used to excite and sense various components of the aircraft for damage detection and nondestructive evaluation purposes. This latter task becomes increasingly important as aging aircraft develop interior cracks, delaminations, et cetera, which can threaten their functional integrity.

The successful development of multiple-component controllers for increasingly complex systems requires detailed modeling of the smart material capabilities and performance. Because none of the components act in isolation, the dynamic interactions between the smart material actuators/sensors and underlying structures must be characterized. Furthermore, interactions between components in the system and coupling with adjacent fields must be fully modeled to utilize the full potential of the controllers. Finally, the control laws themselves must be constructed in a manner which takes advantage of the smart material capabilities and characteristics of the system. Model-based smart material controllers constructed in this manner then have tremendous potential in future applications.

Chapter 2

Modeling Aspects of Shells, Plates and Beams

2.1 Derivation of the Basic Shell Equations

The goal in this chapter is to describe the equations of motion of structures having shell, plate or beam characteristics. We begin by outlining the derivation of the basic shell equations. While this derivation is classical and can be found in numerous references (the exposition in this section follows closely that in [134]), it provides a basis for modeling many structures and illustrates techniques that will be used when including the effects of piezoceramic actuators as discussed in Chapter 3. The resulting equations are sufficiently general so as to include cylindrical shells, plates, and curved and flat beams when judicious choices are made for radii and Lamé parameters. Both the strong and weak forms of the equations of motion for these structures are considered since both are used when employing the models in applications of interest.

2.1.1 Basic Assumptions

The description of shells in a completely general setting involves geometric rather than material considerations. Flügge defines a shell as "an object which, for the purpose of stress analysis, may be considered as the materialization of a curved surface" [85]. From this perspective, soap bubbles, parachutes, and steel containers can be considered shells. For our purpose, it is useful to limit the discussion to elastic materials satisfying linear stress/strain relations. In describing the initial models, we will consider a shell to be a solid bounded by two curved surfaces that are separated by a distance h. The middle surface will then be defined as the locus of points lying midway between the outer bounding surfaces. The foundations of the classical theory of thin shells, described in this manner, were first established by Love [139]. In his formulation, he made the following four assumptions which are basic to what follows.

1. The shell thickness h is very small in comparison with the other dimensions such as radius of curvature and length. This condition is crucial

to thin shell theory and says that the ratio of shell thickness, h, to the smallest radius of curvature, R, is small as compared to unity; that is, $h/R << 1$. In practice, the upper limit for this ratio is set to be of the order of $1/10$ to $1/20$ (see [82, 149]). As discussed in [149, page 2], thickness/radius ratios between $1/1000$ and $1/50$ are common in practical applications and hence thin shell models can be employed in numerous and varied settings.

2. Shell deformations are sufficiently small so as to allow the second and higher powers to be neglected with respect to the first powers. This requirement allows us to refer all kinematic and equilibrium considerations to the original, unperturbed, reference state of the shell and ensures that the differential equations of shell deformation are linear. We note that this hypothesis must be relaxed in order to obtain the nonlinear von Kármán plate model described in Section 2.2.5.

3. Transverse normal stresses (σ_z) are small compared to the other normal stresses $(\sigma_\alpha, \sigma_\beta)$ in the shell and hence can be neglected. In other words, the stress in the direction normal to the thin dimension is taken to be negligible. This assumption, in combination with the fourth, deals with the constitutive properties of thin shells and allows the three-dimensional elasticity problem to be transformed into a two-dimensional one.

4. A line which is originally normal to the shell reference surface will remain normal to the deformed reference surface (in notation to follow, this yields the assumption $\gamma_{\alpha z} = \gamma_{\beta z} = 0$) and will remain unstrained or unextended $(e_z = 0)$. This assumption is analogous to the Euler hypothesis in thin beam theory which states that plane sections remain plane. This is referred to as Kirchhoff's hypothesis or occasionally the "hairbrush hypothesis." We point out that in the case of moderate to thick shells, this hypothesis must be relaxed to allow for rotational effects and shear deformation. As discussed in Section 2.2.5, this leads to theories which include the Mindlin-Reissner plate model and Timoshenko beam model.

We point out that while the latter two assumptions simplify the problem in the sense that a three dimensional elasticity problem is reduced to two dimensions, they lead to models in which the equations for out-of-plane motion involve higher derivatives than do the equations for in-plane motion (fourth derivatives compared with second derivatives in the strong formulation). In many applications, however, the simplification resulting from this drop in dimension fully justifies the imbalance and increased complexity due to the higher derivatives. It should also be noted that Assumptions 3 and 4 lead to a variety of contradictions concerning the vanishing of normal stresses and strains (see [134] for further discussion). In spite of these contradictions, the thin shell models, obtained with these postulates, accurately capture structural physics in a wide range of applications.

2.1.2   Shell Coordinates

When defining shell coordinates, it is convenient (although not universal) to choose the unperturbed middle surface of the shell as the reference surface. On the reference surface, an orthogonal curvilinear coordinate system is established which coincides with the orthogonal lines of principal curvature. The thickness direction which is normal to the reference is taken to be the third coordinate direction. From the fourth shell assumption regarding the preservation of the normal, it follows that the displacements must be linear in the thickness coordinate and thus the behavior of any point on the shell can be determined from the behavior of a corresponding point on the reference surface. Let the reference surface be determined by the vector $\vec{r}(\alpha, \beta)$ where α and β are independent parameters. For an arbitrary point on the shell, a position vector is then defined by

$$\vec{R}(\alpha, \beta, z) = \vec{r}(\alpha, \beta) + z\hat{i}_n$$

where $\hat{i}_n$ is the unit vector normal to the middle surface, and z measures the distance of the point from the corresponding point on the middle surface along $\hat{i}_n$. If we assume that the shell has uniform thickness h, then $-h/2 \leq z \leq h/2$.

With the coordinate system thus established, the fundamental three dimensional shell element of thickness dz at a height z from the middle surface is defined next (see Figure 2.1). First, the magnitude of an arbitrary differential length element is given by

$$
\begin{aligned}
(ds)^2 &= d\vec{R} \cdot d\vec{R} \\
&= (d\vec{r} + z d\hat{i}_n + \hat{i}_n dz) \cdot (d\vec{r} + z d\hat{i}_n + \hat{i}_n dz) \\
&= A^2(1 + z/R_\alpha)^2(d\alpha)^2 + B^2(1 + z/R_\beta)^2(d\beta)^2 + (dz)^2 \\
&= g_1(d\alpha)^2 + g_2(d\beta)^2 + g_3(dz)^2
\end{aligned}
\tag{2.1}
$$

where R_α and R_β are the radii of curvature in the α and β directions and A and B are the Lamé constants which are defined by

$$A^2 = \frac{\partial \vec{r}}{\partial \alpha} \cdot \frac{\partial \vec{r}}{\partial \alpha} \quad , \quad B^2 = \frac{\partial \vec{r}}{\partial \beta} \cdot \frac{\partial \vec{r}}{\partial \beta}$$

(see [82, 134] for further details). The coefficients

$$g_1 = \left[A \left(1 + \frac{z}{R_\alpha} \right) \right]^2$$

$$g_2 = \left[B \left(1 + \frac{z}{R_\beta} \right) \right]^2 \tag{2.2}$$

$$g_3 = 1$$

are a subset of the metric coefficients which provide a link between the length of an element and the differentials $d\alpha$, $d\beta$ and dz (see [163, pages 114-115]).

It follows immediately from (2.1) that the lengths of the edges of the element
are given by

$$ds_\alpha^{(z)} = A(1 + z/R_\alpha)d\alpha$$
$$ds_\beta^{(z)} = B(1 + z/R_\beta)d\beta \tag{2.3}$$

and the corresponding area elements of the faces are

$$dA_\alpha^{(z)} = A(1 + z/R_\alpha)d\alpha dz$$
$$dA_\beta^{(z)} = B(1 + z/R_\beta)d\beta dz \ .$$

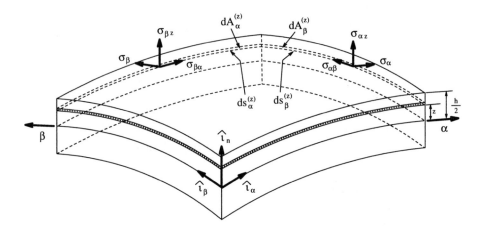

Figure 2.1: The fundamental shell element.

2.1.3   Strain-Displacement Relations

As shown in [82, 134], the strain-displacement equations in orthogonal coordinates which follow from the three dimensional theory of elasticity are

$$e_i = \frac{\partial}{\partial \alpha_i}\left(\frac{U_i}{\sqrt{g_i}}\right) + \frac{1}{2g_i}\sum_{k=1}^{3}\frac{\partial g_i}{\partial \alpha_k}\frac{U_k}{\sqrt{g_k}} \quad , \ i = 1,2,3$$

$$\gamma_{ij} = \frac{1}{\sqrt{g_i g_j}}\left[g_i\frac{\partial}{\partial \alpha_j}\left(\frac{U_i}{\sqrt{g_i}}\right) + g_j\frac{\partial}{\partial \alpha_i}\left(\frac{U_j}{\sqrt{g_j}}\right)\right] \quad , \begin{array}{l} i,j = 1,2,3 \\ i \neq j \end{array} \tag{2.4}$$

where U_i, e_i and γ_{ij} are the displacements, normal strains and shear strains,
respectively, at an arbitrary point in the material. The equations are posed
in shell coordinates by replacing the strain indices 1, 2 and 3 by α, β and
z, respectively, and by letting U, V and W replace U_1, U_2 and U_3 as the

displacements in the α, β and z directions. Finally, the substitution of the metric coefficients in (2.2) yields the general strain displacement equations

$$e_\alpha = \frac{1}{1 + z/R_\alpha} \left(\frac{1}{A} \frac{\partial U}{\partial \alpha} + \frac{V}{AB} \frac{\partial A}{\partial \beta} + \frac{W}{R_\alpha} \right)$$

$$e_\beta = \frac{1}{1 + z/R_\beta} \left(\frac{1}{B} \frac{\partial V}{\partial \beta} + \frac{U}{AB} \frac{\partial B}{\partial \alpha} + \frac{W}{R_\beta} \right)$$

$$e_z = \frac{\partial W}{\partial z}$$

$$\gamma_{\alpha\beta} = \frac{A(1 + z/R_\alpha)}{B(1 + z/R_\beta)} \frac{\partial}{\partial \beta} \left[\frac{U}{A(1 + z/R_\alpha)} \right] \tag{2.5}$$

$$+ \frac{B(1 + z/R_\beta)}{A(1 + z/R_\alpha)} \frac{\partial}{\partial \alpha} \left[\frac{V}{B(1 + z/R_\beta)} \right]$$

$$\gamma_{\alpha z} = \frac{1}{A(1 + z/R_\alpha)} \frac{\partial W}{\partial \alpha} + A(1 + z/R_\alpha) \frac{\partial}{\partial z} \left[\frac{U}{A(1 + z/R_\alpha)} \right]$$

$$\gamma_{\beta z} = \frac{1}{B(1 + z/R_\beta)} \frac{\partial W}{\partial \beta} + B(1 + z/R_\beta) \frac{\partial}{\partial z} \left[\frac{V}{B(1 + z/R_\beta)} \right] .$$

In order to satisfy the fourth (Kirchhoff) hypothesis, the displacements are assumed to be linear in the thickness direction, thus yielding

$$U(\alpha, \beta, z) = u(\alpha, \beta) + z\theta_\alpha(\alpha, \beta)$$

$$V(\alpha, \beta, z) = v(\alpha, \beta) + z\theta_\beta(\alpha, \beta) \tag{2.6}$$

$$W(\alpha, \beta, z) = w(\alpha, \beta)$$

where u, v and w are the displacements of the middle surface in the α, β and z directions, respectively. The quantities θ_α and θ_β are the rotations of the normal to the middle surface which occur during deformation.

In order to determine θ_α and θ_β in terms of the displacements u, v and w, it is noted that the Kirchhoff hypothesis implies that all strain components in the direction of the normal to the reference surface vanish; that is

$$\gamma_{\alpha z} = \gamma_{\beta z} = e_z = 0 .$$

When (2.6) is substituted into (2.5), this constraint on $\gamma_{\alpha z}$ and $\gamma_{\beta z}$ implies that

$$\theta_\alpha = \frac{u}{R_\alpha} - \frac{1}{A} \frac{\partial w}{\partial \alpha}$$

$$\tag{2.7}$$

$$\theta_\beta = \frac{v}{R_\beta} - \frac{1}{B} \frac{\partial w}{\partial \beta} .$$

As noted previously, the Kirchhoff hypothesis must be relaxed in thick shell theory to accommodate the effects of rotation and shear deformation. In that case, the general rotations are retained when (2.6) is substituted into (2.5) to yield equilibrium equations involving θ_α and θ_β. We refer the reader to [171] for details concerning the derivation of equations of motion for thick shells while details concerning an analogous derivation of the Mindlin-Reissner thick plate model are given in Section 2.2.5.

To obtain the strain-displacement equations in terms of u, v and w in the case of thin shells, (2.6) and (2.7) are substituted into (2.5) to obtain

$$e_\alpha = \frac{1}{(1 + z/R_\alpha)}(\varepsilon_\alpha + z\kappa_\alpha)$$

$$e_\beta = \frac{1}{(1 + z/R_\beta)}(\varepsilon_\beta + z\kappa_\beta)$$

$$\gamma_{\alpha\beta} = \frac{1}{(1 + z/R_\alpha)(1 + z/R_\beta)}\left[\left(1 - \frac{z^2}{R_\alpha R_\beta}\right)\varepsilon_{\alpha\beta}\right. \tag{2.8}$$

$$\left. + z\left(1 + \frac{z}{2R_\alpha} + \frac{z}{2R_\beta}\right)\tau\right] .$$

Here ε_α, ε_β and $\varepsilon_{\alpha\beta}$ are the normal and shear strains in the middle surface ($z = 0$), κ_α and κ_β are the midsurface changes in curvature, and τ is the midsurface twist. As shown in [134], these quantities are given by

$$\varepsilon_\alpha = \frac{1}{A}\frac{\partial u}{\partial \alpha} + \frac{v}{AB}\frac{\partial A}{\partial \beta} + \frac{w}{R_\alpha}$$

$$\varepsilon_\beta = \frac{1}{B}\frac{\partial v}{\partial \beta} + \frac{u}{AB}\frac{\partial B}{\partial \alpha} + \frac{w}{R_\beta} \tag{2.9}$$

$$\varepsilon_{\alpha\beta} = \frac{A}{B}\frac{\partial}{\partial \beta}\left(\frac{u}{A}\right) + \frac{B}{A}\frac{\partial}{\partial \alpha}\left(\frac{v}{B}\right)$$

and

$$\kappa_\alpha = \frac{1}{A}\frac{\partial \theta_\alpha}{\partial \alpha} + \frac{\theta_\beta}{AB}\frac{\partial A}{\partial \beta}$$

$$\kappa_\beta = \frac{1}{B}\frac{\partial \theta_\beta}{\partial \beta} + \frac{\theta_\alpha}{AB}\frac{\partial B}{\partial \alpha}$$

$$\tau = \frac{A}{B}\frac{\partial}{\partial \beta}\left(\frac{\theta_\alpha}{A}\right) + \frac{B}{A}\frac{\partial}{\partial \alpha}\left(\frac{\theta_\beta}{B}\right) + \frac{1}{R_\alpha}\left(\frac{1}{B}\frac{\partial u}{\partial \beta} - \frac{v}{AB}\frac{\partial B}{\partial \alpha}\right) \tag{2.10}$$

$$+ \frac{1}{R_\beta}\left(\frac{1}{A}\frac{\partial v}{\partial \alpha} - \frac{u}{AB}\frac{\partial A}{\partial \beta}\right).$$

From (2.8) it can be seen that the total strains at any point can be represented as the sum of two parts, one due to stretching and one due to bending. Note that within the framework of infinitesimal elasticity, these equations are exact and are the strain-displacement equations used in the theories of Byrne, Flügge, Lur'ye, Goldenveizer and Novozhilov.

The existing theories dealing with the deformation of a thin shell differ in how and when the terms z/R_α and z/R_β are to be neglected with respect to unity in setting up the kinematic (strain-displacement), constitutive (Hooke's Law) and equilibrium equations. An overview of the various theories can be found in [134] and [141]. All the theories agree in the expressions for the middle surface strains ε_α, ε_β and $\varepsilon_{\alpha\beta}$ and there is general agreement among the theories about the expressions for the middle surface curvature changes κ_α and κ_β. The greatest disagreement among the theories concerns the proper form of the midsurface change in twist, τ.

In the Donnell-Mushtari theory, one neglects the terms z/R_α and z/R_β in (2.8) and the tangential displacements and their derivatives in (2.10) thus yielding

$$e_\alpha = \varepsilon_\alpha + z\kappa_\alpha$$

$$e_\beta = \varepsilon_\beta + z\kappa_\beta \qquad (2.11)$$

$$\gamma_{\alpha\beta} = \varepsilon_{\alpha\beta} + z\tau$$

and

$$\kappa_\alpha = -\frac{1}{A}\frac{\partial}{\partial\alpha}\left(\frac{1}{A}\frac{\partial w}{\partial\alpha}\right) - \frac{1}{AB^2}\frac{\partial A}{\partial\beta}\frac{\partial w}{\partial\beta}$$

$$\kappa_\beta = -\frac{1}{B}\frac{\partial}{\partial\beta}\left(\frac{1}{B}\frac{\partial w}{\partial\beta}\right) - \frac{1}{A^2B}\frac{\partial B}{\partial\alpha}\frac{\partial w}{\partial\alpha} \qquad (2.12)$$

$$\tau = -\frac{B}{A}\frac{\partial}{\partial\alpha}\left(\frac{1}{B^2}\frac{\partial w}{\partial\beta}\right) - \frac{A}{B}\frac{\partial}{\partial\beta}\left(\frac{1}{A^2}\frac{\partial w}{\partial\alpha}\right)$$

with the midsurface strains given by (2.9). The theories of Love, Timoshenko, Reissner, Naghdi, Berry and Sanders also employ the approximations in (2.11) but retain the expressions in (2.10) for the curvature changes κ_α and κ_β. The expressions for the midsurface change in twist differ among the theories and the reader is directed to Table 1.3 of [134] for a summary of the approximations used in describing τ.

2.1.4   Stress-Strain Relations

When writing down the constitutive properties of the shell, it is assumed that the shell material is elastic and isotropic with Young's modulus E and Poisson ratio ν. In the three dimensional form, Hooke's law is given by

$$e_\alpha = \frac{1}{E}[\sigma_\alpha - \nu(\sigma_\beta + \sigma_z)] \quad , \quad \gamma_{\alpha\beta} = \frac{2(1+\nu)}{E}\sigma_{\alpha\beta}$$

$$e_\beta = \frac{1}{E}[\sigma_\beta - \nu(\sigma_\alpha + \sigma_z)] \quad , \quad \gamma_{\alpha z} = \frac{2(1+\nu)}{E}\sigma_{\alpha z} \qquad (2.13)$$

$$e_z = \frac{1}{E}[\sigma_z - \nu(\sigma_\alpha + \sigma_\beta)] \quad , \quad \gamma_{\beta z} = \frac{2(1+\nu)}{E}\sigma_{\beta z}$$

where σ_α and σ_β are normal stresses acting upon faces of a shell element (see Figure 2.1). The shear strains in the z direction are produced by transverse shear stresses $\sigma_{\alpha z}$ and $\sigma_{\beta z}$ while those in the β and α directions are produced by the tangential stresses $\sigma_{\alpha\beta}$ and $\sigma_{\beta\alpha}$, respectively. To simplify (2.13), it is noted that Kirchhoff's hypothesis yields $e_z = \gamma_{\alpha z} = \gamma_{\beta z} = 0$ whereas from Love's third assumption, σ_z can be taken to be negligible (for contradictions which arise from these assumptions, see [134]). The problem then reduces to one of plane stresses and (2.13) reduces to

$$e_\alpha = \frac{1}{E}(\sigma_\alpha - \nu\sigma_\beta)$$

$$e_\beta = \frac{1}{E}(\sigma_\beta - \nu\sigma_\alpha) \qquad (2.14)$$

$$\gamma_{\alpha\beta} = \frac{2(1+\nu)}{E}\sigma_{\alpha\beta}$$

which yields

$$\sigma_\alpha = \frac{E}{1-\nu^2}(e_\alpha + \nu e_\beta)$$

$$\sigma_\beta = \frac{E}{1-\nu^2}(e_\beta + \nu e_\alpha) \qquad (2.15)$$

$$\sigma_{\alpha\beta} = \sigma_{\beta\alpha} = \frac{E}{2(1+\nu)}\gamma_{\alpha\beta} \ .$$

The relationships (2.14) and (2.15) depend upon the hypothesis that the shell is elastic and contains no internal damping. One way to include the effects of internal damping is to assume a more general constitutive law which posits that stress is proportional to a linear combination of strain and strain rate. In this case, the stress-strain relations are

$$Ee_\alpha + c_D\dot{e}_\alpha = \sigma_\alpha - \nu\sigma_\beta$$

$$Ee_\beta + c_D\dot{e}_\beta = \sigma_\beta - \nu\sigma_\alpha \qquad (2.16)$$

$$E\gamma_{\alpha\beta} + c_D\dot{\gamma}_{\alpha\beta} = 2(1+\nu)\sigma_{\alpha\beta}$$

which implies that

$$\sigma_\alpha = \frac{E}{1 - \nu^2}(e_\alpha + \nu e_\beta) + \frac{c_D}{1 - \nu^2}(\dot{e}_\alpha + \nu \dot{e}_\beta)$$

$$\sigma_\beta = \frac{E}{1 - \nu^2}(e_\beta + \nu e_\alpha) + \frac{c_D}{1 - \nu^2}(\dot{e}_\beta + \nu \dot{e}_\alpha) \tag{2.17}$$

$$\sigma_{\alpha\beta} = \frac{E}{2(1 + \nu)}\gamma_{\alpha\beta} + \frac{c_D}{2(1 + \nu)}\dot{\gamma}_{\alpha\beta} .$$

Here c_D denotes the internal damping coefficient whereas the dot notation indicates differentiation with respect to time.

2.1.5   Force and Moment Resultants

In this section, the force and moment resultants are defined in terms of the stresses which are given in (2.15). Consider first the face of the element in Figure 2.1 which is perpendicular to the α-axis along with the stresses σ_α, $\sigma_{\alpha\beta}$ and $\sigma_{\alpha z}$ which act upon this face. Using (2.3), the intercept of the middle surface with this face has arclength $ds_\beta = Bd\beta$. Similarly, the intercepts of parallel surfaces with this face are found to have arclengths $ds_\beta^{(z)} = B(1 + z/R_\beta)d\beta$. The infinitesimal force acting on the elemental area $ds_\beta^{(z)}dz$ of the face is then given by $\sigma_\alpha ds_\beta^{(z)}dz$. We now want to equate the total force on the face with a force resultant N_α along the middle surface. Doing so yields

$$N_\alpha ds_\beta = \int_{-h/2}^{h/2} \sigma_\alpha ds_\beta^{(z)} dz$$

which implies that

$$N_\alpha = \int_{-h/2}^{h/2} \sigma_\alpha \left(1 + \frac{z}{R_\beta}\right) dz$$

where N_α has units of force per unit length of middle surface. Similar analysis can be used to determine the in-plane and shear forces $N_{\alpha\beta}$ and Q_α, in which case, the three force resultants acting on this face can be expressed as

$$\begin{bmatrix} N_\alpha \\ N_{\alpha\beta} \\ Q_\alpha \end{bmatrix} = \int_{-h/2}^{h/2} \begin{bmatrix} \sigma_\alpha \\ \sigma_{\alpha\beta} \\ \sigma_{\alpha z} \end{bmatrix} \left(1 + \frac{z}{R_\beta}\right) dz . \tag{2.18}$$

Likewise, the force resultants on the face which is perpendicular to the β-axis are

$$\begin{bmatrix} N_\beta \\ N_{\beta\alpha} \\ Q_\beta \end{bmatrix} = \int_{-h/2}^{h/2} \begin{bmatrix} \sigma_\beta \\ \sigma_{\beta\alpha} \\ \sigma_{\beta z} \end{bmatrix} \left(1 + \frac{z}{R_\alpha}\right) dz . \tag{2.19}$$

The sense of the force resultants is shown in Figure 2.2.

The moment resultants can be determined similarly once the moment arm z is included in the integrals. The positive directions of the moment resultants are shown in Figure 2.3 with values given by

$$\begin{bmatrix} M_\alpha \\ M_{\alpha\beta} \end{bmatrix} = \int_{-h/2}^{h/2} \begin{bmatrix} \sigma_\alpha \\ \sigma_{\alpha\beta} \end{bmatrix} \left(1 + \frac{z}{R_\beta}\right) z\, dz \qquad (2.20)$$

and

$$\begin{bmatrix} M_\beta \\ M_{\beta\alpha} \end{bmatrix} = \int_{-h/2}^{h/2} \begin{bmatrix} \sigma_\beta \\ \sigma_{\beta\alpha} \end{bmatrix} \left(1 + \frac{z}{R_\alpha}\right) z\, dz \ . \qquad (2.21)$$

Because the moment resultants are defined with respect to the middle surface, the dimensions are moment per unit length of middle surface. Note that although $\sigma_{\alpha\beta} = \sigma_{\beta\alpha}$, in general $N_{\alpha\beta} \neq N_{\beta\alpha}$ and $M_{\alpha\beta} \neq M_{\beta\alpha}$ unless $R_\alpha = R_\beta$.

In several of the thin shell theories such as those of Donnell, Mushtari, Love, Timoshenko, Reissner, Naghdi, Berry and Sanders, one would at this point neglect z/R_α and z/R_β in comparison with unity thus yielding the force and moment resultants

$$N_\alpha = \frac{Eh}{(1-\nu^2)}(\varepsilon_\alpha + \nu\varepsilon_\beta) \qquad , \qquad M_\alpha = \frac{Eh^3}{12(1-\nu^2)}(\kappa_\alpha + \nu\kappa_\beta)$$

$$N_\beta = \frac{Eh}{(1-\nu^2)}(\varepsilon_\beta + \nu\varepsilon_\alpha) \qquad , \qquad M_\beta = \frac{Eh^3}{12(1-\nu^2)}(\kappa_\beta + \nu\kappa_\alpha) \qquad (2.22)$$

$$N_{\alpha\beta} = N_{\beta\alpha} = \frac{Eh}{2(1+\nu)}\varepsilon_{\alpha\beta} \qquad , \qquad M_{\alpha\beta} = M_{\beta\alpha} = \frac{Eh^3}{24(1+\nu)}\tau \ .$$

To express the force and moment resultants in terms of the displacements u, v and w, the expressions of (2.9) can be substituted for $\varepsilon_\alpha, \varepsilon_\beta$ and $\varepsilon_{\alpha\beta}$. For the theories of Love, Timoshenko, Reissner, Naghdi, Berry and Sanders, κ_α and κ_β in (2.10) are used while the reader is directed to Table 1.3 of [134] for various approximations describing τ. In the Donnell-Mushtari theory, one employs the approximations in (2.12) to describe the midsurface curvature and twist changes and hence the moment and force resultants in terms of u, v and w.

As discussed in [134], higher order approximations to the moment and force resultants can be obtained by replacing the terms $1/(1 + z/R_i), i = \alpha, \beta$, by geometric series which is justified by the assumption that $z/R_i < 1$. In the theory of Byrne, Flügge and Lur'ye, the terms of degree greater than three are discarded before the various integrations are carried out. This yields the force and moment resultants

$$N_\alpha = \frac{Eh}{1-\nu^2}\left[\varepsilon_\alpha + \nu\varepsilon_\beta - \frac{h^2}{12}\left(\frac{1}{R_\alpha} - \frac{1}{R_\beta}\right)\left(\kappa_\alpha - \frac{\varepsilon_\alpha}{R_\alpha}\right)\right]$$

$$N_\beta = \frac{Eh}{1-\nu^2}\left[\varepsilon_\beta + \nu\varepsilon_\alpha - \frac{h^2}{12}\left(\frac{1}{R_\beta} - \frac{1}{R_\alpha}\right)\left(\kappa_\beta - \frac{\varepsilon_\beta}{R_\beta}\right)\right]$$

$$N_{\alpha\beta} = \frac{Eh}{2(1+\nu)}\left[\varepsilon_{\alpha\beta} - \frac{h^2}{12}\left(\frac{1}{R_\alpha} - \frac{1}{R_\beta}\right)\left(\frac{\tau}{2} - \frac{\varepsilon_{\alpha\beta}}{R_\alpha}\right)\right]$$

$$N_{\beta\alpha} = \frac{Eh}{2(1+\nu)}\left[\varepsilon_{\alpha\beta} - \frac{h^2}{12}\left(\frac{1}{R_\beta} - \frac{1}{R_\alpha}\right)\left(\frac{\tau}{2} - \frac{\varepsilon_{\alpha\beta}}{R_\beta}\right)\right]$$

$$M_\alpha = \frac{Eh^3}{12(1-\nu^2)}\left[\kappa_\alpha + \nu\kappa_\beta - \left(\frac{1}{R_\alpha} - \frac{1}{R_\beta}\right)\varepsilon_\alpha\right] \qquad (2.23)$$

$$M_\beta = \frac{Eh^3}{12(1-\nu^2)}\left[\kappa_\beta + \nu\kappa_\alpha - \left(\frac{1}{R_\beta} - \frac{1}{R_\alpha}\right)\varepsilon_\beta\right]$$

$$M_{\alpha\beta} = \frac{Eh^3}{24(1+\nu)}\left[\tau - \frac{\varepsilon_{\alpha\beta}}{R_\alpha}\right]$$

$$M_{\beta\alpha} = \frac{Eh^3}{24(1+\nu)}\left[\tau - \frac{\varepsilon_{\alpha\beta}}{R_\beta}\right].$$

Finally, the midsurface strain, curvature and twist expressions in (2.9) and (2.10) can be used to express these moment and force resultants in terms of the displacements u, v and w.

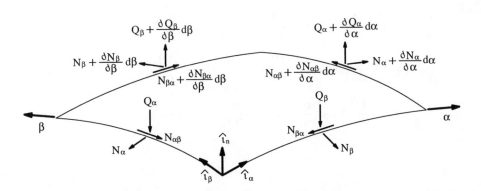

Figure 2.2: The force resultants in shell coordinates.

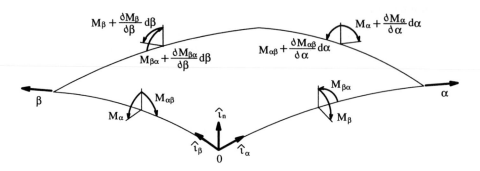

Figure 2.3: The moment resultants in shell coordinates.

2.1.6 Equations of Motion

In order to obtain the strong form of the equations of motion for a thin shell, consider a shell element of thickness h as shown in Figure 2.1. The general equations of the dynamic equilibrium of the element are obtained by balancing the internal force and moment resultants as shown in Figures 2.2 and 2.3 with any externally applied forces and moments. Let

$$\vec{q} = \hat{q}_\alpha \hat{\imath}_\alpha + \hat{q}_\beta \hat{\imath}_\beta + \hat{q}_n \hat{\imath}_n$$

and

$$\vec{m} = \hat{m}_\alpha \hat{\imath}_\alpha + \hat{m}_\beta \hat{\imath}_\beta + \hat{m}_n \hat{\imath}_n$$

denote the external forces and moments which are acting on the middle surface of the shell. Because these forces and moments enter as surface contributions due to an external field, they have units of force and moment per unit area, respectively.

As shown in equations (1.112) and (1.115) of [134], force and moment balancing then yields the equilibrium equations

$$\frac{\partial}{\partial \alpha}(BN_\alpha) + \frac{\partial}{\partial \beta}(AN_{\beta\alpha}) + \frac{\partial A}{\partial \beta}N_{\alpha\beta} - \frac{\partial B}{\partial \alpha}N_\beta + \frac{AB}{R_\alpha}Q_\alpha + AB\hat{q}_\alpha = 0$$

$$\frac{\partial}{\partial \beta}(AN_\beta) + \frac{\partial}{\partial \alpha}(BN_{\alpha\beta}) + \frac{\partial B}{\partial \alpha}N_{\beta\alpha} - \frac{\partial A}{\partial \beta}N_\alpha + \frac{AB}{R_\beta}Q_\beta + AB\hat{q}_\beta = 0 \quad (2.24)$$

$$-\frac{AB}{R_\alpha}N_\alpha - \frac{AB}{R_\beta}N_\beta + \frac{\partial}{\partial \alpha}(BQ_\alpha) + \frac{\partial}{\partial \beta}(AQ_\beta) + AB\hat{q}_n = 0$$

and

$$\frac{\partial}{\partial \alpha}(BM_\alpha) + \frac{\partial}{\partial \beta}(AM_{\beta\alpha}) + \frac{\partial A}{\partial \beta}M_{\alpha\beta} - \frac{\partial B}{\partial \alpha}M_\beta - ABQ_\alpha + AB\hat{m}_\beta = 0$$

$$\frac{\partial}{\partial \beta}(AM_\beta) + \frac{\partial}{\partial \alpha}(BM_{\alpha\beta}) + \frac{\partial B}{\partial \alpha}M_{\beta\alpha} - \frac{\partial A}{\partial \beta}M_\alpha - ABQ_\beta + AB\hat{m}_\alpha = 0 \quad (2.25)$$

$$N_{\alpha\beta} - N_{\beta\alpha} + \frac{M_{\alpha\beta}}{R_\alpha} - \frac{M_{\beta\alpha}}{R_\beta} = 0 \ .$$

A more detailed derivation of these equations is given in the next section where
the specific case of a cylindrical thin shell is considered. It should be noted
that the equations (2.24) and (2.25) are widely accepted among those working
in thin shell theory.

2.1.7   Boundary Conditions

As discussed in [134], appropriate boundary conditions for elastic shells can be
determined both from Newtonian considerations (force and moment balancing) and from a Hamiltonian perspective (energy methods). We assume that
the boundaries lie along the coordinate curves $\alpha_1, \alpha_2, \beta_1$ and β_2. From [134,
page 27], appropriate boundary conditions at the edges α_1 and α_2 are

$$N_\alpha = 0 \qquad \text{or} \quad u = 0$$

$$\left(N_{\alpha\beta} + \frac{M_{\alpha\beta}}{R_\beta}\right) = 0 \quad \text{or} \quad v = 0$$

$$\left(Q_\alpha + \frac{1}{B}\frac{\partial M_{\alpha\beta}}{\partial \beta}\right) = 0 \quad \text{or} \quad w = 0 \qquad (2.26)$$

$$M_\alpha = 0 \qquad \text{or} \quad \theta_\alpha = 0$$

and $M_{\alpha\beta}w|_{\beta_1}^{\beta_2} = 0$. Note that if β is a closed curve (as will be the case with
a cylindrical shell), then this last condition is satisfied identically. Similarly,
appropriate boundary conditions at the edges β_1 and β_2 are

$$\left(N_{\beta\alpha} + \frac{M_{\beta\alpha}}{R_\alpha}\right) = 0 \quad \text{or} \quad u = 0$$

$$N_\beta = 0 \qquad \text{or} \quad v = 0$$

$$\left(Q_\beta + \frac{1}{B}\frac{\partial M_{\beta\alpha}}{\partial \alpha}\right) = 0 \quad \text{or} \quad w = 0 \qquad (2.27)$$

$$M_\beta = 0 \qquad \text{or} \quad \theta_\beta = 0$$

and $M_{\beta\alpha}w|_{\alpha_1}^{\alpha_2} = 0$. A more detailed discussion of the boundary conditions
is given in the next section when the specific case of a cylindrical shell is
considered.

2.2 Equations for Specific Geometries

The discussion of the last section led to the derivation of equations of motion
for general shells in the coordinates α, β with Lamé parameters A, B and radii
R_α, R_β. Through various choices for these parameters, dynamic equations
describing the motion of cylindrical shells, plates and beams will be obtained.
Moreover, the strong forms of the equations that were derived in the last section
through force and moment balancing will be compared with corresponding
weak forms obtained through energy principles. Finally, the thin plate model
derived under the assumption of the four Love postulates will be compared with
the nonlinear von Kármán model which results when the second postulate is
relaxed and the thick plate, Mindlin-Reissner model that is obtained when the
fourth postulate is relaxed.

2.2.1 Cylindrical Shell Equations – Strong Form

We first consider the case in which the thin shell is cylindrical with length ℓ
and a midsurface radius of R as shown in Figure 2.4. If the axial direction is
taken to be along the x-axis, then one can take

$$\begin{aligned}
\alpha &= x & \beta &= \theta \\
A &= 1 & B &= R \\
R_\alpha &= \infty & R_\beta &= R
\end{aligned} \qquad (2.28)$$

in the equations of the previous sections. Note that the choices in (2.28) differ
slightly from those given in [134, page 32] but yield equivalent final results.

From (2.8)-(2.10) and (2.15), it follows that the total strains and corresponding stresses at an arbitrary point on the cylindrical shell are given by

$$e_x = (\varepsilon_x + z\kappa_x) , \qquad\qquad \sigma_x = \frac{E}{1 - \nu^2}(e_x + \nu e_\theta)$$

$$e_\theta = \frac{1}{(1 + z/R)}(\varepsilon_\theta + z\kappa_\theta) , \qquad \sigma_\theta = \frac{E}{1 - \nu^2}(e_\theta + \nu e_x) \qquad (2.29)$$

$$\gamma_{x\theta} = \frac{1}{(1 + z/R)}\left[\varepsilon_{x\theta} + z\left(1 + \frac{z}{2R}\right)\tau\right] , \quad \sigma_{x\theta} = \sigma_{\theta x} = \frac{E}{2(1 + \nu)}\gamma_{x\theta} .$$

Again, it is noted that within the framework of infinitesimal elasticity, the
strain equations in (2.29) are exact and in the Byrne-Flügge-Lur'ye shell theory, these represent the exact form of the kinematic equations. In the Donnell-
Mushtari theory, these strain relations are approximated by neglecting the
underlined terms z/R with respect to unity.

In terms of the axial, circumferential and radial displacements u, v and w,
respectively, the midsurface strains and changes in curvature for the cylindrical

shell are

$$\varepsilon_x = \frac{\partial u}{\partial x} \quad , \qquad\qquad \kappa_x = -\frac{\partial^2 w}{\partial x^2}$$

$$\varepsilon_\theta = \frac{1}{R}\frac{\partial v}{\partial \theta} + \frac{w}{R} \quad , \qquad \kappa_\theta = -\frac{1}{R^2}\frac{\partial^2 w}{\partial \theta^2} + \frac{1}{R^2}\frac{\partial v}{\partial \theta} \qquad (2.30)$$

$$\varepsilon_{x\theta} = \frac{\partial v}{\partial x} + \frac{1}{R}\frac{\partial u}{\partial \theta} \quad , \qquad \tau = -\frac{2}{R}\frac{\partial^2 w}{\partial x \partial \theta} + \frac{2}{R}\frac{\partial v}{\partial x} \ .$$

By substituting the results from (2.30) into (2.23) and applying the relationships in (2.28), the force and moment resultants for a cylindrical shell are found to be

$$N_x = \frac{Eh}{(1-\nu^2)}\left[\frac{\partial u}{\partial x} + \nu\left(\frac{1}{R}\frac{\partial v}{\partial \theta} + \frac{w}{R}\right) - \underline{\frac{h^2}{12R}\frac{\partial^2 w}{\partial x^2}}\right]$$

$$N_\theta = \frac{Eh}{(1-\nu^2)}\left[\frac{1}{R}\frac{\partial v}{\partial \theta} + \frac{w}{R} + \nu\frac{\partial u}{\partial x} + \underline{\frac{h^2}{12R^3}\left(\frac{\partial^2 w}{\partial \theta^2} + w\right)}\right]$$

$$N_{x\theta} = \frac{Eh}{2(1+\nu)}\left[\frac{\partial v}{\partial x} + \frac{1}{R}\frac{\partial u}{\partial \theta} + \underline{\frac{h^2}{12R^2}\left(-\frac{\partial^2 w}{\partial x \partial \theta} + \frac{\partial v}{\partial x}\right)}\right]$$

$$N_{\theta x} = \frac{Eh}{2(1+\nu)}\left[\frac{\partial v}{\partial x} + \frac{1}{R}\frac{\partial u}{\partial \theta} + \underline{\frac{h^2}{12R^2}\left(\frac{\partial^2 w}{\partial x \partial \theta} + \frac{1}{R}\frac{\partial u}{\partial \theta}\right)}\right] \qquad (2.31)$$

$$M_x = \frac{-Eh^3}{12(1-\nu^2)}\left[\frac{\partial^2 w}{\partial x^2} + \frac{\nu}{R^2}\frac{\partial^2 w}{\partial \theta^2} - \underline{\frac{\nu}{R^2}\frac{\partial v}{\partial \theta}} - \underline{\frac{1}{R}\frac{\partial u}{\partial x}}\right]$$

$$M_\theta = \frac{-Eh^3}{12(1-\nu^2)}\left[\frac{1}{R^2}\frac{\partial^2 w}{\partial \theta^2} + \nu\frac{\partial^2 w}{\partial x^2} + \frac{w}{R^2}\right]$$

$$M_{x\theta} = \frac{-Eh^3}{12R(1+\nu)}\left[\frac{\partial^2 w}{\partial x \partial \theta} - \underline{\frac{\partial v}{\partial x}}\right]$$

$$M_{\theta x} = \frac{-Eh^3}{12R(1+\nu)}\left[\frac{\partial^2 w}{\partial x \partial \theta} + \frac{1}{2}\left(-\frac{\partial v}{\partial x} + \underline{\frac{1}{R}\frac{\partial u}{\partial \theta}}\right)\right] \ .$$

The underlined terms are again retained in the Byrne, Flügge and Lur'ye theory and discarded in the Donnell-Mushtari theory.

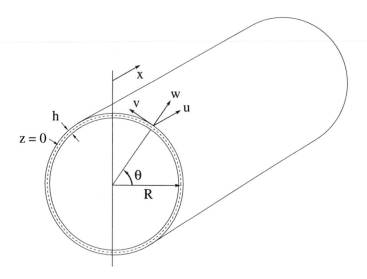

Figure 2.4: The cylindrical thin shell.

For the force and moment orientation shown in Figure 2.5, the equations of equilibrium can be derived as follows.

Force Balancing

Considering first the equilibrium of all forces in the x direction, it follows that

$$\left(N_x + \frac{\partial N_x}{\partial x}dx\right)Rd\theta - N_xRd\theta + \left(N_{\theta x} + \frac{\partial N_{\theta x}}{\partial \theta}d\theta\right)dx - N_{\theta x}dx + \hat{q}_x Rd\theta dx = 0$$

which implies that

$$R\frac{\partial N_x}{\partial x} + \frac{\partial N_{\theta x}}{\partial \theta} + R\hat{q}_x = 0 . \tag{2.32}$$

The equilibrium equations in the θ and z directions can be derived similarly and are given by

$$\frac{\partial N_\theta}{\partial \theta} + R\frac{N_{x\theta}}{\partial x} + Q_\theta + R\hat{q}_\theta = 0 \tag{2.33}$$

and

$$R\frac{\partial Q_x}{\partial x} + \frac{\partial Q_\theta}{\partial \theta} - N_\theta + R\hat{q}_n = 0 . \tag{2.34}$$

We point out that in the Donnell-Mushtari theory, the term Q_θ is considered to be negligible in (2.33) and is neglected in the final equations of motion.

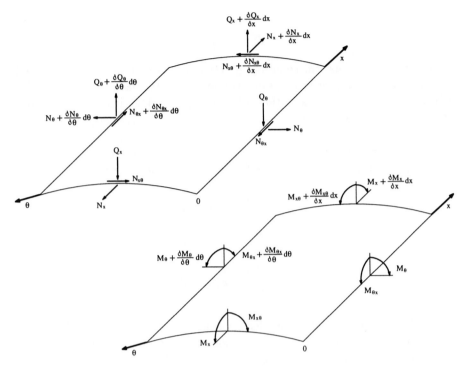

Figure 2.5: Force and moment resultants for the cylindrical shell.

Moment Balancing

With 0 as a reference origin, the balancing of moments with respect to θ yields

$$\left(M_x + \frac{\partial M_x}{\partial x}dx\right)Rd\theta - M_xRd\theta - \left(Q_x + \frac{\partial Q_x}{\partial x}dx\right)Rd\theta dx$$

$$+ \left(M_{\theta x} + \frac{\partial M_{\theta x}}{\partial \theta}d\theta\right)dx - M_{\theta x}dx + Q_\theta \frac{dx}{2}dx - \left(Q_\theta + \frac{\partial Q_\theta}{\partial \theta}d\theta\right)dx\frac{dx}{2}$$

$$+ \hat{q}_n dx Rd\theta \frac{dx}{2} + \hat{m}_\theta Rd\theta dx = 0 \ .$$

When the terms $(dx)^2$ are neglected in accordance with Love's second assumption, the equilibrium equation can be written as

$$R\frac{\partial M_x}{\partial x} + \frac{\partial M_{\theta x}}{\partial \theta} - RQ_x + R\hat{m}_\theta = 0 \ . \tag{2.35}$$

Similarly, the balancing of moments with respect to x and z yields

$$\frac{\partial M_\theta}{\partial \theta} + R\frac{\partial M_{x\theta}}{\partial x} - RQ_\theta + R\hat{m}_x = 0 \tag{2.36}$$

and

$$N_{x\theta} - N_{\theta x} - \frac{M_{\theta x}}{R} = 0 \; , \tag{2.37}$$

respectively. Note that the equations (2.32)–(2.37) are the same as those found in (2.24) and (2.25) once the substitutions in (2.28) have been made. By referring to the integral definitions of $N_{x\theta}, N_{\theta x}$ and $M_{\theta x}$, it can be seen that the expression (2.37) is identically satisfied due to the symmetry of the stress tensor.

Time enters the equilibrium equations through the inertial terms; hence for time dependent problems the body forces $\hat{q}_x$, $\hat{q}_\theta$ and $\hat{q}_n$ are replaced by

$$-\rho h \frac{\partial^2 u}{\partial t^2} + \hat{q}_x(t, x, \theta)$$

$$-\rho h \frac{\partial^2 v}{\partial t^2} + \hat{q}_\theta(t, x, \theta)$$

$$-\rho h \frac{\partial^2 w}{\partial t^2} + \hat{q}_n(t, x, \theta)$$

respectively, where ρ is the mass density (mass per unit volume). When these temporal terms are combined with equations (2.32)–(2.37), the time-dependent Byrne-Flügge-Lur'ye equations for the vibrations of a thin cylindrical shell are given by

$$R\rho h \frac{\partial^2 u}{\partial t^2} - R\frac{\partial N_x}{\partial x} - \frac{\partial N_{\theta x}}{\partial \theta} = R\hat{q}_x$$

$$R\rho h \frac{\partial^2 v}{\partial t^2} - \frac{\partial N_\theta}{\partial \theta} - R\frac{N_{x\theta}}{\partial x} - \frac{1}{R}\frac{\partial M_\theta}{\partial \theta} - \frac{\partial M_{x\theta}}{\partial x} = R\hat{q}_\theta + \hat{m}_x$$

$$R\rho h \frac{\partial^2 w}{\partial t^2} - R\frac{\partial^2 M_x}{\partial x^2} - \frac{1}{R}\frac{\partial^2 M_\theta}{\partial \theta^2} - \frac{\partial^2 M_{x\theta}}{\partial x \partial \theta} - \frac{\partial^2 M_{\theta x}}{\partial x \partial \theta} + N_\theta$$

$$= R\hat{q}_n + R\frac{\partial \hat{m}_\theta}{\partial x} + \frac{\partial \hat{m}_x}{\partial \theta} \; . \tag{2.38}$$

We note that the representation of the external loads as surface moments and forces is convenient when deriving the strong form of the equations of motion. However, in many applications where it is necessary to actually determine expressions for these loads or when using the weak form of the equations, it is advantageous to represent these loads in terms of line forces and moments. To accomplish this, let $\hat{M}_x$, $\hat{M}_\theta$, $\hat{N}_x$, and $\hat{N}_\theta$ denote the external resultants acting on the edge of an infinitesimal element which have the same orientation as the internal resultants depicted in Figure 2.5 (with units of moment and force per unit length of middle surface). Force and moment balancing can be used to write the area moments and in-plane forces in terms of these line moments and forces, thus yielding

$$\hat{q}_x = -\frac{\partial \hat{N}_x}{\partial x} \quad , \quad \hat{q}_\theta = -\frac{1}{R}\frac{\partial \hat{N}_\theta}{\partial \theta}$$

$$\hat{m}_x = -\frac{1}{R}\frac{\partial \hat{M}_\theta}{\partial \theta} \quad , \quad \hat{m}_\theta = -\frac{\partial \hat{M}_x}{\partial x} . \tag{2.39}$$

We point out that the first expression in (2.39) can be obtained from (2.32) simply by replacing N_x by $\hat{N}_x$ and deleting the term $\frac{\partial N_{\theta x}}{\partial \theta}$. Similar analysis leads to the other expressions in (2.39). The use of these line moments and forces in (2.38) is equivalent to including the external resultants directly when determining the equations of moment and force equilibrium for an infinitesimal shell element as done in (2.32)–(2.37).

The substitution of the internal moments and forces in (2.31) and the external resultants from (2.39) then yields

$$\frac{1}{C_L^2}\frac{\partial^2 u}{\partial t^2} - \frac{\partial^2 u}{\partial x^2} - \frac{1-\nu}{2R^2}\frac{\partial^2 u}{\partial \theta^2} - \frac{1+\nu}{2R}\frac{\partial^2 v}{\partial x \partial \theta} - \frac{\nu}{R}\frac{\partial w}{\partial x}$$

$$+ \frac{h^2}{12R^2}\left[R\frac{\partial^3 w}{\partial x^3} - \frac{1-\nu}{2R}\frac{\partial^3 w}{\partial x \partial \theta^2} - \frac{1-\nu}{2R^2}\frac{\partial^2 u}{\partial \theta^2}\right] = -\frac{(1-\nu^2)}{Eh}\frac{\partial \hat{N}_x}{\partial x}$$

$$\frac{1}{C_L^2}\frac{\partial^2 v}{\partial t^2} - \frac{1-\nu}{2}\frac{\partial^2 v}{\partial x^2} - \frac{1}{R^2}\frac{\partial^2 v}{\partial \theta^2} - \frac{1+\nu}{2R}\frac{\partial^2 u}{\partial x \partial \theta} - \frac{1}{R^2}\frac{\partial w}{\partial \theta}$$

$$+ \frac{h^2}{12R^2}\left[\frac{-3(1-\nu)}{2}\frac{\partial^2 v}{\partial x^2} + \frac{3-\nu}{2}\frac{\partial^3 w}{\partial x^2 \partial \theta}\right]$$

$$= -\frac{(1-\nu^2)}{EhR}\frac{\partial \hat{N}_\theta}{\partial \theta} - \frac{(1-\nu^2)}{EhR^2}\frac{\partial \hat{M}_\theta}{\partial \theta} \tag{2.40}$$

$$\frac{1}{C_L^2}\frac{\partial^2 w}{\partial t^2} + \frac{\nu}{R}\frac{\partial u}{\partial x} + \frac{1}{R^2}\frac{\partial v}{\partial \theta} + \frac{1}{R^2}w + \frac{h^2}{12}\nabla^4 w$$

$$+ \frac{h^2}{12R^2}\left[-R\frac{\partial^3 u}{\partial x^3} + \frac{1-\nu}{2R}\frac{\partial^3 u}{\partial x \partial \theta^2} - \frac{3-\nu}{2}\frac{\partial^3 v}{\partial x^2 \partial \theta}\right.$$

$$\left. + \frac{1}{R^2}\left(w + 2\frac{\partial^2 w}{\partial \theta^2}\right)\right] = \frac{(1-\nu^2)}{Eh}\left[\hat{q}_n - \frac{1}{R^2}\frac{\partial^2 \hat{M}_\theta}{\partial \theta^2} - \frac{\partial \hat{M}_x}{\partial x}\right]$$

where $\nabla^4 = \nabla^2 \nabla^2$ with $\nabla^2 \equiv \frac{\partial^2}{\partial x^2} + \frac{1}{R^2}\frac{\partial^2}{\partial \theta^2}$. The constant C_L given by

$$C_L = \left[\frac{E}{\rho(1-\nu^2)}\right]^{\frac{1}{2}}$$

is the phase speed of axial waves in the cylinder wall (see [135]). We reiterate that the external line forces $\hat{N}_x$ and $\hat{N}_\theta$ and moments $\hat{M}_x$ and $\hat{M}_\theta$ have units of force and moment per unit length of middle surface, respectively. As discussed in the next chapter, one means of generating force and moment resultants in this form is through the excitation of piezoceramic patches which are bonded to the shell. The load $\hat{q}_n$ is left as a surface force since this is the form that it usually takes in problems involving the excitation of a shell through normal forces (an example of a normal force in this form is the pressure exerted on the shell due to an exterior or interior acoustic field).

We again emphasize that the resultant expressions in (2.39) (and hence the system (2.40)) were derived for an infinitesimal element; hence certain modifications must be made when considering the *global* form of the resultants and equations (as is necessary when the resultants are generated by a piezoceramic patch). In certain cases (e.g., for certain types of moments and forces), the system (2.40) agrees with the strong form of the global shell equations. In general, however, this is not true, and one must exercise extreme care in determining the form of the global representations for the moments and forces. This point is discussed below and in [44].

As shown in [120, pages 202-203], the system (2.40) can be partially decoupled in terms of the displacements w, u and v thus yielding an eighth order equation for w and fourth order equations for u and v. Given a known loading, one first solves for the radial displacement w and then determines u and v by solving the fourth order equations which depend upon w as well as the external load.

Appropriate boundary conditions can be determined by using the fact that the work performed by the shell responses at the ends must be zero (this follows from energy considerations). The work done during deformation at the point $x = 0$ is given by

$$W = \int_0^{2\pi} \left[N_x u + N_{x\theta} v + Q_x w + M_{x\theta} \theta_\theta + M_x \theta_x \right] R d\theta$$

where the rotations of the normal to the middle surface are

$$\theta_x = -\frac{\partial w}{\partial x} \quad , \quad \theta_\theta = \frac{v}{R} - \frac{1}{R} \frac{\partial w}{\partial \theta}$$

from (2.7) (see [134] for details). Enforcing the condition that no work be done along the boundary and integrating by parts in the obvious terms, we find

$$\int_0^{2\pi} \left[N_x u + N_{x\theta} v + Q_x w + M_{x\theta} \left(\frac{v}{R} - \frac{1}{R} \frac{\partial w}{\partial \theta} \right) - M_x \frac{\partial w}{\partial x} \right] R d\theta = 0$$

$$\Rightarrow \int_0^{2\pi} \left[N_x u + N_{x\theta} v + Q_x w + \frac{v}{R} M_{x\theta} + \frac{w}{R} \frac{\partial M_{x\theta}}{\partial \theta} - M_x \frac{\partial w}{\partial x} \right] R d\theta = 0$$

$$\Rightarrow \int_0^{2\pi} \left[N_x u + \left(N_{x\theta} + \frac{M_{x\theta}}{R} \right) v + \left(Q_x + \frac{1}{R} \frac{\partial M_{x\theta}}{\partial \theta} \right) w - M_x \frac{\partial w}{\partial x} \right] R d\theta = 0 \ .$$

This equation is satisfied if the integrand is set to zero thus yielding the boundary conditions

$$N_x = 0 \qquad \text{or} \quad u = 0$$

$$\left(N_{x\theta} + \frac{M_{x\theta}}{R} \right) = 0 \quad \text{or} \quad v = 0$$

$$\left(Q_x + \frac{1}{R} \frac{\partial M_{x\theta}}{\partial \theta} \right) = 0 \quad \text{or} \quad w = 0$$

$$M_x = 0 \qquad \text{or} \quad \frac{\partial w}{\partial x} = 0$$

at $x = 0$ (see (2.26)). Similar conditions exist at the other end of the shell at $x = \ell$. Some common boundary conditions arising in various settings are

(a) Clamped edge:

$$u = v = w = \frac{\partial w}{\partial x} = 0$$

(b) Free edge:

$$N_x = \left(N_{x\theta} + \frac{M_{x\theta}}{R} \right) = \left(Q_x + \frac{1}{R} \frac{\partial M_{x\theta}}{\partial \theta} \right) = M_x = 0$$

(c) Simply supported edge, not free to move:

$$u = v = w = M_x = 0$$

(d) Simply supported edge, free to move in the x direction:

$$v = w = M_x = N_x = 0 .$$

The condition (d) is also known as a *shear diaphragm* condition, and it is a reasonable approximation to the situation in physical applications where a specially designed thin circular cover is fitted to the end of the shell. This cover prevents movement in the v and w directions but is designed to be flexible enough in the x direction so that it generates only a negligible moment M_x and longitudinal force N_x. This boundary condition is especially popular when analyzing shell structures since the solutions to the shell equations subject to this boundary condition are easily determined (see [134, 141]). In many experimental situations, however, it is perhaps more appropriate to use either condition (a) or (c) since the end-plates are often anchored and very rigid. For example, this might be the case when enforcing hard wall end conditions in coupled structural acoustics problems.

2.2.2   Kelvin-Voigt Damping in the Cylindrical Shell Equations

A Kelvin-Voigt type of internal damping (stress is proportional to a linear combination of strain and strain rate) can be incorporated into the shell model by using the constitutive relations (2.17) when determining the force and moment resultants. In the Donnell-Mushtari theory, this yields the resultants

$$N_x = \frac{Eh}{(1-\nu^2)}\left[\frac{\partial u}{\partial x} + \nu\left(\frac{1}{R}\frac{\partial v}{\partial \theta} + \frac{w}{R}\right)\right] + \frac{c_D h}{(1-\nu^2)}\frac{\partial}{\partial t}\left[\frac{\partial u}{\partial x} + \nu\left(\frac{1}{R}\frac{\partial v}{\partial \theta} + \frac{w}{R}\right)\right]$$

$$N_\theta = \frac{Eh}{(1-\nu^2)}\left[\frac{1}{R}\frac{\partial v}{\partial \theta} + \frac{w}{R} + \nu\frac{\partial u}{\partial x}\right] + \frac{c_D h}{(1-\nu^2)}\frac{\partial}{\partial t}\left[\frac{1}{R}\frac{\partial v}{\partial \theta} + \frac{w}{R} + \nu\frac{\partial u}{\partial x}\right]$$

$$N_{x\theta} = N_{\theta x} = \frac{Eh}{2(1+\nu)}\left[\frac{\partial v}{\partial x} + \frac{1}{R}\frac{\partial u}{\partial \theta}\right] + \frac{c_D h}{2(1+\nu)}\frac{\partial}{\partial t}\left[\frac{\partial v}{\partial x} + \frac{1}{R}\frac{\partial u}{\partial \theta}\right]$$

$$M_x = \frac{-Eh^3}{12(1-\nu^2)}\left[\frac{\partial^2 w}{\partial x^2} + \frac{\nu}{R^2}\frac{\partial^2 w}{\partial \theta^2}\right] - \frac{c_D h^3}{12(1-\nu^2)}\frac{\partial}{\partial t}\left[\frac{\partial^2 w}{\partial x^2} + \frac{\nu}{R^2}\frac{\partial^2 w}{\partial \theta^2}\right]$$

$$M_\theta = \frac{-Eh^3}{12(1-\nu^2)}\left[\frac{1}{R^2}\frac{\partial^2 w}{\partial \theta^2} + \nu\frac{\partial^2 w}{\partial x^2}\right] - \frac{c_D h^3}{12(1-\nu^2)}\frac{\partial}{\partial t}\left[\frac{1}{R^2}\frac{\partial^2 w}{\partial \theta^2} + \nu\frac{\partial^2 w}{\partial x^2}\right]$$

$$M_{x\theta} = M_{\theta x} = \frac{-Eh^3}{12R(1+\nu)}\frac{\partial^2 w}{\partial x \partial \theta} - \frac{c_D h^3}{12R(1+\nu)}\frac{\partial}{\partial t}\left[\frac{\partial^2 w}{\partial x \partial \theta}\right]$$

(compare with the Donnell-Mushtari moment and force terms in (2.31)). Analogous expressions arise in the long shell theory of Byrne, Flügge and Lur'ye. These moment and force terms can then be substituted into (2.38) to yield a damped form of the Donnell-Mushtari shell equations. This type of analysis can also be directly extended to higher order theories by using the constitutive relations (2.17) when determining the higher order force and moment resultants.

2.2.3   Weak Form of the Cylindrical Shell Equations

In order to find the weak form of the shell equations, the kinetic and strain energies of the shell are needed. By combining the Kirchhoff shell hypothesis with the strain results from classical elasticity theory, it follows that the strain energy stored in the shell during deformation is given by

$$U = \frac{1}{2}\int_{-h/2}^{h/2}\int_0^{2\pi}\int_0^\ell \left(\sigma_x e_x + \sigma_\theta e_\theta + \sigma_{x\theta}\gamma_{x\theta}\right)(1+z/R)\,R\,dx\,d\theta\,dz$$

with stresses and strains given by (2.29). Recall that the underlined terms in the strain expressions of (2.29) are retained in the Byrne, Flügge and Lur'ye theory and are discarded in the Donnell-Mushtari theory. Substitution and integration (with $(1 + z/R)^{-1}$ replaced by its geometric series expansion and neglecting powers of z in the integrand which are greater than two) yields

$$U = \frac{1}{2} \int_0^{2\pi} \int_0^{\ell} \frac{Eh}{(1-\nu^2)} \left\{ \left[(\varepsilon_x + \varepsilon_\theta)^2 - 2(1-\nu) \left(\varepsilon_x \varepsilon_\theta - \frac{\varepsilon_{x\theta}^2}{4} \right) \right] \right.$$

$$+ \frac{h^2}{12} \left[(\kappa_x + \kappa_\theta)^2 - 2(1-\nu) \left(\kappa_x \kappa_\theta - \frac{\tau^2}{4} \right) + \frac{2}{R} (\varepsilon_x \kappa_x - \varepsilon_\theta \kappa_\theta) \right.$$

$$\left. \left. - \frac{(1-\nu)}{2} \frac{\varepsilon_{x\theta}}{R} \tau + \frac{\varepsilon_\theta^2}{R^2} + \frac{(1-\nu)}{2} \frac{\varepsilon_{x\theta}^2}{R^2} \right] \right\} R dx d\theta .$$

Expressions relating the midsurface strains and changes in curvature to the displacements u, v and w are given in (2.30).

With the change of variables $s = x/R$, the total strain energy can be written as

$$U = \frac{1}{2} \int_0^{2\pi} \int_0^{\ell/R} \frac{Eh}{(1-\nu^2)} [I_{DM} + k I_{BFL}] \, ds \, d\theta$$

where $k = h^2/(12R^2)$, I_{DM} is the integrand corresponding to the Donnell-Mushtari theory and I_{BFL} denotes the terms which are retained to yield the Byrne, Flügge and Lur'ye strain energy. These two components are given by

$$I_{DM} = \left(\frac{\partial u}{\partial s} + \frac{\partial v}{\partial \theta} + w \right)^2$$

$$- 2(1-\nu) \left[\frac{\partial u}{\partial s} \left(\frac{\partial v}{\partial \theta} + w \right) - \frac{1}{4} \left(\frac{\partial v}{\partial s} + \frac{\partial u}{\partial \theta} \right)^2 \right] \qquad (2.41)$$

$$+ k \left\{ \left(\widetilde{\nabla}^2 w \right)^2 - 2(1-\nu) \left[\frac{\partial^2 w}{\partial s^2} \frac{\partial^2 w}{\partial \theta^2} - \left(\frac{\partial^2 w}{\partial s \partial \theta} \right)^2 \right] \right\}$$

and

$$I_{BFL} = -2\nu \frac{\partial v}{\partial \theta} \frac{\partial^2 w}{\partial s^2} - 3(1-\nu) \frac{\partial v}{\partial s} \frac{\partial^2 w}{\partial s \partial \theta} + \frac{3}{2}(1-\nu) \left(\frac{\partial v}{\partial s} \right)^2$$

$$\qquad (2.42)$$

$$+ (1-\nu) \frac{\partial u}{\partial \theta} \frac{\partial^2 w}{\partial s \partial \theta} + \frac{1}{2}(1-\nu) \left(\frac{\partial u}{\partial \theta} \right)^2 - 2 \frac{\partial u}{\partial s} \frac{\partial^2 w}{\partial s^2} + 2w \frac{\partial^2 w}{\partial \theta^2} + w^2$$

where $\widetilde{\nabla}^2 \equiv \frac{\partial^2}{\partial s^2} + \frac{\partial^2}{\partial \theta^2}$ (hence $\widetilde{\nabla}^2 = R^2 \nabla^2$).

The kinetic energy of the shell is given by

$$T = \frac{1}{2} \int_0^{2\pi} \int_0^{\ell/R} \rho h \left[\left(\frac{\partial u}{\partial t} \right)^2 + \left(\frac{\partial v}{\partial t} \right)^2 + \left(\frac{\partial w}{\partial t} \right)^2 \right] R^2 \, ds \, d\theta .$$

Throughout this development, it is assumed that the shell satisfies *shear diaphragm* boundary conditions at $x = 0, \ell$; that is, it is assumed that

$$v = w = N_x = M_x = 0$$

at the ends. This is done merely to demonstrate the equivalence between the weak form which follows and the strong form already discussed; other boundary conditions can be treated with similar arguments. It should be noted that the conditions $v = w = 0$ at the ends are essential boundary conditions and hence must be enforced on the chosen state space.

For an arbitrary time interval $[t_0, t_1]$, consider the action integral

$$A[\vec{u}] = \int_{t_0}^{t_1} (T - U) dt$$

where $\vec{u} = [u, v, w]$ is considered in the space $V = H_b^1(\Omega) \times H_b^1(\Omega) \times H_b^2(\Omega)$. Here Ω denotes the shell and the subscript b denotes the set of functions satisfying the essential boundary conditions. One then considers variations of the form

$$\hat{u} = \vec{u} + \varepsilon \vec{\Phi} = \begin{bmatrix} u(t, r, \theta, x) \\ v(t, r, \theta, x) \\ w(t, r, \theta, x) \end{bmatrix} + \varepsilon \begin{bmatrix} \eta_1(t)\phi_1(r, \theta, x) \\ \eta_2(t)\phi_2(r, \theta, x) \\ \eta_3(t)\phi_3(r, \theta, x) \end{bmatrix}.$$

Here $\vec{\eta} = [\eta_1, \eta_2, \eta_3]$ and $\vec{\phi} = [\phi_1, \phi_2, \phi_3]$ are chosen so that

(i) $\hat{u}(t, \cdot, \cdot, \cdot) \in V$

(ii) $\hat{u}(t_0, \cdot, \cdot, \cdot) = \hat{u}(t_1, \cdot, \cdot, \cdot)$.

Note that this enforces $\vec{\eta} \in [H^2(0, T)]^3$, $\vec{\eta}(t_0) = \vec{\eta}(t_1)$ and $\vec{\phi} \in V$.

For simplicity of presentation, details concerning the derivation of a weak form of the shell equations will be developed using the Donnell-Mushtari strain expression. A corresponding set of equations can be derived in a similar manner using the Byrne, Flügge and Lur'ye expression, and we summarize the resulting weak form of the equations of motion for this case at the end of the section.

Hamilton's principle states that the motion of the shell must give a stationary value to the action integral when compared to variations in the motion, thus leading to the requirement that for all $\vec{\Phi}$,

$$\frac{d}{d\varepsilon} A\left[\vec{u} + \varepsilon \vec{\Phi}\right]\bigg|_{\varepsilon=0} = 0 \ .$$

Since the action integral for the Donnell-Mushtari case has the form

$$A[\hat{u}] = \int_{t_0}^{t_1} \left\{ \frac{1}{2} \int_0^{2\pi} \int_0^{\ell/R} \rho h \left[\left(\frac{\partial}{\partial t}(u + \varepsilon \eta_1 \phi_1) \right)^2 + \left(\frac{\partial}{\partial t}(v + \varepsilon \eta_2 \phi_2) \right)^2 \right. \right.$$

$$+ \left. \left(\frac{\partial}{\partial t}(w + \varepsilon \eta_3 \phi_3) \right)^2 \right] R^2 \, ds \, d\theta$$

$$- \frac{1}{2} \int_0^{2\pi} \int_0^{\ell/R} \frac{Eh}{(1 - \nu^2)} \left[\left(\frac{\partial}{\partial s}(u + \varepsilon \eta_1 \phi_1) + \frac{\partial}{\partial \theta}(v + \varepsilon \eta_2 \phi_2) + (w + \varepsilon \eta_3 \phi_3) \right)^2 \right.$$

$$- 2(1 - \nu) \left[\frac{\partial}{\partial s}(u + \varepsilon \eta_1 \phi_1) \left(\frac{\partial}{\partial \theta}(v + \varepsilon \eta_2 \phi_2) + (w + \varepsilon \eta_3 \phi_3) \right) \right.$$

$$\left. - \frac{1}{4} \left(\frac{\partial}{\partial s}(v + \varepsilon \eta_2 \phi_2) + \frac{\partial}{\partial \theta}(u + \varepsilon \eta_1 \phi_1) \right)^2 \right]$$

$$+ k \left\{ \left(\tilde{\nabla}^2 (w + \varepsilon \eta_3 \phi_3) \right)^2 \right.$$

$$- 2(1 - \nu) \left[\frac{\partial^2}{\partial s^2}(w + \varepsilon \eta_3 \phi_3) \frac{\partial^2}{\partial \theta^2}(w + \varepsilon \eta_3 \phi_3) \right.$$

$$\left. \left. \left. - \left(\frac{\partial^2}{\partial s \partial \theta}(w + \varepsilon \eta_3 \phi_3) \right)^2 \right] \right\} \right] ds \, d\theta \right\} dt \ ,$$

Hamilton's principle then leads to the condition

$$0 = \left. \frac{\partial}{\partial \varepsilon} A[\hat{u}] \right|_{\varepsilon = 0}$$

$$= \int_{t_0}^{t_1} \int_0^{2\pi} \int_0^{\ell/R} \rho h \left[\frac{\partial u}{\partial t} \frac{\partial \eta_1}{\partial t} \phi_1 + \frac{\partial v}{\partial t} \frac{\partial \eta_2}{\partial t} \phi_2 + \frac{\partial w}{\partial t} \frac{\partial \eta_3}{\partial t} \phi_3 \right] R^2 \, ds \, d\theta \, dt$$

$$- \int_{t_0}^{t_1} \int_0^{2\pi} \int_0^{\ell/R} \frac{Eh}{(1 - \nu^2)} \left\{ \left(\frac{\partial u}{\partial s} + \frac{\partial v}{\partial \theta} + w \right) \left(\eta_1 \frac{\partial \phi_1}{\partial s} + \eta_2 \frac{\partial \phi_2}{\partial \theta} + \eta_3 \phi_3 \right) \right.$$

$$- (1 - \nu) \left[\eta_1 w \frac{\partial \phi_1}{\partial s} + \eta_1 \frac{\partial v}{\partial \theta} \frac{\partial \phi_1}{\partial s} + \eta_2 \frac{\partial u}{\partial s} \frac{\partial \phi_2}{\partial \theta} + \eta_3 \frac{\partial u}{\partial s} \phi_3 \right.$$

$$\left. - \frac{1}{2} \left(\frac{\partial v}{\partial s} + \frac{\partial u}{\partial \theta} \right) \left(\eta_2 \frac{\partial \phi_2}{\partial s} + \eta_1 \frac{\partial \phi_1}{\partial \theta} \right) \right]$$

$$+ k \left\{ \eta_3 \tilde{\nabla}^2 w \tilde{\nabla}^2 \phi_3 - (1 - \nu) \left[\eta_3 \frac{\partial^2 w}{\partial \theta^2} \frac{\partial^2 \phi_3}{\partial s^2} + \eta_3 \frac{\partial^2 w}{\partial s^2} \frac{\partial^2 \phi_3}{\partial \theta^2} \right. \right.$$

$$\left. \left. \left. - 2\eta_3 \frac{\partial^2 w}{\partial s \partial \theta} \frac{\partial^2 \phi_3}{\partial s \partial \theta} \right] \right\} \right\} ds \, d\theta \, dt \ .$$

Note that this must hold for all arbitrary intervals $[t_0, t_1]$ and all admissible perturbations. When integrated by parts, the first integral yields

$$\int_{t_0}^{t_1} \int_0^{2\pi} \int_0^{\ell/R} \rho h \left[\frac{\partial u}{\partial t} \frac{\partial \eta_1}{\partial t} \phi_1 + \frac{\partial v}{\partial t} \frac{\partial \eta_2}{\partial t} \phi_2 + \frac{\partial w}{\partial t} \frac{\partial \eta_3}{\partial t} \phi_3 \right] R^2 \, ds \, d\theta \, dt$$

$$= -\int_{t_0}^{t_1} \int_0^{2\pi} \int_0^{\ell/R} \rho h \left[\frac{\partial^2 u}{\partial t^2} \eta_1 \phi_1 + \frac{\partial^2 v}{\partial t^2} \eta_2 \phi_2 + \frac{\partial^2 w}{\partial t^2} \eta_3 \phi_3 \right] R^2 \, ds \, d\theta \, dt$$

since $\vec{\eta}(t_0) = \vec{\eta}(t_1)$. This yields the coupled system of equations

$$\int_{t_0}^{t_1} \eta_1(t) \int_0^{2\pi} \int_0^{\ell/R} \left\{ -\frac{\rho(1-\nu^2)}{E} \frac{\partial^2 u}{\partial t^2} \phi_1 R^2 - \left(\frac{\partial u}{\partial s} + \frac{\partial v}{\partial \theta} + w \right) \frac{\partial \phi_1}{\partial s} \right.$$

$$\left. +(1-\nu)\left[w \frac{\partial \phi_1}{\partial s} + \frac{\partial v}{\partial \theta} \frac{\partial \phi_1}{\partial s} - \frac{1}{2} \left(\frac{\partial v}{\partial s} + \frac{\partial u}{\partial \theta} \right) \frac{\partial \phi_1}{\partial \theta} \right] \right\} ds \, d\theta \, dt = 0$$

$$\int_{t_0}^{t_1} \eta_2(t) \int_0^{2\pi} \int_0^{\ell/R} \left\{ -\frac{\rho(1-\nu^2)}{E} \frac{\partial^2 v}{\partial t^2} \phi_2 R^2 - \left(\frac{\partial u}{\partial s} + \frac{\partial v}{\partial \theta} + w \right) \frac{\partial \phi_2}{\partial \theta} \right.$$

$$\left. +(1-\nu)\left[\frac{\partial u}{\partial s} \frac{\partial \phi_2}{\partial \theta} - \frac{1}{2} \left(\frac{\partial v}{\partial s} + \frac{\partial u}{\partial \theta} \right) \frac{\partial \phi_2}{\partial s} \right] \right\} ds \, d\theta \, dt = 0$$

$$\int_{t_0}^{t_1} \eta_3(t) \int_0^{2\pi} \int_0^{\ell/R} \left\{ -\frac{\rho(1-\nu^2)}{E} \frac{\partial^2 w}{\partial t^2} \phi_3 R^2 - \left(\frac{\partial u}{\partial s} + \frac{\partial v}{\partial \theta} + w \right) \phi_3 + (1-\nu)\frac{\partial u}{\partial s}\phi_3 \right.$$

$$\left. -k\left\{ \tilde{\nabla}^2 w \tilde{\nabla}^2 \phi_3 - (1-\nu)\left[\frac{\partial^2 w}{\partial \theta^2} \frac{\partial^2 \phi_3}{\partial s^2} + \frac{\partial^2 w}{\partial s^2} \frac{\partial^2 \phi_3}{\partial \theta^2} - 2\frac{\partial^2 w}{\partial s \partial \theta} \frac{\partial^2 \phi_3}{\partial s \partial \theta} \right] \right\} \right\} ds \, d\theta \, dt = 0 \,.$$

The weak form of the equations of motion for the unforced shell is thus

$$\int_0^{2\pi} \int_0^{\ell/R} \left\{ \frac{R^2}{C_L^2} \frac{\partial^2 u}{\partial t^2} \phi_1 + \left(\frac{\partial u}{\partial s} + \nu \frac{\partial v}{\partial \theta} + \nu w \right) \frac{\partial \phi_1}{\partial s} \right.$$

$$\left. + \frac{1}{2}(1-\nu)\left(\frac{\partial v}{\partial s} + \frac{\partial u}{\partial \theta} \right) \frac{\partial \phi_1}{\partial \theta} \right\} ds \, d\theta = 0$$

$$\int_0^{2\pi} \int_0^{\ell/R} \left\{ \frac{R^2}{C_L^2} \frac{\partial^2 v}{\partial t^2} \phi_2 + \left(\nu \frac{\partial u}{\partial s} + \frac{\partial v}{\partial \theta} + w \right) \frac{\partial \phi_2}{\partial \theta} \right.$$

$$\left. + \frac{1}{2}(1-\nu)\left(\frac{\partial v}{\partial s} + \frac{\partial u}{\partial \theta} \right) \frac{\partial \phi_2}{\partial s} \right\} ds \, d\theta = 0$$

$$\int_0^{2\pi} \int_0^{\ell/R} \left\{ \frac{R^2}{C_L^2} \frac{\partial^2 w}{\partial t^2} \phi_3 + \left(\nu \frac{\partial u}{\partial s} + \frac{\partial v}{\partial \theta} + w \right) \phi_3 + k\left\{ \tilde{\nabla}^2 w \tilde{\nabla}^2 \phi_3 \right. \right.$$

$$\left. \left. -(1-\nu)\left[\frac{\partial^2 w}{\partial \theta^2} \frac{\partial^2 \phi_3}{\partial s^2} + \frac{\partial^2 w}{\partial s^2} \frac{\partial^2 \phi_3}{\partial \theta^2} - 2\frac{\partial^2 w}{\partial s \partial \theta} \frac{\partial^2 \phi_3}{\partial s \partial \theta} \right] \right\} \right\} ds \, d\theta = 0$$

for all $\vec{\phi} \in V$. Again, the constant $C_L = \left[\frac{E}{\rho(1-\nu^2)}\right]^{1/2}$ is the phase speed of axial waves in the cylinder wall.

In terms of the moment and force resultants (see the Donnell-Mushtari terms in (2.31)) and the original axial variable x, the weak form of the Donnell-Mushtari shell equations is

$$\int_0^{2\pi} \int_0^\ell \left\{ R\rho h \frac{\partial^2 u}{\partial t^2} \phi_1 + RN_x \frac{\partial \phi_1}{\partial x} + N_{\theta x} \frac{\partial \phi_1}{\partial \theta} \right\} dx d\theta = 0$$

$$\int_0^{2\pi} \int_0^\ell \left\{ R\rho h \frac{\partial^2 v}{\partial t^2} \phi_2 + N_\theta \frac{\partial \phi_2}{\partial \theta} + RN_{x\theta} \frac{\partial \phi_2}{\partial x} \right\} dx d\theta = 0 \tag{2.43}$$

$$\int_0^{2\pi} \int_0^\ell \left\{ R\rho h \frac{\partial^2 w}{\partial t^2} \phi_3 + N_\theta \phi_3 - RM_x \frac{\partial^2 \phi_3}{\partial x^2} \right.$$

$$\left. - \frac{1}{R} M_\theta \frac{\partial^2 \phi_3}{\partial \theta^2} - 2M_{x\theta} \frac{\partial^2 \phi_3}{\partial x \partial \theta} \right\} dx d\theta = 0 \; .$$

The derivation thus far has been for the unforced shell. To include the contributions of applied external forces and moments which do nonconservative work on the shell, one can appeal to an extended form of Hamilton's principle or more formally include these contributions directly in the system (2.43). Both techniques yield identical final equations and for ease of presentation, we will take the latter approach.

The inclusion of the applied line forces and moments $\hat{N}_x, \hat{N}_\theta$ and $\hat{M}_x, \hat{M}_\theta$ and the surface load $\hat{q}_n$ in the system then yields

$$\int_0^{2\pi} \int_0^\ell \left\{ R\rho h \frac{\partial^2 u}{\partial t^2} \phi_1 + RN_x \frac{\partial \phi_1}{\partial x} + N_{\theta x} \frac{\partial \phi_1}{\partial \theta} - R\hat{N}_x \frac{\partial \phi_1}{\partial x} \right\} dx d\theta = 0$$

$$\int_0^{2\pi} \int_0^\ell \left\{ R\rho h \frac{\partial^2 v}{\partial t^2} \phi_2 + N_\theta \frac{\partial \phi_2}{\partial \theta} + RN_{x\theta} \frac{\partial \phi_2}{\partial x} - \hat{N}_\theta \frac{\partial \phi_2}{\partial \theta} \right\} dx d\theta = 0 \tag{2.44}$$

$$\int_0^{2\pi} \int_0^\ell \left\{ R\rho h \frac{\partial^2 w}{\partial t^2} \phi_3 + N_\theta \phi_3 - RM_x \frac{\partial^2 \phi_3}{\partial x^2} - \frac{1}{R} M_\theta \frac{\partial^2 \phi_3}{\partial \theta^2} \right.$$

$$\left. - 2M_{x\theta} \frac{\partial^2 \phi_3}{\partial x \partial \theta} - R\hat{q}_n \phi_3 + R\hat{M}_x \frac{\partial^2 \phi_3}{\partial x^2} + \frac{1}{R} \hat{M}_\theta \frac{\partial^2 \phi_3}{\partial \theta^2} \right\} dx d\theta = 0$$

for all $\vec{\phi} \in V$ as the weak form of the Donnell-Mushtari equations for the forced shell.

With the assumption of sufficient smoothness, the weak solution in this form is consistent with the strong solution in (2.38). The vanishing of several of the boundary terms which arise during integration by parts is a result of the choice $V = H_b^1(\Omega) \times H_b^1(\Omega) \times H_b^2(\Omega)$ for the function space since the state variables and test functions are required to satisfy the essential boundary conditions

$$v = w = 0$$

at $x = 0, \ell$. It can also be noted that these boundary terms will also vanish with the enforcement of other essential boundary conditions such as the *fixed end* conditions

$$u = v = w = \frac{\partial w}{\partial x} = 0$$

at $x = 0, \ell$.

We point out that in the weak form (2.44), one is not required to differentiate the applied force and moment resultants $\hat{N}_x, \hat{N}_\theta, \hat{M}_x$ and $\hat{M}_\theta$ as is required in the strong form (2.40). This proves to be very beneficial when these terms are generated by finite piezoceramic patches as discussed in the next section.

2.2.4   Accuracy of the Donnell-Mushtari Equations

A discussion concerning the accuracy of the Donnell-Mushtari equations is given in [120, pages 221-229] as well as in [104, 145]. In these references, results obtained with the Donnell-Mushtari theory are compared to those obtained via the theory of Byrne-Flügge-Lur'ye (which is considered to be perhaps the most accurate of the shell theories) as well as results obtained via the Morley theory which is developed by retaining only selected Byrne-Flügge-Lur'ye terms in the final displacement equations. Two of the Morley equations are identical to those of the Donnell-Mushtari theory with the third containing more of the Byrne-Flügge-Lur'ye contributions. The general conclusion for cylindrical shells under distributed loads is that the Donnell-Mushtari theory is accurate for short shells but that the Morley or another more accurate theory should be used for shells which are long in comparison to the radius. As discussed in [104], the ratio of length to radius for which the Donnell-Mushtari equations are accurate is somewhat dependent on the type of loading but in general, Hoff concludes that the use of the Donnell equations is not recommended unless the cylinder is relatively short. It should be noted most of the shell theories give marginal results for very short shells due to the end effects. In this case, one usually must resort back to general elasticity theory in order to obtain accurate results.

As a final note concerning the accuracy of the thin shell theories, we mention briefly the consistency of the theories in describing the dynamics of a shell which is subjected to rigid body translations. As discussed in [120, 134], the equations of Byrne, Flügge and Lur'ye are consistent with regard to rigid body motions (no changes are introduced in the strain-displacement equations), whereas changes in curvature and twist occur in the Donnell-Mushtari theory when a rigid body translation is applied to the shell. This inconsistency is eliminated in the Sander's theory (in which the equations of motion are derived from the principle of virtual work), and hence this latter theory is preferable to that of Donnell and Mushtari in some settings. Like the Donnell-Mushtari theory, however, the Sander's shell theory is first order in nature and in some situations is less accurate than the higher order Byrne-Flügge-Lur'ye theory. A more detailed discussion concerning the merits and internal consistency of the various thin shell theories can be found in [82, 120, 134, 141].

2.2.5 Plate Equations

In this section, plate models derived under a variety of assumptions and in rectangular and circular geometries are considered. For the first model, Love's four hypotheses are retained and the previously discussed shell results are used to describe the longitudinal and transverse motion of a rectangular plate through appropriate choices for Lamé parameters and radii. This yields a model in which the longitudinal and transverse vibrations of a uniform homogeneous plate are decoupled and includes the well known Love-Kirchhoff plate model. To obtain a more accurate model which includes the coupling between in-plane and out-of-plane vibrations, Love's second postulate is relaxed and moments and force are considered in the deformed state rather than along the neutral surface as done previously. We briefly outline this second approach and compare the resulting nonlinear von Kármán equation with that obtained under the assumption of all four postulates. Thirdly, the Mindlin-Reissner model for thick rectangular plates is briefly discussed. Here the fourth Love postulate is relaxed to allow for rotational effects and shear deformation. The final model considered here describes the transverse motion of a thin circular plate. In this case, geometrical coupling when balancing moments and forces leads to equations which differ from those describing rectangular plate dynamics.

Thin Rectangular Plates – Uncoupled Motion

The equations for a rectangular plate which result from the previously described shell equations (derived under the assumption of Love's four postulates) are described first. With the axial direction being taken along the x-axis, the choices

$$\alpha = x \qquad \beta = y$$
$$A = 1 \qquad B = 1$$
$$R_\alpha = \infty \qquad R_\beta = \infty$$

for Lamé constants and radii can be made in the general shell equations in the first section. From (2.8)–(2.10) and (2.15), it follows directly that the strain-displacement equations and stress-strain relations are given by

$$e_x = \varepsilon_x + z\kappa_x \quad , \qquad \sigma_x = \frac{E}{1-\nu^2}(e_x + \nu e_y)$$

$$e_y = \varepsilon_y + z\kappa_y \quad , \qquad \sigma_y = \frac{E}{1-\nu^2}(e_y + \nu e_x) \qquad (2.45)$$

$$\gamma_{xy} = \varepsilon_{xy} + z\tau \quad , \qquad \sigma_{xy} = \frac{E}{2(1+\nu)}\gamma_{xy}$$

where

$$\varepsilon_x = \frac{\partial u}{\partial x} \quad , \qquad \kappa_x = -\frac{\partial^2 w}{\partial x^2}$$

$$\varepsilon_y = \frac{\partial v}{\partial y} \quad , \qquad \kappa_y = -\frac{\partial^2 w}{\partial y^2} \qquad (2.46)$$

$$\varepsilon_{xy} = \frac{\partial v}{\partial x} + \frac{\partial u}{\partial y} \quad , \qquad \tau = -2\frac{\partial^2 w}{\partial x \partial y} \; .$$

The combination of the constitutive equations and the resultant expressions

$$\begin{bmatrix} N_x \\ N_{xy} \end{bmatrix} = \int_{-h/2}^{h/2} \begin{bmatrix} \sigma_x \\ \sigma_{xy} \end{bmatrix} dz \quad , \qquad \begin{bmatrix} M_x \\ M_{xy} \end{bmatrix} = \int_{-h/2}^{h/2} \begin{bmatrix} \sigma_x \\ \sigma_{xy} \end{bmatrix} z\,dz$$

$$\begin{bmatrix} N_y \\ N_{yx} \end{bmatrix} = \int_{-h/2}^{h/2} \begin{bmatrix} \sigma_y \\ \sigma_{yx} \end{bmatrix} dz \quad , \qquad \begin{bmatrix} M_y \\ M_{yx} \end{bmatrix} = \int_{-h/2}^{h/2} \begin{bmatrix} \sigma_y \\ \sigma_{yx} \end{bmatrix} z\,dz$$

then yields

$$N_x = \frac{Eh}{(1-\nu^2)}(\varepsilon_x + \nu\varepsilon_y) \quad , \qquad M_x = \frac{Eh^3}{12(1-\nu^2)}(\kappa_x + \nu\kappa_y)$$

$$N_y = \frac{Eh}{(1-\nu^2)}(\varepsilon_y + \nu\varepsilon_x) \quad , \qquad M_y = \frac{Eh^3}{12(1-\nu^2)}(\kappa_y + \nu\kappa_x) \qquad (2.47)$$

$$N_{xy} = N_{yx} = \frac{Eh}{2(1+\nu)}\varepsilon_{xy} \quad , \qquad M_{xy} = M_{yx} = \frac{Eh^3}{24(1+\nu)}\tau \; .$$

Moreover, force and moment balancing equations in a manner analogous to that summarized in (2.24) and (2.25) and detailed in the discussion of thin cylindrical shells yields the relations

$$\rho h \frac{\partial^2 u}{\partial t^2} = \frac{\partial N_x}{\partial x} + \frac{\partial N_{yx}}{\partial y} + \hat{q}_x$$

$$\rho h \frac{\partial^2 v}{\partial t^2} = \frac{\partial N_y}{\partial y} + \frac{N_{xy}}{\partial x} + \hat{q}_y \qquad (2.48)$$

$$\rho h \frac{\partial^2 w}{\partial t^2} = \frac{\partial Q_x}{\partial x} + \frac{\partial Q_y}{\partial y} + \hat{q}_n$$

and

$$\frac{\partial M_x}{\partial x} + \frac{\partial M_{yx}}{\partial y} - Q_x + \hat{m}_y = 0$$

$$\frac{\partial M_y}{\partial y} + \frac{\partial M_{xy}}{\partial x} - Q_y + \hat{m}_x = 0$$

(2.49)

where again, $\vec{q} = \hat{q}_x \hat{i}_x + \hat{q}_y \hat{i}_y + \hat{q}_n \hat{i}_n$ and $\vec{m} = \hat{m}_x \hat{i}_x + \hat{m}_y \hat{i}_y$ denote the forces and moments due to an external field which is considered to be acting on the *undeformed* middle surface. Hence $\vec{q}$ and $\vec{m}$ have units of force and moment per unit area, respectively.

Eliminating Q_x and Q_y in (2.48) and (2.49) yields the three equilibrium equations

$$\rho h \frac{\partial^2 u}{\partial t^2} - \frac{\partial N_x}{\partial x} - \frac{\partial N_{yx}}{\partial y} = \hat{q}_x$$

$$\rho h \frac{\partial^2 v}{\partial t^2} - \frac{\partial N_y}{\partial y} - \frac{N_{xy}}{\partial x} = \hat{q}_y$$

(2.50)

$$\rho h \frac{\partial^2 w}{\partial t^2} - \frac{\partial^2 M_x}{\partial x^2} - \frac{\partial^2 M_y}{\partial y^2} - \frac{\partial^2 M_{xy}}{\partial x \partial y} - \frac{\partial^2 M_{yx}}{\partial y \partial x} = \hat{q}_n + \frac{\partial \hat{m}_x}{\partial y} + \frac{\partial \hat{m}_y}{\partial x}$$

with the moment and force resultants given by (2.47). In the case of a plate with constant material parameters, the equations of motion can be written as

$$\frac{1}{C_L^2} \frac{\partial^2 u}{\partial t^2} - \frac{\partial^2 u}{\partial x^2} - \frac{1}{2}(1-\nu)\frac{\partial^2 u}{\partial y^2} - \frac{1}{2}(1+\nu)\frac{\partial^2 v}{\partial x \partial y} = \frac{1-\nu^2}{Eh}\hat{q}_x$$

$$\frac{1}{C_L^2} \frac{\partial^2 v}{\partial t^2} - \frac{\partial^2 v}{\partial y^2} - \frac{1}{2}(1-\nu)\frac{\partial^2 v}{\partial x^2} - \frac{1}{2}(1+\nu)\frac{\partial^2 u}{\partial x \partial y} = \frac{1-\nu^2}{Eh}\hat{q}_y$$

$$\frac{1}{C_L^2} \frac{\partial^2 w}{\partial t^2} + \frac{h^2}{12}\left(\frac{\partial^4 w}{\partial x^4} + 2\frac{\partial^4 w}{\partial x^2 \partial y^2} + \frac{\partial^4 w}{\partial y^4}\right)$$

$$= \frac{1-\nu^2}{Eh}\left[\hat{q}_n + \frac{\partial \hat{m}_x}{\partial y} + \frac{\partial \hat{m}_y}{\partial x}\right]$$

(2.51)

with $C_L = \left[\frac{E}{\rho(1-\nu^2)}\right]^{\frac{1}{2}}$. The first two equations describe the longitudinal movement of the plate (see [123, 125, 139]). The transverse (third) equation is the familiar equation describing the bending of a thin plate. Note that internal damping may be incorporated by using a more general constitutive law such as the Kelvin-Voigt relation in (2.17). We also point out that, under the assumption of all four Love postulates, the equations modeling longitudinal and transverse displacement of a uniform, homogeneous plate are uncoupled even though there is coupling in the physical plate. This issue is pursued in the next subsection where the nonlinear von Kármán plate model is discussed.

To find the weak form of the equations, the vector $\vec{u} = [u, v, w]$ containing the displacements in the x, y and normal directions is considered in the space

$V = H_b^1(\Omega) \times H_b^1(\Omega) \times H_b^2(\Omega)$ where Ω denotes the plate and the subscript b denotes the set of functions satisfying essential boundary conditions for a specific problem. By using analysis similar to that just described for cylindrical shells, the weak form of the equations of motion for the plate can be found to be

$$\int_0^a \int_0^\ell \left\{ \rho h \frac{\partial^2 u}{\partial t^2} \phi_1 + N_x \frac{\partial \phi_1}{\partial x} + N_{yx} \frac{\partial \phi_1}{\partial y} - \hat{N}_x \frac{\partial \phi_1}{\partial x} \right\} dx dy = 0$$

$$\int_0^a \int_0^\ell \left\{ \rho h \frac{\partial^2 v}{\partial t^2} \phi_2 + N_y \frac{\partial \phi_2}{\partial y} + N_{xy} \frac{\partial \phi_2}{\partial x} - \hat{N}_y \frac{\partial \phi_2}{\partial y} \right\} dx dy = 0$$

(2.52)

$$\int_0^a \int_0^\ell \left\{ \rho h \frac{\partial^2 w}{\partial t^2} \phi_3 - M_x \frac{\partial^2 \phi_3}{\partial x^2} - M_y \frac{\partial^2 \phi_3}{\partial y^2} - M_{xy} \frac{\partial^2 \phi_3}{\partial x \partial y} - M_{yx} \frac{\partial^2 \phi_3}{\partial x \partial y} \right.$$

$$\left. - \hat{q}_n \phi_3 + \hat{M}_x \frac{\partial^2 \phi_3}{\partial x^2} + \hat{M}_y \frac{\partial^2 \phi_3}{\partial y^2} \right\} dx dy = 0$$

for all $\vec{\phi} = [\phi_1, \phi_2, \phi_3] \in V$. As in the case of the thin shell, the external line forces and moments $\hat{N}_x, \hat{N}_y, \hat{M}_x$ and $\hat{M}_y$ are related in an infinitesimal sense to the corresponding area forces and moments $\hat{q}_x, \hat{q}_y, \hat{m}_y$ and $\hat{m}_x$ appearing in the strong form of the equations by the relations

$$\hat{q}_x = -\frac{\partial \hat{N}_x}{\partial x} \quad , \quad \hat{q}_y = -\frac{\partial \hat{N}_y}{\partial y}$$

(2.53)

$$\hat{m}_x = -\frac{\partial \hat{M}_y}{\partial y} \quad , \quad \hat{m}_y = -\frac{\partial \hat{M}_x}{\partial x} \ .$$

If the solution has sufficient smoothness, integration by parts can be used to show that the weak solution is consistent with the strong solution in (2.50).

von Kármán Plate Equations

The balancing of force and moments leading to the equilibrium equations (2.48) and (2.49) was performed under the assumption of Love's second postulate; thus all resultants were considered in the *undeformed* middle surface. For plates that are subjected to both in-plane and out-of-plane forces (and thus exhibit both longitudinal and transverse vibrations) or plates in which deformations are large with respect to the thickness, it is common to relax the assumption of linearity and consider forces and moments in the *deformed* middle surface. As depicted in Figure 2.6, deformation of the middle surface leads to rotations of the four corners of the infinitesimal element. Under the assumption of small displacements, the *sine* of these rotation angles can be approximated by changes in slopes.

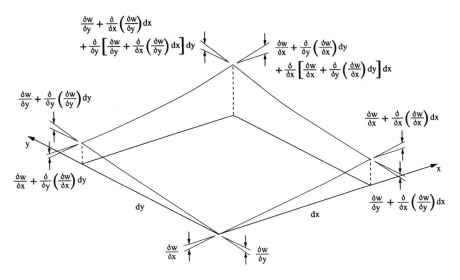

Figure 2.6: Deformed middle surface of the thin plate.

The balancing of transverse forces yields the *nonlinear* relations

$$\rho h \frac{\partial^2 w}{\partial t^2} = \frac{\partial Q_x}{\partial x} + \frac{\partial Q_y}{\partial y} + \frac{\partial}{\partial x}\left(N_x \frac{\partial w}{\partial x}\right) + \frac{\partial}{\partial y}\left(N_y \frac{\partial w}{\partial y}\right)$$
$$+ \frac{\partial}{\partial x}\left(N_{xy} \frac{\partial w}{\partial y}\right) + \frac{\partial}{\partial y}\left(N_{xy} \frac{\partial w}{\partial x}\right) + \hat{q}_n \tag{2.54}$$

when third order differential terms are neglected (see [133, 174] for details). Similarly, force balancing in the x and y directions yields

$$\rho h \frac{\partial^2 u}{\partial t^2} = \frac{\partial N_x}{\partial x} + \frac{\partial N_{xy}}{\partial y} - \frac{\partial}{\partial x}\left(Q_x \frac{\partial w}{\partial x}\right) - \frac{\partial}{\partial y}\left(Q_y \frac{\partial w}{\partial x}\right) + \hat{q}_x$$
$$\rho h \frac{\partial^2 v}{\partial t^2} = \frac{\partial N_y}{\partial y} + \frac{\partial N_{xy}}{\partial x} - \frac{\partial}{\partial y}\left(Q_y \frac{\partial w}{\partial y}\right) - \frac{\partial}{\partial x}\left(Q_x \frac{\partial w}{\partial y}\right) + \hat{q}_y \ . \tag{2.55}$$

It can be noted that if the in-plane forces are small as compared with transverse forces and slopes are small, the equilibrium equation (2.54) reduces to those in (2.48). Similarly, the in-plane equations (2.55) reduce to those in (2.48) when transverse shearing forces are small relative to in-plane forces.

The balancing of moments for an infinitesimal element yields

$$\frac{\rho h^3}{12} \frac{\partial^3 w}{\partial x \partial t^2} = Q_x - \frac{\partial M_x}{\partial x} - \frac{\partial M_{yx}}{\partial y} - \hat{m}_y$$
$$\frac{\rho h^3}{12} \frac{\partial^3 w}{\partial y \partial t^2} = Q_y - \frac{\partial M_y}{\partial y} - \frac{\partial M_{xy}}{\partial x} - \hat{m}_x \tag{2.56}$$

when the effects of rotational inertia are retained and higher-order terms are neglected. It should be noted that in many applications involving the use of this model, the inertial contributions due to rotation are neglected, in which case, the moment relations (2.56) reduce to those considered in the linear theory (2.49).

To obtain the final equations of motion, the transverse shear terms are eliminated and force and moments resultants are replaced by appropriate expressions. In the case of large deformations, one retains quadratic terms in the strain-displacement equations

$$\varepsilon_x = \frac{\partial u}{\partial x} + \frac{1}{2}\left(\frac{\partial w}{\partial x}\right)^2 \quad , \quad \varepsilon_y = \frac{\partial v}{\partial y} + \frac{1}{2}\left(\frac{\partial w}{\partial y}\right)^2 \quad , \quad \varepsilon_{xy} = \frac{\partial v}{\partial x} + \frac{\partial u}{\partial y} + \frac{\partial w}{\partial x}\frac{\partial w}{\partial y}$$

which yields the force resultants

$$N_x = \frac{Eh}{1-\nu^2}\left[\frac{\partial u}{\partial x} + \nu\frac{\partial v}{\partial y} + \frac{1}{2}\left(\frac{\partial w}{\partial x}\right)^2 + \frac{\nu}{2}\left(\frac{\partial w}{\partial y}\right)^2\right]$$

$$N_y = \frac{Eh}{1-\nu^2}\left[\frac{\partial v}{\partial y} + \nu\frac{\partial u}{\partial x} + \frac{1}{2}\left(\frac{\partial w}{\partial y}\right)^2 + \frac{\nu}{2}\left(\frac{\partial w}{\partial x}\right)^2\right] \qquad (2.57)$$

$$N_{xy} = \frac{Eh}{2(1+\nu)}\left[\frac{\partial u}{\partial y} + \frac{\partial v}{\partial x} + \frac{\partial w}{\partial x}\frac{\partial w}{\partial y}\right]$$

(compare with the linear expressions in (2.46) and (2.47)). Typically the shear forces are assumed to be small in the longitudinal equilibrium equations, in which case, longitudinal motion is described by the u and v equations in (2.50) with N_x, N_y and N_{xy} given in (2.57). In the case of constant material parameters, the plate bending is modeled by the equation

$$\rho h w_{tt} - \frac{\rho h^3}{12}\nabla^2 w_{tt} + \frac{Eh^3}{12(1-\nu^2)}\nabla^4 w$$

$$= \left(N_x\frac{\partial^2 w}{\partial x^2} + 2N_{xy}\frac{\partial^2 w}{\partial x\partial y} + N_y\frac{\partial^2 w}{\partial y^2}\right) + \hat{q}_n + \frac{\partial \hat{m}_x}{\partial y} + \frac{\partial \hat{m}_y}{\partial x} \qquad (2.58)$$

with N_x, N_y, N_{xy} defined in (2.57) and $\nabla^2 \equiv \frac{\partial^2}{\partial x^2} + \frac{\partial^2}{\partial y^2}, \nabla^4 = \nabla^2\nabla^2$. These nonlinear equations describing the longitudinal and transverse vibrations are typically referred to as the von Kármán system for a thin plate. We point out that the longitudinal and transverse equations are coupled through the products of the force resultants and the curvature components. This provides a nonlinear model which captures the coupling between the in-plane and out-of-plane vibrations of a uniform homogeneous plate (see [133] for further details concerning this model and [123] for a derivation of the equations using energy principles).

Mindlin-Reissner Plate Model

A third plate model under consideration includes the Mindlin-Reissner equations for transverse plate vibrations in conjunction with previously discussed equations modeling longitudinal motion. To obtain these equations, the first three Love postulates are assumed whereas the fourth is relaxed to permit transverse shear deformation and rotational effects (it is still assumed that the plate filaments remain straight and unstrained during deformation).

With $\alpha = x, A = 1, R_\alpha = \infty, \beta = y, B = 1$ and $R_\beta = \infty$, the general displacement relations in (2.6) can be written as

$$U(x,y,z) = u(x,y) + z\theta_x(x,y)$$
$$V(x,y,z) = v(x,y) + z\theta_y(x,y)$$
$$W(x,y,z) = w(x,y)$$

where θ_x and θ_y are rotations of the middle surface in the x and y directions. It follows immediately from (2.5) that the strain displacement relations are given by

$$e_x = \frac{\partial u}{\partial x} + z\frac{\partial \theta_x}{\partial x} \quad , \quad \gamma_{xz} = \frac{\partial w}{\partial x} + \theta_x$$

$$e_y = \frac{\partial v}{\partial y} + z\frac{\partial \theta_y}{\partial y} \quad , \quad \gamma_{yz} = \frac{\partial w}{\partial y} + \theta_y$$

$$\gamma_{xy} = \left(\frac{\partial v}{\partial x} + \frac{\partial u}{\partial y}\right) + z\left(\frac{\partial \theta_y}{\partial x} + \frac{\partial \theta_x}{\partial y}\right).$$

It can be noted that if one assumes the absence of shear deformations, one obtains the relations $\theta_x = -\frac{\partial w}{\partial x}$ and $\theta_y = -\frac{\partial w}{\partial y}$ from (2.7) which lead to the previous expressions in (2.45) and (2.46). Moreover, it can be seen that while Love's fourth postulate leads to vanishing shear strains γ_{xz} and γ_{yz}, these strains are nonzero in the model which results when this postulate is relaxed to allow shear deformations. This eliminates one of the contradictions which arise under the assumption of all four postulates (see [134]).

Expressions for the moment and force resultants are obtained in the usual manner by integrating stresses through the thickness of the plate. With the notation

$$D = \frac{Eh^3}{12(1-\nu^2)} \quad , \quad G = \frac{E}{2(1+\nu)} ,$$

this yields the expressions

$$M_x = D\left(\frac{\partial \theta_x}{\partial x} + \nu\frac{\partial \theta_y}{\partial y}\right) , \qquad N_x = \frac{Eh}{1-\nu^2}\left(\frac{\partial u}{\partial x} + \nu\frac{\partial v}{\partial y}\right)$$

$$M_y = D\left(\frac{\partial \theta_y}{\partial y} + \nu\frac{\partial \theta_x}{\partial x}\right) , \qquad N_y = \frac{Eh}{1-\nu^2}\left(\frac{\partial v}{\partial y} + \nu\frac{\partial u}{\partial x}\right)$$

$$M_{xy} = \frac{D(1-\nu)}{2}\left(\frac{\partial \theta_y}{\partial x} + \frac{\partial \theta_x}{\partial y}\right) , \qquad N_{xy} = \frac{Eh}{2(1+\nu)}\left(\frac{\partial u}{\partial y} + \frac{\partial v}{\partial x}\right)$$

$$Q_x = K^2 Gh\left(\frac{\partial w}{\partial x} + \theta_x\right) , \qquad Q_y = K^2 Gh\left(\frac{\partial w}{\partial y} + \theta_y\right).$$

The constant K^2 in the two shear expressions is used to compensate for the fact that the outer surfaces of the plate cannot support a shear stress. Hence an averaging technique is used to obtain values at the neutral surface (see [133] for details). Reissner proposed the choice 5/6 whereas $K^2 = 2/3$ is used if one assumes a parabolic shear stress distribution. In applications, this parameter can be treated as an unknown to be recovered when estimating physical parameters using fit-to-data techniques (this was the approach of Mindlin).

Force balancing yields expressions identical to those considered previously in (2.48), namely

$$\rho h \frac{\partial^2 u}{\partial t^2} = \frac{\partial N_x}{\partial x} + \frac{\partial N_{yx}}{\partial y} + \hat{q}_x$$

$$\rho h \frac{\partial^2 v}{\partial t^2} = \frac{\partial N_y}{\partial y} + \frac{N_{xy}}{\partial x} + \hat{q}_y \qquad (2.59)$$

$$\rho h \frac{\partial^2 w}{\partial t^2} = \frac{\partial Q_x}{\partial x} + \frac{\partial Q_y}{\partial y} + \hat{q}_n .$$

The effects of rotational inertia become significant for geometries and frequencies when shear deformation must be considered. The inclusion of these contributions when balancing moments yields

$$\frac{\rho h^3}{12} \frac{\partial^2 \theta_x}{\partial t^2} = -Q_x + \frac{\partial M_x}{\partial x} + \frac{\partial M_{xy}}{\partial y} + \hat{m}_y$$

$$\frac{\rho h^3}{12} \frac{\partial^2 \theta_y}{\partial t^2} = -Q_y + \frac{\partial M_y}{\partial y} + \frac{\partial M_{xy}}{\partial x} + \hat{m}_x .$$

From (2.59), it can be seen that the equations describing longitudinal motion are identical to those obtained previously under the assumption of all four Love postulates. In the case of constant physical parameters, the system of equations describing transverse motion are

$$\rho h \frac{\partial^2 w}{\partial t^2} = GK^2 h \left(\nabla^2 w + \frac{\partial \theta_x}{\partial x} + \frac{\partial \theta_y}{\partial y} \right) + \hat{q}_n$$

$$\frac{\rho h^3}{12} \frac{\partial^2 \theta_x}{\partial t^2} = -K^2 Gh \left(\frac{\partial w}{\partial x} + \theta_x \right)$$

$$+ \frac{D}{2} \left[(1 - \nu) \nabla^2 \theta_x + (1 + \nu) \frac{\partial}{\partial x} \left(\frac{\partial \theta_x}{\partial x} + \frac{\partial \theta_y}{\partial y} \right) \right] + \hat{m}_y$$

$$\frac{\rho h^3}{12} \frac{\partial^2 \theta_y}{\partial t^2} = -K^2 Gh \left(\frac{\partial w}{\partial y} + \theta_y \right)$$

$$+ \frac{D}{2} \left[(1 - \nu) \nabla^2 \theta_y + (1 + \nu) \frac{\partial}{\partial y} \left(\frac{\partial \theta_x}{\partial x} + \frac{\partial \theta_y}{\partial y} \right) \right] + \hat{m}_x .$$

As discussed in [133, 171], the inclusion of the rotational inertia increases the mass effect while the shear deformation tends to decrease the stiffness. Both effects will lower the structural frequencies, and in a variety of applications (including thick structures and systems with multiple frequencies), this model more accurately captures the physics of the structure than does the thin plate Kirchhoff model presented at the beginning of this section.

Thin Circular Plates – Transverse Motion

In the discussions regarding the previous three plate models, the plates were assumed to be rectangular. Here we consider a circular plate having radius a and thickness h. All four of Love's postulates are assumed to hold, and emphasis is placed on describing the equation which models the transverse plate motion since this is the case considered in parameter estimation and control examples of later chapters. Throughout the discussion, r and θ are used to denote the radial and angular coordinates, respectively. We point out that due to geometrical differences, the angular terms in the circular plate discussion differ significantly from those arising in the shell and curved beam analyses. A more reasonable comparison is to equate the y component in the rectangular plate with $r\theta$ in the present case.

Since the constitutive properties do not depend upon the geometry, a change of variables in the expressions for M_x and M_y of (2.47) and use of the Laplacian expression $\nabla^2 = \frac{\partial^2}{\partial r^2} + \frac{1}{r}\frac{\partial}{\partial r} + \frac{1}{r^2}\frac{\partial^2}{\partial \theta^2}$ for polar coordinates yields

$$M_r = -D\left(\frac{\partial^2 w}{\partial r^2} + \frac{\nu}{r}\frac{\partial w}{\partial r} + \frac{\nu}{r^2}\frac{\partial^2 w}{\partial \theta^2}\right)$$

$$M_\theta = -D\left(\frac{1}{r}\frac{\partial w}{\partial r} + \frac{1}{r^2}\frac{\partial^2 w}{\partial \theta^2} + \nu\frac{\partial^2 w}{\partial r^2}\right) \tag{2.60}$$

$$M_{r\theta} = M_{\theta r} = -D(1-\nu)\left(\frac{1}{r}\frac{\partial^2 w}{\partial r\partial \theta} - \frac{1}{r^2}\frac{\partial w}{\partial \theta}\right)$$

where again, $D = \frac{Eh^3}{12(1-\nu^2)}$. Here M_r and M_θ denote the internal radial and circumferential bending moments and $M_{r\theta}$ is the twisting moment. As was the case with the rectangular plate model, internal damping can be incorporated by using a more general constitutive law such as a Kelvin-Voigt relation of the form (2.17).

For this discussion, we will assume that the external load to the plate consists of a normal surface force $\hat{q}_n$ and applied line moments $\hat{M}_r$ and $\hat{M}_\theta$. The representation in terms of line moments rather than surface moments $\hat{m}_r$ and $\hat{m}_\theta$ facilitates the discussion in later chapters when the external load to the plate is generated by piezoceramic patches. Contributions due to applied twisting moments are not considered since the patches do not generate this kind of load.

As detailed in [37], balancing of moments with respect to θ and r yields

$$-M_r - r\frac{\partial M_r}{\partial r} + rQ_r + M_\theta + \frac{\partial M_{\theta r}}{\partial \theta} = -\hat{M}_r - r\frac{\partial \hat{M}_r}{\partial r} + \hat{M}_\theta$$

$$M_{r\theta} + r\frac{\partial M_{r\theta}}{\partial r} + \frac{\partial M_\theta}{\partial \theta} - M_{\theta r} - rQ_\theta = \frac{\partial \hat{M}_\theta}{\partial \theta} \tag{2.61}$$

respectively, when higher order terms are eliminated. Here Q_r and Q_θ denote the shear resultants with orientations similar to those in the rectangular case. Similarly, force balancing yields the dynamic equilibrium equation

$$\rho h r\frac{\partial^2 w}{\partial t^2} - Q_r - r\frac{\partial Q_r}{\partial r} - \frac{\partial Q_\theta}{\partial \theta} = r\hat{q}_n \ . \tag{2.62}$$

The synthesis of the moment and force relations (2.61) and (2.62) then yields the dynamic equation

$$\rho h\frac{\partial^2 w}{\partial t^2} - \frac{\partial^2 M_r}{\partial r^2} - \frac{2}{r}\frac{\partial M_r}{\partial r} + \frac{1}{r}\frac{\partial M_\theta}{\partial r} - \frac{2}{r}\frac{\partial^2 M_{r\theta}}{\partial r\partial\theta} - \frac{2}{r^2}\frac{\partial M_{r\theta}}{\partial \theta} - \frac{1}{r^2}\frac{\partial^2 M_\theta}{\partial\theta^2}$$

$$= \hat{q}(t,r,\theta) - \frac{\partial^2 \hat{M}_r}{\partial r^2} - \frac{2}{r}\frac{\partial \hat{M}_r}{\partial r} + \frac{1}{r}\frac{\partial \hat{M}_\theta}{\partial r} - \frac{1}{r^2}\frac{\partial^2 \hat{M}_\theta}{\partial\theta^2} \ , \tag{2.63}$$

with moments given by (2.60), as a model for the transverse vibrations of the thin circular plate. This expression can be compared with the final equation in (2.50) which models the transverse vibrations of a thin rectangular plate.

In the case of truly clamped edges, zero displacement and slope are maintained at the plate perimeter and the fixed-edge boundary conditions

$$w(t,a,\theta) = \frac{\partial w}{\partial r}(t,a,\theta) = 0 \tag{2.64}$$

accurately model the edge behavior of the plate.

In applications or experimental situations, however, perfect clamps modeled by fixed-edge boundary conditions usually are difficult to attain. This results in frequencies that are lower than predicted by perfectly clamped boundary models [126, 147, 159, 160]. As discussed in [45], one means of modeling small boundary displacements and rotations is through boundary moment conditions of the form

$$\frac{1}{a}M_r(t,a,\theta) + \frac{\partial M_r}{\partial r}(t,a,\theta)$$

$$= -k_t w(t,a,\theta) - c_t\frac{\partial w}{\partial t}(t,a,\theta) - \rho h\frac{\partial^2 w}{\partial t^2}(t,a,\theta) \tag{2.65}$$

$$M_r(t,a,\theta) = k_p\frac{\partial w}{\partial r}(t,a,\theta) + c_p\frac{\partial^2 w}{\partial r\partial t}(t,a,\theta) \ .$$

The parameters k_t and c_t denote stiffness and damping contributions when boundary displacement occurs with similar interpretations for k_p and c_p when accounting for boundary rotation. It can be noted that if one divides through by k_t and k_p and takes the limits $k_p \to \infty$ and $k_t \to \infty$, the imperfectly clamped boundary conditions converge to the fixed-edge conditions (2.64). This should be expected since increased stiffness in displacement and rotation implies that one is approaching an ideal clamp in which displacement and slope are truly zero. Advantages and limitations associated with using the "imperfectly clamped" boundary condition (2.65) to model the edge movement for a circular plate with slightly loosened boundary clamps can be found in [5].

The states, state spaces and weak forms of the modeling equations depend upon the assumed boundary conditions. For a plate having perfectly clamped edges and hence boundary conditions (2.64), the state for the problem is taken to be the transverse displacement w in the state space $H = L^2(\Omega)$ where $\Omega = \{(r, \theta) | 0 \leq r \leq a, 0 \leq \theta \leq 2\pi\}$ again denotes the region occupied by the unstrained neutral surface of the plate. Motivated by the energy considerations discussed in [37], we also define the space of test functions $V = H_b^2(\Omega) \equiv \{\eta \in H^2(\Omega) \mid \eta(a, \theta) = \frac{\partial \eta}{\partial r}(a, \theta) = 0\}$.

A weak or variational form of the equation describing the transverse motion of a thin circular plate having perfectly clamped edges is then

$$\int_\Omega \rho h \frac{\partial^2 w}{\partial t^2} \phi_3 d\omega - \int_\Omega M_r \frac{\partial^2 \phi_3}{\partial r^2} d\omega - \int_\Omega \frac{1}{r^2} M_\theta \left[r \frac{\partial \phi_3}{\partial r} + \frac{\partial^2 \phi_3}{\partial \theta^2} \right] d\omega$$

$$-2 \int_\Omega \frac{1}{r^2} M_{r\theta} \left[r \frac{\partial^2 \phi_3}{\partial r \partial \theta} - \frac{\partial \phi_3}{\partial \theta} \right] d\omega \qquad (2.66)$$

$$= - \int_\Omega \hat{M}_r \frac{\partial^2 \phi_3}{\partial r^2} d\omega - \int_\Omega \frac{1}{r^2} \hat{M}_\theta \left[r \frac{\partial \phi_3}{\partial r} + \frac{\partial^2 \phi_3}{\partial \theta^2} \right] d\omega + \int_\Omega \hat{q}_n \phi_3 d\omega$$

for all $\phi_3 \in V$. The differential here is $d\omega = r d\theta dr$.

If the "imperfectly clamped" boundary conditions (2.65) are assumed, the state is taken to be $z = (w(a, \cdot), w(\cdot, \cdot))$ in the state space $H \equiv L^2(0, 2\pi) \times L^2(\Omega)$ (see [45] for details). A suitable space for test functions is $V = \{\Psi = (\xi(\cdot), \eta(\cdot, \cdot)) \in H \mid \eta \in H^2(\Omega), \eta(a, \theta) = \xi(\theta)\}$. As detailed in [45], the usual integration by parts arguments (starting from the integrated strong form) then yields the weak form

$$\int_\Omega \rho h \frac{\partial^2 w}{\partial t^2} \phi_3 d\omega - \int_\Omega M_r \frac{\partial^2 \phi_3}{\partial r^2} d\omega - \int_\Omega \frac{1}{r^2} M_\theta \left[r \frac{\partial \phi_3}{\partial r} + \frac{\partial^2 \phi_3}{\partial \theta^2} \right] d\omega$$

$$-2 \int_\Omega \frac{1}{r^2} M_{r\theta} \left[r \frac{\partial^2 \phi_3}{\partial r \partial \theta} - \frac{\partial \phi_3}{\partial \theta} \right] d\omega$$

$$+ \int_0^{2\pi} a \left[k_t w(t,a,\theta) + c_t \frac{\partial w}{\partial t}(t,a,\theta) + \rho h \frac{\partial^2 w}{\partial t^2}(t,a,\theta) \right] \phi_3(a,\theta) d\theta \quad (2.67)$$

$$+ \int_0^{2\pi} a \left[k_p \frac{\partial w}{\partial r}(t,a,\theta) + c_p \frac{\partial^2 w}{\partial r \partial t}(t,a,\theta) \right] \frac{\partial \phi_3}{\partial r}(a,\theta) d\theta$$

$$= - \int_\Omega \hat{M}_r \frac{\partial^2 \phi_3}{\partial r^2} d\omega - \int_\Omega \frac{1}{r^2} \hat{M}_\theta \left[r \frac{\partial \phi_3}{\partial r} + \frac{\partial^2 \phi_3}{\partial \theta^2} \right] d\omega + \int_\Omega \hat{q}_n \phi_3 d\omega$$

for all $\Psi = (\phi_3(a,\cdot), \phi_3(\cdot,\cdot)) \in V$. The weak form of the modeling equations proves to be advantageous in many applications since the transfer of derivatives onto test functions lowers smoothness requirements and eliminates deleterious effects due to discontinuous parameters and input terms.

2.2.6 Beam Equations

Curved-Beam Equations

From the cylindrical shell equations discussed in the first section, one can easily obtain modeling equations of motion for a thin ring, curved beam or arch as depicted in Figure 2.7. In this case, motion in the axial direction (across the width) is assumed to be negligible with respect to that in the circumferential and transverse directions. Hence direct axial motion as well as Poisson effects are neglected.

Considering only vibrations in the v and w directions, the Donnell-Mushtari thin shell equations (2.38) reduce to

$$R\rho h b \frac{\partial^2 v}{\partial t^2} - b \frac{\partial N_\theta}{\partial \theta} = Rb\hat{q}_\theta$$
$$R\rho h b \frac{\partial^2 w}{\partial t^2} - \frac{b}{R} \frac{\partial^2 M_\theta}{\partial \theta^2} + bN_\theta = Rb\hat{q}_n + b \frac{\partial \hat{m}_x}{\partial \theta} \quad (2.68)$$

where

$$bN_\theta = Ehb \left(\frac{1}{R} \frac{\partial v}{\partial \theta} + \frac{w}{R} \right)$$
$$bM_\theta = \frac{-Eh^3 b}{12} \frac{1}{R^2} \frac{\partial^2 w}{\partial \theta^2} = -EI \frac{1}{R^2} \frac{\partial^2 w}{\partial \theta^2} ,$$

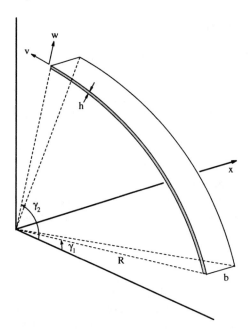

Figure 2.7: Curved beam subject to circumferential and transverse vibrations.

b denotes the width of the beam, and I is the moment of inertia. In the case of constant parameters, this reduces to

$$\rho h b \frac{\partial^2 v}{\partial t^2} - \frac{E h b}{R^2} \frac{\partial^2 v}{\partial \theta^2} - \frac{E h b}{R^2} \frac{\partial w}{\partial \theta} = b \hat{q}_\theta$$
$$\rho h b \frac{\partial^2 w}{\partial t^2} + \frac{E h b}{R^2} \frac{\partial v}{\partial \theta} + \frac{E h b}{R^2} w + \frac{E h^3 b}{12 R^4} \frac{\partial^4 w}{\partial \theta^4} = b \hat{q}_n + \frac{b}{R} \frac{\partial \hat{m}_x}{\partial \theta} \ .$$

(2.69)

We point out that (2.68) and (2.69) retain the geometric coupling between in-plane and out-of-plane vibrations due to curvature.

It was noted in the discussion concerning the accuracy of the Donnell-Mushtari shell equations that basic elasticity equations must be used when considering very short shells due to inaccuracies resulting from end effects. The behavior of the rings, arches and curved beams discussed here differs from that of the short shells presented previously both in the types of motion considered and in effects due to end conditions. In the case of a closed ring with no external mounts, rigid body motion is observed whereas clamped, free, simply supported or other analogous boundary conditions are applied at the ends of the curved beams or arch when it is mounted to an underlying structure.

A corresponding weak or variational form of the equations can be determined by choosing $V = H_b^1(\Omega) \times H_b^2(\Omega)$ for the space of trial functions where Ω denotes the beam and the subscript b again denotes the set of functions which must satisfy any essential boundary conditions. Through either an energy derivation such as that given for the thin shell, or simply integration by parts arguments, one arrives at the variational form

$$\int_{-1}^{-2} \left\{ R\rho h b \frac{\partial^2 v}{\partial t^2} \phi_1 + b N_\theta \frac{\partial \phi_1}{\partial \theta} - b\hat{N}_\theta \frac{\partial \phi_1}{\partial \theta} \right\} d\theta = 0$$

$$\int_{-1}^{-2} \left\{ R\rho h b \frac{\partial^2 w}{\partial t^2} \phi_2 + b N_\theta \phi_2 - \frac{b}{R} M_\theta \frac{\partial^2 \phi_2}{\partial \theta^2} - R b\hat{q}_n \phi_2 + \frac{b}{R} \hat{M}_\theta \frac{\partial^2 \phi_2}{\partial \theta^2} \right\} d\theta = 0$$

for all $(\phi_1, \phi_2) \in V$. Again, from (2.39) it can be seen that the external line resultants $\hat{N}_\theta$ and $\hat{M}_\theta$ are related to the corresponding area force and moment terms through the relations $\hat{q}_\theta = -\frac{1}{R}\frac{\partial \hat{N}_\theta}{\partial \theta}$ and $\hat{m}_x = -\frac{1}{R}\frac{\partial \hat{M}_\theta}{\partial \theta}$. As was the case with the thin shells and plate, the formulation of the problem in weak form transfers the derivatives onto the test functions which proves to be advantageous in problems involving discontinuous parameter and inputs.

Flat-Beam Equations

The motion of an undamped flat thin beam of length ℓ and width b can be determined from either the dynamics of thin plate theory (2.50) or through a change in radii and Lamé parameters in the curved beam model. In the former case, one simply ignores vibrations in the x direction as well as Poisson effects to obtain equations modeling motion in the y (longitudinal) and z (transverse) directions. To obtain these equations from those used to describe curved beam dynamics, the parameters $\beta = \theta, B = R, R_\beta = \infty$ are replaced by $\beta = y, B = 1, R_\beta = \infty$ which implies that differential increments $R d\theta$ in the curved beam model are replaced by dy. Simplification in both cases yields the in-plane (rod) and out-of-plane (transverse or bending) equations

$$\rho h b \frac{\partial^2 v}{\partial t^2} - b\frac{\partial N_y}{\partial y} = b\hat{q}_y ,$$

$$\rho h b \frac{\partial^2 w}{\partial t^2} - b\frac{\partial^2 M_y}{\partial y^2} = b\hat{q}_n + b\frac{\partial \hat{m}_x}{\partial y}$$

(2.70)

where

$$bN_y = Ehb\frac{\partial v}{\partial y}$$

$$bM_y = -\frac{Eh^3 b}{12}\frac{\partial^2 w}{\partial y^2} = -EI\frac{\partial^2 w}{\partial y^2} .$$

Here b and I again denote the width and moment of inertia of the beam, respectively. This model can also be extended to include internal damping by using a more general constitutive law in which it is assumed that the stress is proportional to a linear combination of strain and strain rate.

Arguments analogous to those used in the curved beam case can then be used to obtain the weak form of the equations

$$\int_0^\ell \left\{ \rho h b \frac{\partial^2 v}{\partial t^2} \phi_1 + b N_y \frac{\partial \phi_1}{\partial y} - b \hat{N}_y \frac{\partial \phi_1}{\partial y} \right\} dy = 0 \quad \text{for all } \phi_1 \in H_b^1(\Omega)$$

$$\int_0^\ell \left\{ \rho h b \frac{\partial^2 w}{\partial t^2} \phi_2 - b M_y \frac{\partial^2 \phi_2}{\partial y^2} - b \hat{q}_n \phi_2 + b \hat{M}_y \frac{\partial^2 \phi_2}{\partial y^2} \right\} dy = 0 \tag{2.71}$$

$$\text{for all } \phi_2 \in H_b^2(\Omega) .$$

We reiterate that in this form, one is not required to differentiate the external force or moment resultants, $\hat{N}_y$ and $\hat{M}_y$, which proves to be very useful when these terms are generated by the activation of finite piezoceramic patches.

Chapter 3

Patch Contributions to Structural Equations

The crux of this chapter concerns the modeling of piezoceramic patch interactions with the underlying structures described in Chapter 2. The contributions due to the patches can be categorized into two types, namely, internal (material) and external moments and forces. The internal moments and forces account for the material changes in the structure due to the presence of the patches and are present even when no voltage is being applied to the patches. The external contributions are due to the strains induced by the patches when voltage is applied, and they enter the equations of motion as external loads.

Throughout the analysis, care is taken to produce models which are sufficiently accurate so as to account for potential coupling between extension and bending due to geometry and/or the manner in which the patches are excited. This provides structure/patch interaction models which can be used in various structural and structural acoustics control settings.

3.1 Patch Contributions to Thin Shell Equations

Several techniques are employed when designing and bonding piezoceramic patches to curved structures. In many applications, the patches are molded so as to have the same curvature as the underlying structure, and it is this case which is considered in detail in this chapter. In structures having large radii of curvature, flat patches are occasionally bent using a vacuum pump and then bonded to the structure. This pre-stresses the underlying structure and in some cases, leads to nonlinear dynamics. On other occasions, flat regions are sanded into curved structures with sufficient thickness and flat patches are again used as actuators. While this eliminates the difficulties due to pre-stresses, it does change the structural dynamics in a manner which may be difficult to characterize.

In the previous chapter, the strong and weak forms of the equations of motion for a homogeneous, thin cylindrical shell having uniform thickness were presented (see (2.38) and (2.44)). When piezoceramic patches are bonded to the shell, these basic equations are affected in two ways. First, the presence of the patches and manner of bonding to the shell significantly alters the material properties and thickness of the structure in regions covered by the patches; this must be taken into account when determining the internal (material) moment and force resultants to be used in (2.38) and (2.44). Moreover, when a voltage is applied, mechanical strains are induced in the patches. This leads to external moments and forces as loads in the shell model. Both contributions are discussed here, and a general model describing the structural dynamics when patches are bonded to the shell is presented.

3.1.1   Internal Forces and Moments – Curved Piezoceramic Patches

We assume for now that a pair of *curved* piezoceramic patches having thickness h_{pe} and radius of curvature R are perfectly bonded to a cylindrical shell of thickness h with midsurface radius R (see Figure 3.1). Throughout the shell discussion, the contributions due to the bonding layer are not directly considered. This is done to simplify the exposition and is justified in the circular plate analysis of Section 3.2.2 where it is demonstrated that bonding layer contributions can be combined with patch contributions to yield single coefficients which, in applications, must be estimated using fit-to-data techniques.

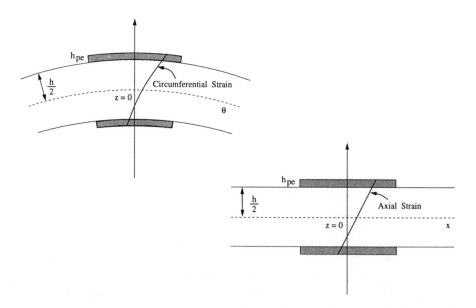

Figure 3.1: Strain distribution for the composite structure.

As shown in Figure 3.2, the patches are assumed to be situated so that their edges are parallel to lines of constant x and θ. Because the patches and shell are assumed to have the same radius of curvature, the bonding of the patches to the shell does not pre-stress the structure. The stresses which result when flat patches are bent and bonded to the cylindrical shell are discussed in the next section.

We remark here that the assumptions that the patches have edges parallel to the coordinate axes are only for convenience. Indeed the derivations and arguments below hold for rather arbitrarily shaped patches, with the shape affecting only the characteristic function χ_{pe} which has a value of one for coordinates that the patch covers and zero elsewhere (e.g., see (3.4) below). Thus, if one has a plate (rectangular, circular, etc.), our formulations and models can be used to treat patches that are shaped as a general triangle or quadrangle as well as circular and rectangular shaped patches.

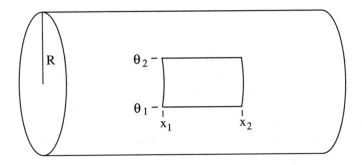

Figure 3.2: Piezoceramic patch placement.

As noted in (2.29), the infinitesimally exact strain relationships for a cylindrical shell having midsurface radius R are given by

$$e_x = (\varepsilon_x + z\kappa_x)$$

$$e_\theta = \frac{1}{(1 + z/R)}(\varepsilon_\theta + z\kappa_\theta)$$

$$\gamma_{x\theta} = \frac{1}{(1 + z/R)}\left[\varepsilon_{x\theta} + z\left(1 + \frac{z}{2R}\right)\tau\right]$$

with $\varepsilon_x, \varepsilon_\theta, \kappa_x, \kappa_\theta, \varepsilon_{x\theta}$ and τ described in (2.30). To simplify the discussion which follows, we will neglect the term z/R with respect to unity as is done in the Donnell-Mushtari theory. We emphasize that this is done merely for brevity of presentation and the infinitesimally exact terms can be used in a manner corresponding to that of the Byrne-Flügge-Lur'ye theory and should be used if this latter theory is used to model the underlying structural dynamics. Moreover, it is reasonable to assume that this relationship is maintained throughout the combined thickness $h + 2h_{pe}$ as shown in Figure 3.1 (see [118]). Hence we will take $e_x = \varepsilon_x + z\kappa_x$, $e_\theta = \varepsilon_\theta + z\kappa_\theta$, and $\gamma_{x\theta} = \varepsilon_{x\theta} + z\tau$ throughout the combined thickness of the structure. Note that this assumption implies that the strains at the interface are continuous and that the centers for the radii of curvature for the shell and patch are concurrent.

Although the same strain distribution is assumed throughout the patch and shell, the stress changes since the Young's modulus and Poisson ratio for the patch will, in general, differ from those of the shell. For an undamped shell with E_{pe_1}, ν_{pe_1} and E_{pe_2}, ν_{pe_2} denoting the Young's modulus and Poisson ratio for the outer and inner patches, respectively, the stress component σ_x is given by

$$
\sigma_x = \begin{cases}
\dfrac{E}{1 - \nu^2}(e_x + \nu e_\theta) \ , & \text{shell} \\[2ex]
\dfrac{E_{pe_1}}{1 - \nu_{pe_1}^2}(e_x + \nu_{pe_1} e_\theta) \ , & \text{outer patch} \\[2ex]
\dfrac{E_{pe_2}}{1 - \nu_{pe_2}^2}(e_x + \nu_{pe_2} e_\theta) \ , & \text{inner patch}
\end{cases}
\tag{3.1}
$$

with similar expressions for σ_θ and $\sigma_{x\theta} = \sigma_{\theta x}$ (see (2.15)). The subscripts 1 and 2 will be used throughout this discussion to denote outer and inner patch properties, respectively. We point out that the stresses in bonding layers can be described in a similar manner, and the reader is directed to Section 3.2.2 for such a discussion in the context of the circular plate model.

The moment and force resultants are obtained by integrating the stresses across the thickness of the structure. This yields the expressions

$$
\begin{bmatrix} N_x \\ N_{x\theta} \end{bmatrix} = \int_{-h/2 - h_{pe}}^{h/2 + h_{pe}} \begin{bmatrix} \sigma_x \\ \sigma_{x\theta} \end{bmatrix} \left(1 + \frac{z}{R} \right) dz
$$

$$
\begin{bmatrix} N_\theta \\ N_{\theta x} \end{bmatrix} = \int_{-h/2 - h_{pe}}^{h/2 + h_{pe}} \begin{bmatrix} \sigma_\theta \\ \sigma_{\theta x} \end{bmatrix} dz
$$

$$
\begin{bmatrix} M_x \\ M_{x\theta} \end{bmatrix} = \int_{-h/2 - h_{pe}}^{h/2 + h_{pe}} \begin{bmatrix} \sigma_x \\ \sigma_{x\theta} \end{bmatrix} \left(1 + \frac{z}{R} \right) z\, dz
\tag{3.2}
$$

$$
\begin{bmatrix} M_\theta \\ M_{\theta x} \end{bmatrix} = \int_{-h/2 - h_{pe}}^{h/2 + h_{pe}} \begin{bmatrix} \sigma_\theta \\ \sigma_{\theta x} \end{bmatrix} z\, dz
$$

in regions of the structure covered by the patches and the previously discussed expressions (2.18)-(2.21) (with $\alpha = x, \beta = \theta, R_\alpha = \infty$ and $R_\beta = R$) in those regions of the structure consisting solely of shell material. In accordance with the Donnell-Mushtari assumptions, the curvature terms z/R appearing in the integrals are neglected with respect to unity, but again, this is done for ease of presentation and higher order results can be obtained by retaining these terms. For the *undamped combined* structure when *both patches* in a pair are present and have *potentially differing material properties*, this yields the force and moment resultants

$$N_x = \frac{Eh}{1-\nu^2}(\varepsilon_x + \nu\varepsilon_\theta)$$

$$+ \frac{E_{pe_1}}{1-\nu_{pe_1}^2}\left[(\varepsilon_x + \nu_{pe_1}\varepsilon_\theta)h_{pe} + \frac{a_2}{2}(\kappa_x + \nu_{pe_1}\kappa_\theta)\right]\chi_{pe}(x,\theta)$$

$$+ \frac{E_{pe_2}}{1-\nu_{pe_2}^2}\left[(\varepsilon_x + \nu_{pe_2}\varepsilon_\theta)h_{pe} - \frac{a_2}{2}(\kappa_x + \nu_{pe_2}\kappa_\theta)\right]\chi_{pe}(x,\theta)$$

$$N_\theta = \frac{Eh}{1-\nu^2}(\varepsilon_\theta + \nu\varepsilon_x)$$

$$+ \frac{E_{pe_1}}{1-\nu_{pe_1}^2}\left[(\varepsilon_\theta + \nu_{pe_1}\varepsilon_x)h_{pe} + \frac{a_2}{2}(\kappa_\theta + \nu_{pe_1}\kappa_x)\right]\chi_{pe}(x,\theta)$$

$$+ \frac{E_{pe_2}}{1-\nu_{pe_2}^2}\left[(\varepsilon_\theta + \nu_{pe_2}\varepsilon_x)h_{pe} - \frac{a_2}{2}(\kappa_\theta + \nu_{pe_2}\kappa_x)\right]\chi_{pe}(x,\theta)$$

$$N_{x\theta} = N_{\theta x} = \frac{Eh}{2(1+\nu)}\varepsilon_{x\theta}$$

$$+ E_{pe_1}\left[\frac{h_{pe}}{2(1+\nu_{pe_1})}\varepsilon_{x\theta} + \frac{a_2}{4(1+\nu_{pe_1})}\tau\right]\chi_{pe}(x,\theta)$$

$$+ E_{pe_2}\left[\frac{h_{pe}}{2(1+\nu_{pe_2})}\varepsilon_{x\theta} - \frac{a_2}{4(1+\nu_{pe_2})}\tau\right]\chi_{pe}(x,\theta)$$

$$\quad (3.3)$$

$$M_x = \frac{Eh^3}{12(1-\nu^2)}(\kappa_x + \nu\kappa_\theta)$$

$$+ \frac{E_{pe_1}}{1-\nu_{pe_1}^2}\left[(\varepsilon_x + \nu_{pe_1}\varepsilon_\theta)\frac{a_2}{2} + (\kappa_x + \nu_{pe_1}\kappa_\theta)\frac{a_3}{3}\right]\chi_{pe}(x,\theta)$$

$$+ \frac{E_{pe_2}}{1-\nu_{pe_2}^2}\left[-(\varepsilon_x + \nu_{pe_2}\varepsilon_\theta)\frac{a_2}{2} + (\kappa_x + \nu_{pe_2}\kappa_\theta)\frac{a_3}{3}\right]\chi_{pe}(x,\theta)$$

$$M_\theta = \frac{Eh^3}{12(1-\nu^2)}(\kappa_\theta + \nu\kappa_x)$$

$$+ \frac{E_{pe_1}}{1-\nu_{pe_1}^2}\left[(\varepsilon_\theta + \nu_{pe_1}\varepsilon_x)\frac{a_2}{2} + (\kappa_\theta + \nu_{pe_1}\kappa_x)\frac{a_3}{3}\right]\chi_{pe}(x,\theta)$$

$$+ \frac{E_{pe_2}}{1-\nu_{pe_2}^2}\left[-(\varepsilon_\theta + \nu_{pe_2}\varepsilon_x)\frac{a_2}{2} + (\kappa_\theta + \nu_{pe_2}\kappa_x)\frac{a_3}{3}\right]\chi_{pe}(x,\theta)$$

$$M_{x\theta} = M_{\theta x} = \frac{Eh^3}{24(1+\nu)}\tau$$

$$+ E_{pe_1}\left[\frac{a_2}{4(1+\nu_{pe_1})}\varepsilon_{x\theta} + \frac{a_3}{6(1+\nu_{pe_1})}\tau\right]\chi_{pe}(x,\theta)$$

$$+ E_{pe_2}\left[-\frac{a_2}{4(1+\nu_{pe_2})}\varepsilon_{x\theta} + \frac{a_3}{6(1+\nu_{pe_2})}\tau\right]\chi_{pe}(x,\theta).$$

The constants a_2 and a_3 are given by

$$a_2 = \left(\frac{h}{2} + h_{pe}\right)^2 - \frac{h^2}{4}$$

$$a_3 = \left(\frac{h}{2} + h_{pe}\right)^3 - \frac{h^3}{8}$$

while the characteristic function $\chi_{pe}(x,\theta)$ has the definition

$$\chi_{pe}(x,\theta) = \begin{cases} 1 \quad , \quad x_1 \leq x \leq x_2 \,, \theta_1 \leq \theta \leq \theta_2 \\ 0 \quad , \quad \text{otherwise} \ . \end{cases} \tag{3.4}$$

The midsurface characteristics $\varepsilon_x, \varepsilon_\theta, \kappa_x, \kappa_\theta, \varepsilon_{x\theta}$ and τ are described in (2.30). We point out that the resultant values in (3.3) for those regions of the structure not covered by patches are identical to the Donnell-Mushtari components in (2.31). If greater accuracy is required, high-order terms can be retained and integrated through the combined structure to yield a composite model having the accuracy of the Byrne-Flügge-Lur'ye equations.

In the case where *both patches* have *identical material properties* ($E_{pe_2} = E_{pe_1} = E_{pe}$ and $\nu_{pe_2} = \nu_{pe_1} = \nu_{pe}$), these expressions simplify to yield

$$N_x = \frac{Eh}{1-\nu^2}\left(\varepsilon_x + \nu\varepsilon_\theta\right) + \frac{2E_{pe}h_{pe}}{1-\nu_{pe}^2}\left(\varepsilon_x + \nu_{pe}\varepsilon_\theta\right)\chi_{pe}(x,\theta)$$

$$N_\theta = \frac{Eh}{1-\nu^2}\left(\varepsilon_\theta + \nu\varepsilon_x\right) + \frac{2E_{pe}h_{pe}}{1-\nu_{pe}^2}\left(\varepsilon_\theta + \nu_{pe}\varepsilon_x\right)\chi_{pe}(x,\theta)$$

$$N_{x\theta} = N_{\theta x} = \frac{Eh}{2(1+\nu)}\varepsilon_{x\theta} + \frac{E_{pe}h_{pe}}{(1+\nu_{pe})}\varepsilon_{x\theta}\chi_{pe}(x,\theta)$$

$$M_x = \frac{Eh^3}{12(1-\nu^2)}\left(\kappa_x + \nu\kappa_\theta\right) + \frac{2E_{pe}a_3}{3(1-\nu_{pe}^2)}\left(\kappa_x + \nu_{pe}\kappa_\theta\right)\chi_{pe}(x,\theta) \tag{3.5}$$

$$M_\theta = \frac{Eh^3}{12(1-\nu^2)}\left(\kappa_\theta + \nu\kappa_x\right) + \frac{2E_{pe}a_3}{3(1-\nu_{pe}^2)}\left(\kappa_\theta + \nu_{pe}\kappa_x\right)\chi_{pe}(x,\theta)$$

$$M_{x\theta} = M_{\theta x} = \frac{Eh^3}{24(1+\nu)}\tau + \frac{E_{pe}a_3}{3(1+\nu_{pe})}\tau\chi_{pe}(x,\theta) \ .$$

If only *one* patch is present, the internal force and moment resultants for the structure can be determined from (3.3) by omitting the contributions from the missing patch. For example, if only an outer patch is bonded to the shell, one can obtain the internal resultants for the structure by deleting those terms in (3.3) which are multiplied by E_{pe_2}.

Finally, internal Kelvin-Voigt damping can be incorporated in the model by assuming a more general constitutive relation in which stress is taken to be proportional to a linear combination of strain and strain rate. Let c_D, c_{Dpe_1} and c_{Dpe_2} denote the damping coefficients in the shell, outer, and inner patches, respectively. The stress component σ_x in (3.1) is then replaced by the more general expression

$$
\sigma_x = \begin{cases}
\dfrac{E}{1-\nu^2}(e_x + \nu e_\theta) + \dfrac{c_D}{1-\nu^2}(\dot{e}_x + \nu\dot{e}_\theta) \ , & \text{shell} \\[3mm]
\dfrac{E_{pe_1}}{1-\nu^2_{pe_1}}(e_x + \nu_{pe_1} e_\theta) + \dfrac{c_{Dpe_1}}{1-\nu^2_{pe_1}}(\dot{e}_x + \nu_{pe_1}\dot{e}_\theta) \ , & \text{outer patch} \\[3mm]
\dfrac{E_{pe_2}}{1-\nu^2_{pe_2}}(e_x + \nu_{pe_2} e_\theta) + \dfrac{c_{Dpe_2}}{1-\nu^2_{pe_2}}(\dot{e}_x + \nu_{pe_2}\dot{e}_\theta) \ , & \text{inner patch}
\end{cases}
$$

with analogous expressions for σ_θ and $\sigma_{x\theta}$. The substitution of these stress terms into (3.2) then yields moment and force expressions analogous to those in (3.3) and (3.5) but which now include damping contributions containing the temporal derivatives of the terms $\varepsilon_x, \varepsilon_\theta, \kappa_x, \kappa_\theta, \varepsilon_{x\theta}$ and τ given in (2.30).

The internal moments and forces determined by (3.3) or (3.5) are then substituted into (2.38) if one is using the strong form of the shell equations or (2.44) if employing the weak form of the equations. As noted in the resultant expressions (3.3) or (3.5), the bonding of patches to the shell leads to discontinuous stiffness, Poisson and damping constants in the equations of motion. Similarly, the density ρ in (2.38) and (2.44) must be assumed to be piecewise constant in order to account for the differing geometrical and material properties of the structure in regions of the patches (see [42, 49] for experimental results related to this point). In this manner, the material contributions due to the presence of the piezoceramic patches are incorporated in the dynamic equations of motion. An important advantage of the weak form of the equations of motion for models of structures involving piezoceramic patches is that the problems associated with differentiating discontinuous functions are eliminated since the derivatives are transferred onto the test functions.

3.1.2 Flat Patches

In the previous section, moment and force resultants for the composite cylindrical shell were obtained under the assumption that the patches were molded so as to have the same curvature as the shell. For shells having a large radius of curvature, it is also possible to bend flat patches, using a vacuum pump, and then bond them to the underlying curved shell. The advantage to this technique is due to the availability of pre-molded flat patches of varying dimensions. The bending of the patches, however, leads to pre-stresses which can adversely affect system dynamics. The manner in which these stresses enter the system model depends upon the exact techniques used to bond the patches. In general, the effects tend to be nonlinear due to the large displacements involved in bending the patches as well as the potential breakdown in the linear constitutive relations. Hence the previously discussed equations may not adequately describe the system dynamics when flat patches are bonded to the structure due to the effects of unmodeled internal pre-stresses.

3.1.3　External Moments and Forces

The second contribution from the piezoceramic patches is the generation of external moments and forces which results from the property that when a voltage is applied, mechanical strains are induced in the x and θ directions (see Figure 3.3). Here we assume that when the patch is activated, in accordance with basic shell theory, equal strains are induced in the x and θ directions and the radius of curvature is not changed in either direction. Patches satisfying this assumption could be made, for example, by taking a portion of a thin-walled tubular piezoceramic element.

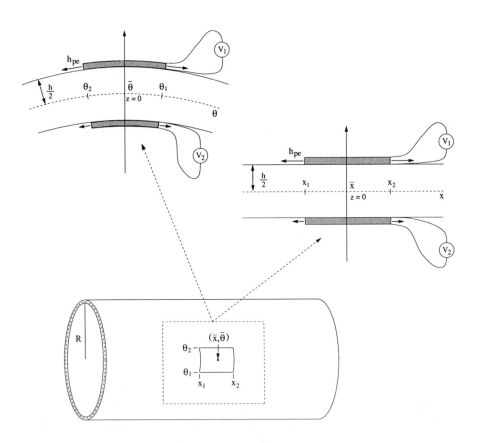

Figure 3.3: The activation of patches bonded to a thin cylindrical shell.

The magnitude of the induced free strains is taken to be

$$e_{pe_1} = (e_x)_{pe_1} = (e_\theta)_{pe_1} = \frac{d_{31}}{h_{pe}} V_1$$

$$e_{pe_2} = (e_x)_{pe_2} = (e_\theta)_{pe_2} = \frac{d_{31}}{h_{pe}} V_2$$

where d_{31} is a piezoceramic strain constant, and V_1 and V_2 are the applied voltages into the outer and inner patches, respectively. We point out that when a voltage is applied to a patch with edge coordinates x_1, x_2, θ_1 and θ_2, as depicted in Figure 3.3, the point $(\bar{x}, \bar{\theta}) = ((x_1 + x_2)/2, R(\theta_1 + \theta_2)/2)$ will not move whereas the axially symmetric points on either side will move an equal amount in opposite directions. This observation is important when determining the sense of the force resultants and it motivates the use of indicator functions in several of the following expressions.

With E_{pe_1}, ν_{pe_1} and E_{pe_2}, ν_{pe_2} again denoting the Young's modulus and Poisson ratio for the outer and inner patch, respectively, the induced external stress distribution in the individual patches is taken to be

$$(\sigma_x)_{pe_1} = (\sigma_\theta)_{pe_1} = -\frac{E_{pe_1}}{1 - \nu_{pe_1}} e_{pe_1}$$

$$(\sigma_x)_{pe_2} = (\sigma_\theta)_{pe_2} = -\frac{E_{pe_2}}{1 - \nu_{pe_2}} e_{pe_2} \; .$$

(3.6)

The negative signs result from the conservation of forces when balancing the material and induced stresses in the patch.

By integrating the stresses over the face of a fundamental element, it follows that the external moment and force resultants due to the activation of the individual patches can be expressed as

$$(M_x)_{pe_1} = \int_{h/2}^{h/2+h_{pe}} (\sigma_x)_{pe_1} \left(1 + \frac{z}{R}\right) z\, dz$$

$$(M_\theta)_{pe_1} = \int_{h/2}^{h/2+h_{pe}} (\sigma_\theta)_{pe_1} z\, dz$$

$$(N_x)_{pe_1} = \int_{h/2}^{h/2+h_{pe}} (\sigma_x)_{pe_1} \left(1 + \frac{z}{R}\right) dz$$

$$(N_\theta)_{pe_1} = \int_{h/2}^{h/2+h_{pe}} (\sigma_\theta)_{pe_1} dz$$

with analogous expression for $(M_x)_{pe_2}, (M_\theta)_{pe_2}, (N_x)_{pe_2}$ and $(N_\theta)_{pe_2}$. The units here are moment per unit length and force per unit length. Integration then yields the external forces and moments

$$(M_x)_{pe_1} = \frac{-E_{pe_1}}{1 - \nu_{pe_1}} \left[\frac{1}{8} \left(4\left(\frac{h}{2} + h_{pe}\right)^2 - h^2 \right) \right.$$

$$\left. + \frac{1}{R}\frac{1}{24} \left(8\left(\frac{h}{2} + h_{pe}\right)^3 - h^3 \right) \right] e_{pe_1}$$

$$(M_x)_{pe_2} = \frac{E_{pe_2}}{1 - \nu_{pe_2}} \left[\frac{1}{8} \left(4\left(\frac{h}{2} + h_{pe}\right)^2 - h^2 \right) \right.$$

$$\left. - \frac{1}{R}\frac{1}{24} \left(8\left(\frac{h}{2} + h_{pe}\right)^3 - h^3 \right) \right] e_{pe_2}$$

$$(M_\theta)_{pe_1} = \frac{-E_{pe_1}}{1 - \nu_{pe_1}} \left[\frac{1}{8} \left(4\left(\frac{h}{2} + h_{pe}\right)^2 - h^2 \right) \right] e_{pe_1}$$

$$(M_\theta)_{pe_2} = \frac{E_{pe_2}}{1 - \nu_{pe_2}} \left[\frac{1}{8} \left(4\left(\frac{h}{2} + h_{pe}\right)^2 - h^2 \right) \right] e_{pe_2}$$

$$(N_x)_{pe_1} = \frac{-E_{pe_1}}{1 - \nu_{pe_1}} \left[h_{pe} + \frac{1}{R}\frac{1}{8} \left(4\left(\frac{h}{2} + h_{pe}\right)^2 - h^2 \right) \right] e_{pe_1}$$

$$(N_x)_{pe_2} = \frac{-E_{pe_2}}{1 - \nu_{pe_2}} \left[h_{pe} - \frac{1}{R}\frac{1}{8} \left(4\left(\frac{h}{2} + h_{pe}\right)^2 - h^2 \right) \right] e_{pe_2}$$

$$(N_\theta)_{pe_1} = \frac{-E_{pe_1}}{1 - \nu_{pe_1}} h_{pe} e_{pe_1}$$

$$(N_\theta)_{pe_2} = \frac{-E_{pe_2}}{1 - \nu_{pe_2}} h_{pe} e_{pe_2} \ .$$

(3.7)

We note that in evaluating these integrals, we have retained the terms z/R which result from the curvature of the shell. Although this yields external terms having slightly more accuracy than the internal resultants obtained via the Donnell-Mushtari assumptions, it provides expressions for the loads due to the excitation of the individual patches which can be directly used in higher order theories (e.g., the Byrne-Flügge-Lur'ye theory) without alteration.

We also emphasize that the expressions in (3.7) admit differing voltages into the patches including the possibility of letting one patch remain passive with no voltage being applied. This provides a great deal of flexibility in applying various types of loads though the activation of the patches.

Thus far in the development of the external forces and moments due to the activation of the patches, edge effects have been ignored, and hence the expressions in (3.7) apply to patches covering the full circumference of the shell and having infinite axial length. The equations can be modified for finite patches in the following manner. For a patch with bounding values x_1, x_2, θ_1 and θ_2 as shown in Figures 3.2 and 3.3, the total line moments and forces are

$$(M_x)_{pe} = [(M_x)_{pe_1} + (M_x)_{pe_2}] \chi_{pe}(x, \theta)$$

$$(M_\theta)_{pe} = [(M_\theta)_{pe_1} + (M_\theta)_{pe_2}] \chi_{pe}(x, \theta)$$

$$(N_x)_{pe} = [(N_x)_{pe_1} + (N_x)_{pe_2}] \chi_{pe}(x, \theta) S_{1,2}(x)\hat{S}_{1,2}(\theta)$$

$$(N_\theta)_{pe} = [(N_\theta)_{pe_1} + (N_\theta)_{pe_2}] \chi_{pe}(x, \theta) S_{1,2}(x)\hat{S}_{1,2}(\theta)$$

(3.8)

where $\chi_{pe}(x, \theta)$ is the characteristic function defined in (3.4). The presence of the indicator functions

$$S_{1,2}(x) = \begin{cases} 1 & , \quad x < (x_1 + x_2)/2 \\ 0 & , \quad x = (x_1 + x_2)/2 \\ -1 & , \quad x > (x_1 + x_2)/2 \end{cases}$$

$$\hat{S}_{1,2}(\theta) = \begin{cases} 1 & , \quad \theta < (\theta_1 + \theta_2)/2 \\ 0 & , \quad \theta = (\theta_1 + \theta_2)/2 \\ -1 & , \quad \theta > (\theta_1 + \theta_2)/2 \end{cases}$$

(3.9)

derives from the property that for homogeneous patches having uniform thickness, opposite but equal strains are generated about the point $(\bar{x}, \bar{\theta}) = ((x_1 + x_2)/2, R(\theta_1 + \theta_2)/2)$ in the two coordinate directions (see Figure 3.3).

If the weak form (2.44) is used, the external line moments and forces are

$$\hat{M}_x = (M_x)_{pe} \quad , \quad \hat{M}_\theta = (M_\theta)_{pe}$$
$$\hat{N}_x = (N_x)_{pe} \quad , \quad \hat{N}_\theta = (N_\theta)_{pe}$$

(3.10)

where $(M_x)_{pe}, (M_\theta)_{pe}, (N_x)_{pe}, (N_\theta)_{pe}$ given in (3.8) are the respective moments and in-plane forces which are generated by the input of voltage to the patches.

However, if one is using the strong form (2.38) of the equations of motion with piezoceramic actuators, the surface moments and forces to be used in (2.38) are given by

$$\hat{m}_x = -\frac{1}{R}\frac{\partial(M_\theta)_{pe}}{\partial\theta} \quad , \qquad \hat{m}_\theta = -\frac{\partial(M_x)_{pe}}{\partial x}$$

$$\hat{q}_x = -S_{1,2}(x)\hat{S}_{1,2}(\theta)\frac{\partial(N_x)_{pe}}{\partial x} \quad , \quad \hat{q}_\theta = -S_{1,2}(x)\hat{S}_{1,2}(\theta)\frac{1}{R}\frac{\partial(N_\theta)_{pe}}{\partial\theta} \; .$$

(3.11)

We point out that the differences between the external surface force expressions in (2.39) and (3.11) are due to the fact that the former were derived for an infinitesimal element whereas the latter are global expressions which preserve the overall signs of the forces generated by the patches as well as reflect the discontinuities due to changes in sign. These differences result from the property that the sense of the forces is highly dependent on the specified location of the axis origin on the neutral surface. Hence the direction of forces throughout the patch differs in some locations from those observed in the infinitesimal element. This necessitates the inclusion of the indicator functions in (3.11).

Unlike the forces, the action of the moments is specified with respect to a fixed point on the neutral surface (the point 0 for the element in Figure 2.5 or a point on the left edge of the shell in Figure 2.4). As long as the orientation of the infinitesimal element and full shell with patches are the same, the line moments derived for the infinitesimal element will be consistent with those of the full structure. Thus the expressions for the general infinitesimal moments in (2.39) need no modifications when describing the surface moments generated by the patches as given in (3.11).

3.2   Patch Contributions to Plate Equations

Analysis similar to that used for the thin cylindrical shells can be used to determine the forces and moments which are due to the presence and activation of piezoceramic patches which have been bonded to a flat rectangular or circular plate.

3.2.1   Rectangular Plate Model

By repeating the analysis used in the last section for determining the internal moments and forces for the structure consisting of piezoceramic patches bonded to a thin shell, it is straightforward to show that the internal moments and forces for a *damped* rectangular plate having a pair of identical patches with edges at x_1, x_2, y_1 and y_2 are given by

$$
\begin{aligned}
N_x &= \left[\frac{Eh}{1-\nu^2}(\varepsilon_x + \nu\varepsilon_y) + \frac{c_D h}{1-\nu^2}(\dot{\varepsilon}_x + \nu\dot{\varepsilon}_y) \right] \\
&\quad + \left[\frac{2E_{pe}h_{pe}}{1-\nu_{pe}^2}(\varepsilon_x + \nu_{pe}\varepsilon_y) + \frac{2c_{Dpe}h_{pe}}{1-\nu_{pe}^2}(\dot{\varepsilon}_x + \nu_{pe}\dot{\varepsilon}_y) \right] \chi_{pe}(x,y)
\end{aligned}
$$

$$
\begin{aligned}
N_y &= \left[\frac{Eh}{1-\nu^2}(\varepsilon_y + \nu\varepsilon_x) + \frac{c_D h}{1-\nu^2}(\dot{\varepsilon}_y + \nu\dot{\varepsilon}_x) \right] \\
&\quad + \left[\frac{2E_{pe}h_{pe}}{1-\nu_{pe}^2}(\varepsilon_y + \nu_{pe}\varepsilon_x) + \frac{2c_{Dpe}h_{pe}}{1-\nu_{pe}^2}(\dot{\varepsilon}_y + \nu_{pe}\dot{\varepsilon}_x) \right] \chi_{pe}(x,y)
\end{aligned}
$$

$$
\begin{aligned}
N_{xy} = N_{yx} &= \left[\frac{Eh}{2(1+\nu)}\varepsilon_{xy} + \frac{c_D h}{2(1+\nu)}\dot{\varepsilon}_{xy} \right] \\
&\quad + \left[\frac{E_{pe}h_{pe}}{2(1+\nu_{pe})}\varepsilon_{xy} + \frac{c_{Dpe}h_{pe}}{2(1+\nu_{pe})}\dot{\varepsilon}_{xy} \right] \chi_{pe}(x,y)
\end{aligned}
$$

and

$$M_x = \left[\frac{Eh^3}{12(1-\nu^2)} (\kappa_x + \nu\kappa_y) + \frac{c_D h^3}{12(1-\nu^2)} (\dot{\kappa}_x + \nu\dot{\kappa}_y) \right]$$

$$+ \left[\frac{2E_{pe}a_3}{3(1-\nu_{pe}^2)} (\kappa_x + \nu_{pe}\kappa_y) + \frac{2c_{Dpe}a_3}{3(1-\nu_{pe}^2)} (\dot{\kappa}_x + \nu_{pe}\dot{\kappa}_y) \right] \chi_{pe}(x,y)$$

$$M_y = \left[\frac{Eh^3}{12(1-\nu^2)} (\kappa_y + \nu\kappa_x) + \frac{c_D h^3}{12(1-\nu^2)} (\dot{\kappa}_y + \nu\dot{\kappa}_x) \right]$$

$$\text{(3.12)}$$

$$+ \left[\frac{2E_{pe}a_3}{3(1-\nu_{pe}^2)} (\kappa_y + \nu_{pe}\kappa_x) + \frac{2c_{Dpe}a_3}{3(1-\nu_{pe}^2)} (\dot{\kappa}_y + \nu_{pe}\dot{\kappa}_x) \right] \chi_{pe}(x,y)$$

$$M_{xy} = M_{yx} = \left[\frac{Eh^3}{24(1+\nu)} \tau + \frac{c_D h^3}{24(1+\nu)} \dot{\tau} \right]$$

$$+ \left[\frac{E_{pe}a_3}{3(1+\nu_{pe})} \tau + \frac{c_{Dpe}a_3}{3(1+\nu_{pe})} \dot{\tau} \right] \chi_{pe}(x,y)$$

(compare to (3.5)). The characteristic function here is given by

$$\chi_{pe}(x,y) = \begin{cases} 1 & , \quad x_1 \le x \le x_2 \,, y_1 \le y \le y_2 \\ 0 & , \quad \text{otherwise} \,, \end{cases}$$

whereas $a_3 = (h/2 + h_{pe})^3 - h^3/8$, and the midsurface characteristics $\varepsilon_x, \varepsilon_y, \kappa_x,$ $\kappa_y, \varepsilon_{xy}$ and τ are defined in (2.46). As before, E, c_D, ν and $E_{pe}, c_{Dpe}, \nu_{pe}$ are the Young's modulus, damping coefficient and Poisson ratio of the plate and patches, respectively.

If the patches have differing material properties or if only one patch is present, the moments and forces can be determined from (3.3) with θ replaced by y.

These internal (material) moments and forces are then used in the strong form (2.50) or weak form (2.52) of the plate equations with the choice depending on the application of interest. In either case, the use of these moments and forces incorporates the structural contributions due to the presence of the piezoceramic patches. Finally, the density ρ in both (2.50) and (2.52) is assumed to be piecewise continuous to account for the differing structural density in regions in which patches are bonded.

The external moments and forces due to the activation of the patches are also found in a manner analogous to that used in the shell analysis. The induced stresses

$$(\sigma_x)_{pe_1} = (\sigma_\theta)_{pe_1} = \frac{-E_{pe_1}}{1-\nu_{pe_1}} e_{pe_1} = \frac{-E_{pe_1}}{1-\nu_{pe_1}} \frac{d_{31}}{h_{pe}} V_1$$

$$(\sigma_x)_{pe_2} = (\sigma_\theta)_{pe_2} = \frac{-E_{pe_2}}{1-\nu_{pe_2}} e_{pe_2} = \frac{-E_{pe_2}}{1-\nu_{pe_2}} \frac{d_{31}}{h_{pe}} V_2$$

are integrated through the thickness of the respective patches thus yielding
the external moments and forces

$$(M_x)_{pe_1} = (M_y)_{pe_1} = -\frac{1}{8}\frac{E_{pe_1}}{1 - \nu_{pe_1}}\left(4\left(\frac{h}{2} + h_{pe}\right)^2 - h^2\right)e_{pe_1}$$

$$(M_x)_{pe_2} = (M_y)_{pe_2} = \frac{1}{8}\frac{E_{pe_2}}{1 - \nu_{pe_2}}\left(4\left(\frac{h}{2} + h_{pe}\right)^2 - h^2\right)e_{pe_2}$$

$$(N_x)_{pe_1} = (N_y)_{pe_1} = \frac{-E_{pe_1}h_{pe}}{1 - \nu_{pe_1}}e_{pe_1}$$

$$(N_x)_{pe_2} = (N_y)_{pe_2} = \frac{-E_{pe_2}h_{pe}}{1 - \nu_{pe_2}}e_{pe_2} \ .$$

The total external moments and forces generated by the patches are then
given by

$$(M_x)_{pe} = (M_y)_{pe} = [(M_x)_{pe_1} + (M_x)_{pe_2}]\chi_{pe}(x,y)$$

$$(N_x)_{pe} = (N_y)_{pe} = [(N_x)_{pe_1} + (N_x)_{pe_2}]\chi_{pe}(x,y)S_{1,2}(x)\tilde{S}_{1,2}(y)$$

$$(3.13)$$

where $S_{1,2}(x)$ denotes the indicator function described in (3.9) with a similar
definition for $\tilde{S}_{1,2}(y)$.

These loads can be substituted directly into the weak form of the plate
equations (2.52) as the load on the system (with $\hat{q}_n = 0$ and $\hat{N}_x = (N_x)_{pe}$,
$\hat{N}_y = (N_y)_{pe}$, $\hat{M}_x = (M_x)_{pe}$, $\hat{M}_y = (M_y)_{pe}$). If the strong form of the plate
equations is being used, the surface loads can be determined via the expressions

$$\hat{q}_x = -S_{1,2}(x)\tilde{S}_{1,2}(y)\frac{\partial(N_x)_{pe}}{\partial x} \quad , \quad \hat{q}_y = -S_{1,2}(x)\tilde{S}_{1,2}(y)\frac{\partial(N_y)_{pe}}{\partial y}$$

$$\hat{m}_x = -\frac{\partial(M_y)_{pe}}{\partial y} \quad , \qquad\qquad \hat{m}_y = -\frac{\partial(M_x)_{pe}}{\partial x} \quad ,$$

and these latter values can be substituted into the equilibrium equations (2.50).

As in the case of the shells, the use of the strong form results in up to
two derivatives of the characteristic function whereas the use of the weak form
alleviates this problem by transferring the derivatives onto the test functions.

It should be noted that the voltage choice $e_{pe} = e_{pe_1} = e_{pe_2}$ causes pure
extension (patch pairs excited "in-phase") in the plate while pure bending
occurs with the choice $e_{pe} = -e_{pe_1} = e_{pe_2}$ ("out-of-phase" excitation).

3.2.2 Circular Plate Model

We also consider the case in which s piezoceramic patch pairs are bonded to a
thin circular plate having radius a as depicted in Figure 3.4. To simplify exposition, we will assume that all patches have thickness h_{pe} and Young's modulus,

Poisson ratio and Kelvin-Voigt damping coefficients E_{pe}, ν_{pe} and c_{Dpe}, respectively (individual patches or patches with differing thickness and/or material properties can be handled in a manner analogous to that discussed for the shell). To illustrate the contributions due to bonding layers, we consider layers of adhesive having thickness $h_{b\ell}$ with material properties $E_{b\ell}, \nu_{b\ell}, c_{Db\ell}$. Finally, we will consider only transverse vibrations of the plate since this is the case in applications discussed in later chapters; extensions to include coupled transverse and longitudinal vibrations follow directly using the techniques discussed for shells.

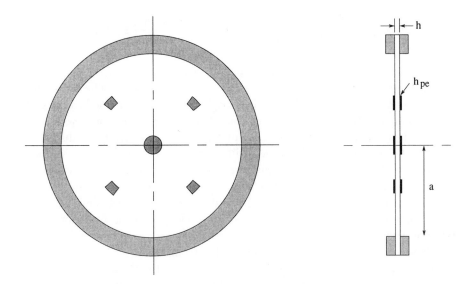

Figure 3.4: Clamped thin circular plate with surface-mounted piezoceramic patches.

From arguments similar to those used to obtain the moment expressions (3.12) for the rectangular plate, it is determined that the moments for the circular plate with s pairs of surface-mounted piezoceramic patches are

$$M_r = D K_r + \tilde{D} K_\theta + C \dot{K}_r + \tilde{C} \dot{K}_\theta$$

$$M_\theta = D K_\theta + \tilde{D} K_r + C \dot{K}_\theta + \tilde{C} \dot{K}_r \qquad (3.14)$$

$$M_{r\theta} = M_{\theta r} = \frac{D}{2}\tau - \frac{\tilde{D}}{2}\tau + \frac{C}{2}\dot{\tau} - \frac{\tilde{C}}{2}\dot{\tau}$$

where

$$K_r = -\frac{\partial^2 w}{\partial r^2} \quad , \quad K_\theta = -\frac{1}{r}\frac{\partial w}{\partial r} - \frac{1}{r^2}\frac{\partial^2 w}{\partial \theta^2} \quad , \quad \tau = -\frac{2}{r}\frac{\partial^2 w}{\partial r \partial \theta} + \frac{2}{r^2}\frac{\partial w}{\partial \theta} \ .$$

The parameters $D, \tilde{D}, C$ and $\tilde{C}$ are given by

$$D(r, \theta) = \frac{Eh^3}{12(1 - \nu^2)} + \frac{2}{3} \sum_{i=1}^{s} \left[\frac{E_{pe} a_3}{1 - \nu_{pe}^2} + \frac{E_{b\ell} a_{3b\ell}}{1 - \nu_{b\ell}^2} \right] \chi_{pe_i}(r, \theta)$$

$$\tilde{D}(r, \theta) = \frac{Eh^3 \nu}{12(1 - \nu^2)} + \frac{2}{3} \sum_{i=1}^{s} \left[\frac{E_{pe} a_3 \nu_{pe}}{1 - \nu_{pe}^2} + \frac{E_{b\ell} a_{3b\ell} \nu_{b\ell}}{1 - \nu_{b\ell}^2} \right] \chi_{pe_i}(r, \theta)$$

$$\hspace{10cm} (3.15)$$

$$C(r, \theta) = \frac{c_D h^3}{12(1 - \nu^2)} + \frac{2}{3} \sum_{i=1}^{s} \left[\frac{c_{Dpe} a_3}{1 - \nu_{pe}^2} + \frac{c_{Db\ell} a_{3b\ell}}{1 - \nu_{b\ell}^2} \right] \chi_{pe_i}(r, \theta)$$

$$\tilde{C}(r, \theta) = \frac{c_D h^3 \nu}{12(1 - \nu^2)} + \frac{2}{3} \sum_{i=1}^{s} \left[\frac{c_{Dpe} a_3 \nu_{pe}}{1 - \nu_{pe}^2} + \frac{c_{Db\ell} a_{3b\ell} \nu_{b\ell}}{1 - \nu_{b\ell}^2} \right] \chi_{pe_i}(r, \theta) \, .$$

Here $a_{3b\ell} = (h/2 + h_{b\ell})^3 - (h/2)^3$, $a_3 = (h/2 + h_{b\ell} + h_{pe})^3 - (h/2 + h_{b\ell})^3$ and $\chi_{pe_i}(r, \theta)$ denotes the characteristic function over the ith patch. The expressions in (3.15) can be adopted to the case of a single patch that is bonded to the plate by replacing the 2/3 by 1/3. We point out that the expression for a_3 reduces to that used previously for shells and flat plates when the contribution due to the bonding layer is ignored. While $h_{b\ell}$ will never be zero, the Poisson, stiffness, damping and density contributions from the bonding layer can be combined with those of the patches and plate to yield a single coefficient for each region covered by a patch that must be estimated using fit-to-data techniques. This is the reason for simplifying discussion in previous sections by ignoring the bonding layers.

The moments (3.14) are then used in the strong form (2.63) or weak forms (2.66) or (2.67) to obtain models which incorporate the Poisson, stiffness and damping effects due to the patches and bonding layer. The density changes in regions of the plate covered by patches are taken into account by modeling ρ as piecewise constant with regions delineated by the characteristic functions χ_{pe_i}.

External moments are generated when out-of-phase voltages generating out-of-phase strains $e_{pe_i}(t) = e_{pe_{i1}}(t) = -e_{pe_{i2}}(t) = d_{31}V_i(t)/h_{pe}$ are applied to the ith patch pair. The total external moments created in this manner are given by

$$(M_r)_{pe} = (M_\theta)_{pe} = -\sum_{i=1}^{s} \mathcal{K}_i^B V_i(t) \chi_{pe_i}(r, \theta)$$

where

$$\mathcal{K}_i^B = \frac{E_{pe}}{1 - \nu_{pe}} d_{31}(h + 2h_{b\ell} + h_{pe}) \hspace{2cm} (3.16)$$

(compare with (3.13)). By taking $\hat{q}_n = 0$, $\hat{M}_r = (M_r)_{pe}$, $\hat{M}_\theta = (M_\theta)_{pe}$ in the strong form (2.63) or weak forms (2.66) or (2.67) of the circular plate equations, one obtains dynamic models describing the transverse motion of a circular plate driven by piezoceramic patch pairs. As was the case with the shell

and rectangular plate, use of the weak form of the modeling equations eliminates the difficulties associated with the differentiation of piecewise constant
parameters and input terms.

3.3 Beam/Patch Interactions

The patch contributions to the dynamics of curved and flat beams are considered here. In both cases, the equations describing transverse and in-plane
vibrations are retained. For curved beams, the equations are coupled due to
curvature whereas nonsymmetric patch configurations can lead to coupling
between in-plane and transverse vibrations in the flat beam.

3.3.1 Curved Beam Model

In Chapter 2, the strong form

$$R\rho h b \frac{\partial^2 v}{\partial t^2} - b\frac{\partial N_\theta}{\partial \theta} = Rb\hat{q}_\theta$$

$$R\rho h b \frac{\partial^2 w}{\partial t^2} - \frac{b}{R}\frac{\partial^2 M_\theta}{\partial \theta^2} + bN_\theta = Rb\hat{q}_n + b\frac{\partial \hat{m}_x}{\partial \theta}$$

(3.17)

and weak form

$$\int_{\gamma_1}^{\gamma_2} \left\{ R\rho h b \frac{\partial^2 v}{\partial t^2}\phi_1 + bN_\theta\frac{\partial \phi_1}{\partial \theta} - b\hat{N}_\theta\frac{\partial \phi_1}{\partial \theta} \right\} d\theta = 0 \ , \ \phi_1 \in H_b^1(\Omega)$$

$$\int_{\gamma_1}^{\gamma_2} \left\{ R\rho h b \frac{\partial^2 w}{\partial t^2}\phi_2 + bN_\theta\phi_2 - \frac{b}{R}M_\theta\frac{\partial^2 \phi_2}{\partial \theta^2} \right.$$

$$\left. - Rb\hat{q}_n\phi_2 + \frac{b}{R}\hat{M}_\theta\frac{\partial^2 \phi_2}{\partial \theta^2} \right\} d\theta = 0 \ , \ \phi_2 \in H_b^2(\Omega)$$

(3.18)

of the equations describing the circumferential and transverse motion of a
curved beam having thickness h, radius of curvature R, and width b, were
derived (see (2.68) and Figure 3.5). We now extend that analysis to include the
effects due to piezoceramic patches that are bonded to the beam. The internal
contributions to bN_θ and bM_θ as well as the external loads $b\hat{q}_n$, $b\hat{q}_\theta$, $b\hat{m}_x$ and
$b\hat{M}_\theta$ can be derived directly from the thin shell equations described previously
if motion is restricted to the circumferential and transverse directions and
Poisson effects are ignored. If one assumes that identical patches ($E_{pe_1} =
E_{pe_2} = E_{pe}, c_{Dpe_1} = c_{Dpe_2} = c_{Dpe}$) are bonded as a pair to the beam, as
depicted in Figure 3.5, then reduction of the shell expressions in (3.3) and use

of the linear strain-displacement relations (2.31) yields

$$bN_\theta = [Ehb + 2E_{pe}h_{pe}b\chi_{pe}(\theta)]\left(\frac{1}{R}\frac{\partial v}{\partial\theta} + \frac{w}{R}\right)$$

$$+ [c_D hb + 2c_{Dpe}h_{pe}b\chi_{pe}(\theta)]\left(\frac{1}{R}\frac{\partial^2 v}{\partial\theta\partial t} + \frac{1}{R}\frac{\partial w}{\partial t}\right)$$

$$bM_\theta = -\left[\frac{Eh^3 b}{12R^2} + \frac{2E_{pe}ba_3}{3R^2}\chi_{pe}(\theta)\right]\frac{\partial^2 w}{\partial\theta^2} - \left[\frac{c_D h^3 b}{12R^2} + \frac{2c_{Dpe}ba_3}{3R^2}\chi_{pe}(\theta)\right]\frac{\partial^3 w}{\partial\theta^2\partial t}$$

where χ_{pe} again denotes the characteristic function and $a_3 \equiv (h/2 + h_{pe})^3 - h^3/8$. This representation of the internal resultants illustrates the piecewise constant nature of the physical stiffness and damping parameters. Analogous expressions can also be determined from (3.3) in the more general case of patches with differing material properties. As in the cases of the shell and plate, the density ρ in (3.17) and (3.18) is taken to be piecewise constant to account for the density discontinuity in the region of the patches.

The external moments and forces generated by the patches in response to an applied voltage can be determined in a similar manner from the corresponding shell expressions. For patches with identical material properties and piezoelectric constant d_{31}, the external forces and moments generated by the patches are

$$(bN_\theta)_{pe} = \left[(bN_\theta)_{pe_1} + (bN_\theta)_{pe_2}\right]\chi_{pe}(\theta)\hat{S}_{1,2}(\theta)$$

$$= -E_{pe}d_{31}b(V_1 + V_2)\chi_{pe}(\theta)\hat{S}_{1,2}(\theta)$$

$$(bM_\theta)_{pe} = \left[(bM_\theta)_{pe_1} + (bM_\theta)_{pe_2}\right]\chi_{pe}(\theta)$$

$$= -\frac{E_{pe}d_{31}b}{2}(h + h_{pe})(V_1 - V_2)\chi_{pe}(\theta)$$

where V_1 and V_2 are the voltages generated by the outer and inner patches, respectively, and $\hat{S}_{1,2}(\theta)$ is the indicator function defined in (3.9).

In the weak form of the equations (3.18), these expressions can be substituted directly as loads on the beam with $b\hat{q}_n = 0, b\hat{N}_\theta = (bN_\theta)_{pe}$ and $b\hat{M}_\theta = (bM_\theta)_{pe}$. For the strong form of the beam equations, corresponding surface loads are determined from the expressions

$$b\hat{q}_\theta = (bq_\theta)_{pe} = -\frac{1}{R}\hat{S}_{1,2}(\theta)\frac{\partial(bN_\theta)_{pe}}{\partial\theta}$$

$$b\hat{m}_x = (bm_x)_{pe} = -\frac{1}{R}\frac{\partial(bM_\theta)_{pe}}{\partial\theta}\,.$$

As in previous cases, the use of the weak form of the equations eliminates the problem of differentiating the discontinuous characteristic function.

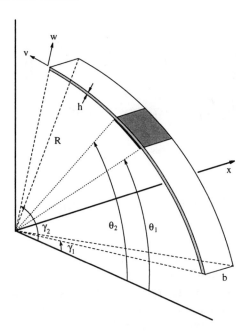

Figure 3.5: Curved beam with curved patches subject to circumferential and transverse vibrations.

3.3.2   Flat Beam Model

The patch contributions to the dynamics of a thin flat beam can be determined directly from either the plate/patch interaction model if one considers only vibrations in the y-direction along with the usual transverse vibrations, or the curved beam model if one considers the radius of curvature to be infinite (hence the differential increment $R d\theta$ is replaced by dy). For a *damped* flat beam of thickness h and width b having a pair of bonded patches of thickness h_{pe} with edges at y_1 and y_2, both approaches yield the internal force and moment

$$
\begin{aligned}
b N_y &= \left[E h b + \left(E_{pe_1} h_{pe} b + E_{pe_2} h_{pe} b \right) \chi_{pe}(y) \right] \frac{\partial v}{\partial y} \\[2mm]
&+ \left[c_D h b + \left(c_{Dpe_1} h_{pe} b + c_{Dpe_2} h_{pe} b \right) \chi_{pe}(y) \right] \frac{\partial^2 v}{\partial y \partial t} \\[2mm]
&+ \frac{1}{2} \left(E_{pe_1} a_2 b - E_{pe_2} a_2 b \right) \chi_{pe}(y) \frac{\partial^2 w}{\partial y^2} \\[2mm]
&+ \frac{1}{2} \left(c_{Dpe_1} a_2 b - c_{Dpe_2} a_2 b \right) \chi_{pe}(y) \frac{\partial^3 w}{\partial y^2 \partial t}
\end{aligned}
\tag{3.19}
$$

and

$$bM_y = -\left[E\frac{h^3b}{12} + \frac{1}{3}\left(E_{pe_1}a_3b + E_{pe_2}a_3b\right)\chi_{pe}(y)\right]\frac{\partial^2 w}{\partial y^2}$$

$$-\left[c_D\frac{h^3b}{12} + \frac{1}{3}\left(c_{Dpe_1}a_3b + c_{Dpe_2}a_3b\right)\chi_{pe}(y)\right]\frac{\partial^3 w}{\partial y^2\partial t}$$

$$-\frac{1}{2}\left(E_{pe_1}a_2b - E_{pe_2}a_2b\right)\chi_{pe}(y)\frac{\partial v}{\partial y}$$

$$-\frac{1}{2}\left(c_{Dpe_1}a_2b - c_{Dpe_2}a_2b\right)\chi_{pe}(y)\frac{\partial^2 v}{\partial y\partial t}$$

(3.20)

where again, $a_2 = (h/2 + h_{pe})^2 - h^2/4$ and $a_3 = (h/2 + h_{pe})^3 - h^3/8$. Also, E_{pe_1}, c_{Dpe_1} and E_{pe_2}, c_{Dpe_2} denote the Young's modulus and Kelvin-Voigt damping parameter for the outer and inner patches, respectively. We note that in obtaining these expressions for the internal forces and moments, we have assumed that the patches also have width b. This was done for clarity of presentation and more general expressions for the case when the patches are narrower than the beam can be obtained in a similar fashion.

Two special cases of (3.19) and (3.20) are worth mentioning in more detail since they occur quite commonly in applications. If both patches have the same material properties ($E_{pe_1} = E_{pe_2} = E_{pe}$ and $c_{Dpe_1} = c_{Dpe_2} = c_{Dpe}$), then

$$bN_y = [Ehb + 2E_{pe}h_{pe}b\chi_{pe}(y)]\frac{\partial v}{\partial y}$$

$$+ [c_Dhb + 2c_{Dpe}h_{pe}b\chi_{pe}(y)]\frac{\partial^2 v}{\partial y\partial t}$$

$$bM_y = -\left[E\frac{h^3b}{12} + \frac{2}{3}E_{pe}a_3b\chi_{pe}(y)\right]\frac{\partial^2 w}{\partial y^2}$$

$$- \left[c_D\frac{h^3b}{12} + \frac{2}{3}c_{Dpe}a_3b\chi_{pe}(y)\right]\frac{\partial^3 w}{\partial y^2\partial t} .$$

(3.21)

If only one patch is present, the expressions reduce to

$$bN_y = [Ehb + E_{pe}h_{pe}b\chi_{pe}(y)]\frac{\partial v}{\partial y}$$

$$+ [c_Dhb + c_{Dpe}h_{pe}b\chi_{pe}(y)]\frac{\partial^2 v}{\partial y\partial t}$$

$$+ \left[\frac{1}{2}E_{pe}a_2b\frac{\partial^2 w}{\partial y^2} + \frac{1}{2}c_{Dpe}a_2b\frac{\partial^3 w}{\partial y^2\partial t}\right]\chi_{pe}(y)$$

and

$$bM_y = -\left[\frac{Eh^3b}{12} + \frac{1}{3}E_{pe}a_3b\chi_{pe}(y)\right]\frac{\partial^2 w}{\partial y^2}$$

$$-\left[\frac{c_D h^3 b}{12} + \frac{1}{3}c_{Dpe}a_3b\chi_{pe}(y)\right]\frac{\partial^3 w}{\partial y^2 \partial t}$$

$$-\left[\frac{1}{2}E_{pe}a_2b\frac{\partial v}{\partial y} + \frac{1}{2}c_{Dpe}a_2b\frac{\partial^2 v}{\partial y \partial t}\right]\chi_{pe}(y) \, .$$

When the latter expressions for the internal force and moment resultants are substituted into the strong form (2.70) or weak form (2.71) of the beam equations, it is apparent that the longitudinal and transverse vibrations are coupled as a result of the asymmetry due to the single patch. Finally, we remind the reader that ρ in (2.70) or (2.71) should be assumed to be piecewise constant to account for the discontinuities in density in regions to which patches are bonded.

The external forces and moments generated by the activation of the patches follows directly from the expressions obtained in the case of the plate. Summarizing from those results, we see that the total external forces and moments are

$$(bM_y)_{pe} = [(bM_y)_{pe_1} + (bM_y)_{pe_2}]\chi_{pe}(y)$$

$$(bN_y)_{pe} = [(bN_y)_{pe_1} + (bN_y)_{pe_2}]\chi_{pe}(y)\tilde{S}_{1,2}(y)$$

(3.22)

where

$$(bM_y)_{pe_1} = -\frac{1}{8}E_{pe_1}b\left(4\left(\frac{h}{2}+h_{pe}\right)^2 - h^2\right)e_{pe_1} = -\frac{1}{2}E_{pe_1}b(h+h_{pe})d_{31}V_1$$

$$(bM_y)_{pe_2} = \frac{1}{8}E_{pe_2}b\left(4\left(\frac{h}{2}+h_{pe}\right)^2 - h^2\right)e_{pe_2} = \frac{1}{2}E_{pe_2}b(h+h_{pe})d_{31}V_2$$

$$(bN_y)_{pe_1} = -E_{pe_1}h_{pe}be_{pe_1} = -E_{pe_1}bd_{31}V_1$$

$$(bN_y)_{pe_2} = -E_{pe_2}h_{pe}be_{pe_2} = -E_{pe_2}bd_{31}V_2 \, .$$

These expressions can then be substituted directly into the weak equations (2.71) as loads on the beam (with $b\hat{q}_n = 0$ and $b\hat{N}_y = (bN_y)_{pe}, b\hat{M}_y = (bM_y)_{pe}$). In order to determine the patch loads for the strong form of the beam equations, the corresponding surface moments and forces are found via the relationships

$$b\hat{q}_y = (bq_y)_{pe} = -\tilde{S}_{1,2}(y)\frac{\partial(bN_y)_{pe}}{\partial y} \quad , \quad b\hat{m}_x = (bm_x)_{pe} = -\frac{\partial(bM_y)_{pe}}{\partial y}$$

and these latter values are used in (2.70). We again point out that this results in the need to differentiate the characteristic function (once for the force and

twice for the moment) whereas this problem is avoided in the weak formulation
since the derivatives are transferred on the test functions. In fact, the effect of
the characteristic functions in the latter case is to simply restrict the integrals
to the region covered by the patches.

Chapter 4

Well-Posedness of Abstract Structural Models

4.1 Motivating Example

In this chapter we present existence, uniqueness and continuous dependence results for an abstract class of damped second-order (in time) partial differential equation models with input or control operators that are unbounded in the natural state space for these systems. The models for shells, plates and beams with piezoceramic actuators, developed in Chapters 2 and 3, can be included in the abstract framework presented below if one makes appropriate choices of state spaces and operators or sesquilinear forms.

We shall discuss both a weak or variational formulation (equivalent to the weak form of the systems discussed in Chapters 2 and 3) and a semigroup formulation. The variational formulation will be advantageous when developing approximation and system identification or parameter estimation techniques in subsequent chapters whereas a semigroup formulation (with an associated "variation-of-parameters" representation for solutions to nonhomogeneous systems) will be desirable for our discussions of control problems in later chapters.

To motivate and illustrate the abstract formulation, we recall the example from Chapters 2 and 3 involving a flat beam subject to transverse and longitudinal vibrations (see Sections 2.2.6 and 3.3.2). The beam, depicted in Figure 4.1, is assumed to have length ℓ, width b, thickness h and cantilever end conditions with the fixed end at $y = 0$ and free end at $y = \ell$. It is assumed that the beam is homogeneous and constructed from a material which essentially satisfies the Euler-Bernoulli hypothesis for displacements and Kelvin-Voigt damping hypothesis. This latter posits that damping is proportional to strain rate. Finally, it is assumed that a pair of identical piezoceramic patches are bonded to opposite sides of the beam over the region $y_1 < y < y_2$. The Young's moduli, linear mass densities and damping coefficients for the beam and patch are denoted by $E, E_{pe}, \rho, \rho_{pe}$ and c_D, c_{Dpe}, respectively, with the subscript pe again denoting patch properties. The patch thickness is denoted h_{pe} while, for simplicity, the bonding layer is taken to be negligible (as discussed in Chapter 3, the bonding layer contributions enter the equations in exactly the same manner as the passive patch contributions).

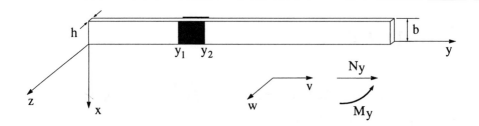

Figure 4.1: Cantilever beam with piezoceramic patches.

For such a beam, subject only to forces and moments generated by actuating the patches, we have seen that force and moment balancing lead to the dynamic system of equations

$$\tilde{\rho}(y)\frac{\partial^2 v}{\partial t^2} - \frac{\partial(bN_y)}{\partial y} = -\tilde{S}_{1,2}(y)\frac{\partial(bN_y)_{pe}}{\partial y}$$

$$\tilde{\rho}(y)\frac{\partial^2 w}{\partial t^2} - \frac{\partial^2(bM_y)}{\partial y^2} = -\frac{\partial^2(bM_y)_{pe}}{\partial y^2}$$

(4.1)

for the longitudinal displacement $v = v(t,y)$ and transverse displacement $w = w(t,y)$ (see (2.70)). Due to the presence of the patches, the linear mass density $\tilde{\rho}(y) = \rho h b + 2b\rho_{pe}h_{pe}\chi_{pe}(y)$ is piecewise constant with the characteristic function $\chi_{pe}(y)$ used to isolate the patch contributions. Similar material discontinuities appear in the *internal* force and moment resultants

$$bN_y = \widetilde{Eb}(y)\frac{\partial v}{\partial y} + \widetilde{c_Db}(y)\frac{\partial^2 v}{\partial y \partial t}$$

$$bM_y = -\widetilde{EI}(y)\frac{\partial^2 w}{\partial y^2} - \widetilde{c_DI}(y)\frac{\partial^3 w}{\partial y^2 \partial t}$$

(4.2)

where

$$\widetilde{Eb}(y) = Ehb + 2bE_{pe}h_{pe}\chi_{pe}(y)$$

$$\widetilde{c_Db}(y) = c_Dhb + 2bc_{Dpe}h_{pe}\chi_{pe}(y)$$

$$\widetilde{EI}(y) = E\frac{h^3b}{12} + \frac{2b}{3}E_{pe}a_3\chi_{pe}(y)$$

$$\widetilde{c_DI}(y) = c_D\frac{h^3b}{12} + \frac{2b}{3}c_{Dpe}a_3\chi_{pe}(y)$$

(4.3)

follow from (3.21) and $a_3 = (h/2 + h_{pe})^3 - h^3/8$.

The *external* forces and moments $(bN_y)_{pe}$ and $(bM_y)_{pe}$ depend on the voltages supplied to each of the two patches (see (3.22)). If the time-dependent voltages to the outer and inner patches are denoted by $V_1(t)$ and $V_2(t)$, respectively, these forces and moments are given by

$$(bN_y)_{pe} = \mathcal{K}^A \tilde{S}_{1,2}(y)\chi_{pe}(y)[V_1(t) + V_2(t)]$$

$$(bM_y)_{pe} = -\mathcal{K}^B \chi_{pe}(y)[V_1(t) - V_2(t)],$$

(4.4)

where $\mathcal{K}^A = -E_{pe}bd_{31}$, $\mathcal{K}^B = -\frac{1}{2}E_{pe}bd_{31}(h + h_{pe})$ are constants depending on the piezoceramic material properties. The indicator function $\tilde{S}_{1,2}(y)$ is defined in a manner similar to $S_{1,2}(x)$ of (3.9), with y_1, y_2 in place of x_1, x_2, and serves to indicate the sense of the in-plane forces generated by the patches.

Coupled to the system (4.1)–(4.4) are appropriate boundary conditions

$$v(t,0) = 0, \quad \frac{\partial v}{\partial y}(t, \ell) = 0$$

$$w(t,0) = \frac{\partial w}{\partial y}(t,0) = 0, \quad M_y(t,\ell) = \frac{\partial}{\partial y}M_y(t,\ell) = 0,$$

(4.5)

and initial conditions

$$v(0,y) = v_0(y), \quad \frac{\partial v}{\partial t}(0,y) = v_1(y),$$

$$w(0,y) = w_0(y), \quad \frac{\partial w}{\partial t}(0,y) = w_1(y) .$$

(4.6)

For a beam containing a pair of identical patches which are bonded symmetrically about the middle surface, the differential equations (4.1)–(4.4), under the first order Euler-Bernoulli assumptions, describe vibrations in the longitudinal and transverse directions that are uncoupled. As we have seen earlier (see Section 3.3.2), if one has only a single patch bonded to the beam, or if the patches are not identical, then one obtains a set of equations for longitudinal and transverse vibrations that are *coupled* (which is not surprising since the structure consisting of beam plus patch or patches is no longer symmetric). We note that even though the longitudinal and transverse motions are uncoupled in the case of symmetric patches, both motions are present unless we have $V_1(t) = V_2(t)$ (in-phase excitation) which results in longitudinal motion only or $V_1(t) = -V_2(t)$ (out-of-phase excitation) which results in transverse motion only.

The strong form (4.1) of the equations involves irregularities which can (and have) led to computational difficulties for estimation and control efforts found in the literature. Retention of such irregular terms as the discontinuous coefficients in (4.3) and the impulse derivatives resulting from differentiation of the force and moment terms in (4.4) has been shown (see [49, 50] and the references therein) to be of great importance (indeed, essential) when using such models with experimental data from actual structures. It is therefore useful for both theoretical analysis and computational techniques to formulate the system (4.1)–(4.6) in a weak or variational form.

Recalling (2.71), we see that (4.1)–(4.5) in weak form can be written as

$$\int_0^\ell \left\{ \tilde{\rho}\frac{\partial^2 v}{\partial t^2}\phi_1 + \widetilde{Eb}\frac{\partial v}{\partial y}\frac{\partial \phi_1}{\partial y} + \widetilde{c_D b}\frac{\partial^2 v}{\partial y \partial t}\frac{\partial \phi_1}{\partial y} - (bN_y)_{pe}\frac{\partial \phi_1}{\partial y} \right\} dy = 0$$

$$\int_0^\ell \left\{ \tilde{\rho}\frac{\partial^2 w}{\partial t^2}\phi_2 + \widetilde{EI}\frac{\partial^2 w}{\partial y^2}\frac{\partial^2 \phi_2}{\partial y^2} + \widetilde{c_D I}\frac{\partial^3 w}{\partial y^2 \partial t}\frac{\partial^2 \phi_2}{\partial y^2} + (bM_y)_{pe}\frac{\partial^2 \phi_2}{\partial y^2} \right\} dy = 0$$

for all $\phi_1 \in H_L^1(0,\ell)$ and $\phi_2 \in H_L^2(0,\ell)$, where

$$H_L^1(0,\ell) = \{\phi \in H^1(0,\ell) \,|\, \phi(0) = 0\}$$

$$H_L^2(0,\ell) = \{\phi \in H^2(0,\ell) \,|\, \phi(0) = \phi'(0) = 0\} \,.$$

We then seek solutions (v,w) in $L^2(0,\ell) \times L^2(0,\ell)$ having appropriate smoothness. Letting $\langle\cdot,\cdot\rangle$ denote the inner product in $L^2(0,\ell)$ and using the notation $\dot{v} = \frac{\partial v}{\partial t}$ and $\phi' = \frac{\partial \phi}{\partial y}$, we see that these uncoupled equations can be written as follows. Find $v \in H_L^1(0,\ell)$ satisfying the initial conditions in (4.6) and

$$\langle \tilde{\rho}\ddot{v}, \phi_1 \rangle + \langle \widetilde{Eb}v', \phi_1' \rangle + \langle \widetilde{c_D b}\dot{v}', \phi_1' \rangle = \langle (bN_y)_{pe}, \phi_1' \rangle \qquad (4.7)$$

for all $\phi_1 \in H_L^1(0,\ell)$, and find $w \in H_L^2(0,\ell)$ satisfying the initial conditions in (4.6) and

$$\langle \tilde{\rho}\ddot{w}, \phi_2 \rangle + \langle \widetilde{EI}w'', \phi_2'' \rangle + \langle \widetilde{c_D I}\dot{w}'', \phi_2'' \rangle = \langle -(bM_y)_{pe}, \phi_2'' \rangle \qquad (4.8)$$

for all $\phi_2 \in H_L^2(0,\ell)$.

Each of these equations can be written in a concise operator-theoretic form in an appropriate space V^*. For example, consider (4.7) and define the Hilbert space $V = H_L^1(0,\ell)$ with its conjugate dual space V^* (i.e., the linear space of all conjugate linear continuous functionals on V). Define operators A_1, A_2 that are bounded linear maps from V to V^* (denoted by $A_i \in \mathcal{L}(V,V^*)$) by

$$\begin{aligned} (A_1\psi)(\phi) &= \langle \widetilde{Eb}\psi', \phi' \rangle \\ (A_2\psi)(\phi) &= \langle \widetilde{c_D b}\psi', \phi' \rangle \end{aligned}$$

for $\phi, \psi \in V$. Moreover, define $f \in V^*$ by

$$f(\phi) = \langle (bN_y)_{pe}, \phi' \rangle$$

for $\phi \in V$. Then equation (4.7) may be equivalently written

$$\tilde{\rho}\ddot{v}(t) + A_1 v(t) + A_2 \dot{v}(t) = f \quad \text{in } V^*.$$

A similar reformulation can be readily made for equation (4.8) and indeed, for all the weak forms of the various systems discussed in Chapters 2 and 3.

We consider, in subsequent sections, a general class of systems that include models such as (4.7), (4.8) as special cases. In Section 4.2 below, we consider a weak or variational form of these systems which, as noted above, is most useful for development of identification methods and general computational methods (e.g., Galerkin methods). For feedback control theoretic developments on the other hand (e.g., an LQR or MinMax theory), it is most advantageous to have a semigroup formulation (with the usual variation-of-parameters representation of solutions to forced systems). This approach is developed in Section 4.3. Questions related to the equivalence of weak formulations and semigroup formulations are then addressed in Section 4.4.

4.2 Variational or Weak Form

Motivated by the examples given in Section 4.1, we consider the abstract second-order (in time) system

$$\ddot{w}(t) + A_2 \dot{w}(t) + A_1 w(t) = f(t) \quad \text{in } V^*,$$
$$w(0) = w_0, \qquad \dot{w}(0) = w_1. \tag{4.9}$$

More precisely, let V and H be complex Hilbert spaces forming a Gelfand triple $V \hookrightarrow H \cong H^* \hookrightarrow V^*$ with pivot space H and duality pairing $\langle \cdot, \cdot \rangle_{V^*,V}$ (see [22, pages 43-44] [172, page 25], [185, pages 165, 261]). That is, V is continuously and densely embedded in H with $|\phi|_H \le c|\phi|_V$, and we identify H^* with H through the Riesz map. Here V^*, H^* are the (conjugate) dual spaces to V, H, respectively. The duality pairing $\langle \cdot, \cdot \rangle_{V^*,V}$ is the extension by continuity of the inner product $\langle \cdot, \cdot \rangle_H$ from $V \times H$ to $V^* \times H$. Hence, elements $v^* \in V^*$ have the representation $v^*(v) = \langle v^*, v \rangle_{V^*,V}$.

In this context, we will assume that the operators A_1 and A_2 of (4.9) are generated by sesquilinear forms σ_1 and σ_2. Specifically, we assume that $\sigma_1 : V \times V \to \mathbb{C}$ is a sesquilinear form on V that is symmetric, continuous and V-elliptic. That is, σ_1 satisfies the following conditions.

(H1) For all $\phi, \psi \in V$, $\sigma_1(\phi, \psi) = \overline{\sigma_1(\psi, \phi)}$.

(H2) There exists a constant c_1 such that for all $\phi, \psi \in V$

$$|\sigma_1(\phi, \psi)| \le c_1 |\phi|_V |\psi|_V .$$

(H3) There exists a positive constant k_1 such that for all $\phi \in V$

$$\text{Re } \sigma_1(\phi, \phi) = \sigma_1(\phi, \phi) \ge k_1 |\phi|_V^2 .$$

We note that hypothesis (H1) implies that $\sigma_1(\phi, \phi)$ is real for all $\phi \in V$. Under these assumptions we then have that there exists $A_1 \in \mathcal{L}(V, V^*)$ such that $\sigma_1(\phi, \psi) = \langle A_1\phi, \psi \rangle_{V^*,V}$ for all $\phi, \psi \in V$.

To allow for a wide class of damping operators of physical interest, we introduce a sesquilinear form σ_2 defined on a Hilbert space V_2 that can be the same as V, the same as H, or somewhere between V and H; that is, we assume that V_2 is a complex Hilbert space satisfying $V \subset V_2 \subset H$. We can then consider V_2 and H in a Gelfand setting with duality pairing $\langle \cdot, \cdot \rangle_{V_2^*, V_2}$ and can write $V \hookrightarrow V_2 \hookrightarrow H \cong H^* \hookrightarrow V_2^* \hookrightarrow V^*$. We note that $\langle \phi, \psi \rangle_{V^*, V} = \langle \phi, \psi \rangle_{V_2^*, V_2}$ if $\phi \in V_2^*$, $\psi \in V$ and these quantities both reduce to $\langle \phi, \psi \rangle_H$ if $\phi \in H$.

We assume that $\sigma_2 : V_2 \times V_2 \to \mathbb{C}$ satisfies the following continuity and ellipticity conditions.

(H4) There exists a constant c_2 such that for all ϕ, $\psi \in V_2$

$$|\sigma_2(\phi, \psi)| \leq c_2 |\phi|_{V_2} |\psi|_{V_2} .$$

(H5) There exist constants $k_2 > 0$, $\lambda_0 \geq 0$ such that for all $\phi \in V_2$

$$\operatorname{Re} \sigma_2(\phi, \phi) + \lambda_0 |\phi|_H^2 \geq k_2 |\phi|_{V_2}^2 .$$

Under these assumptions we have the existence of $A_2 \in \mathcal{L}(V_2, V_2^*)$ such that $\sigma_2(\phi, \psi) = \langle A_2 \phi, \psi \rangle_{V_2^*, V_2}$ for all ϕ, $\psi \in V_2$.

In subsequent discussions, we shall need some regularity on f in (4.9). Thus we assume

(H6) The input function f satisfies $f \in L^2((0, T), V_2^*)$.

Given the above hypotheses and formulations, we consider the weak or variational form of our system given by

$$\langle \ddot{w}(t), \phi \rangle + \sigma_2(\dot{w}(t), \phi) + \sigma_1(w(t), \phi) = \langle f(t), \phi \rangle \quad \text{for all } \phi \in V,$$

$$w(0) = w_0, \qquad \dot{w}(0) = w_1. \tag{4.10}$$

We point out that (4.9) and (4.10) are equivalent equations if we interpret $\langle \cdot, \cdot \rangle$ as $\langle \cdot, \cdot \rangle_{V^*, V}$ and note that $\langle f, \phi \rangle_{V^*, V} = \langle f, \phi \rangle_{V_2^*, V_2}$ since $f \in L^2((0, T), V_2^*)$. We can then establish the following fundamental existence and uniqueness results.

Theorem 4.1 *Suppose that σ_1, σ_2 and f satisfy (H1)–(H6) and that $w_0 \in V$, $w_1 \in H$. Then there exists a unique solution w of (4.10) (equivalently (4.9)) with $w \in L^2((0, T), V)$, $\dot{w} \in L^2((0, T), V_2)$ and $\ddot{w} \in L^2((0, T), V^*)$. Moreover, solutions of (4.10) depend continuously on the data (w_0, w_1, f) in that the map $(w_0, w_1, f) \to (w, \dot{w})$ is continuous from $V \times H \times L^2((0, T), V_2^*)$ to $L^2((0, T), V) \times L^2((0, T), V_2)$.*

This result differs from analogous results and arguments for undamped systems [137, pages 272-278], [185, pages 439-442] in that the contributions due to the damping sesquilinear form are introduced and treated here. Before sketching the proof to this theorem, we point out that all of the structural models of Chapters 2 and 3 (beams, plates and shells) with a variety of damping models (Kelvin-Voigt, viscous or air, square-root or "structural" and spatial hysteresis) can be treated within this framework. Some of these examples will be discussed in more detail in subsequent sections.

Proof: Let $\{\xi_i\}_{i=1}^{\infty}$ denote a linearly independent total subset of V. For each m, let $V^m = \text{span}\,\{\xi_1, \ldots, \xi_m\}$ and let $w_{0m}, w_{1m} \in V^m$ be chosen so that $w_{0m} \to w_0$ in V, $w_{1m} \to w_1$ in H as $m \to \infty$. Let $w_m(t) \equiv \sum_{i=1}^{m} \eta_{im}(t)\xi_i$ be the unique solution to the m dimensional linear system

$$\langle \ddot{w}_m(t), \xi_j \rangle_H + \sigma_2(\dot{w}_m(t), \xi_j) + \sigma_1(w_m(t), \xi_j) = \langle f(t), \xi_j \rangle_{V_2^*, V_2}$$

$$w_m(0) = w_{0m} \tag{4.11}$$

$$\dot{w}_m(0) = w_{1m}\,,$$

where $j = 1, 2, \ldots, m$. Multiplying the equation in (4.11) by $\dot{\bar{\eta}}_{jm}(t)$ and summing over j, we obtain

$$\langle \ddot{w}_m(t), \dot{w}_m(t) \rangle_H + \sigma_2(\dot{w}_m(t), \dot{w}_m(t)) + \sigma_1(w_m(t), \dot{w}_m(t)) = \langle f(t), \dot{w}_m(t) \rangle_{V_2^*, V_2}.$$

Since $\frac{d}{dt}\sigma_1(w_m(t), w_m(t)) = 2\,\mathrm{Re}\,\sigma_1(w_m(t), \dot{w}_m(t))$, this equality implies that

$$\frac{d}{dt}\left\{|\dot{w}_m(t)|_H^2 + \sigma_1(w_m(t), w_m(t))\right\} \;+\; 2\,\mathrm{Re}\,\sigma_2(\dot{w}_m(t), \dot{w}_m(t))$$
$$= 2\,\mathrm{Re}\,\langle f(t), \dot{w}_m(t) \rangle_{V_2^*, V_2}.$$

Upon integrating this equality, we obtain

$$|\dot{w}_m(t)|_H^2 \;+\; \sigma_1(w_m(t), w_m(t)) + \int_0^t 2\,\mathrm{Re}\,\sigma_2(\dot{w}_m(s), \dot{w}_m(s))ds$$
$$= |\dot{w}_m(0)|_H^2 + \sigma_1(w_m(0), w_m(0)) + \int_0^t 2\,\mathrm{Re}\,\langle f(s), \dot{w}_m(s) \rangle_{V_2^*, V_2}ds.$$

Using (H3), (H5) and for arbitrary $\epsilon > 0$, the inequality

$$|\langle f(s), \dot{w}_m(s) \rangle_{V_2^*, V_2}| \leq \frac{1}{4\epsilon}|f(s)|_{V_2^*}^2 + \epsilon|\dot{w}_m(s)|_{V_2}^2\,, \tag{4.12}$$

in the above equality, we find that

$$|\dot{w}_m(t)|_H^2 + k_1|w_m(t)|_V^2 + \int_0^t 2(k_2 - \epsilon)|\dot{w}_m(s)|_{V_2}^2 ds$$
$$\leq |w_{1m}|_H^2 + c_1|w_{0m}|_V^2 + 2\lambda_0 \int_0^t |\dot{w}_m(s)|_H^2 ds + \int_0^t \frac{1}{2\epsilon}|f(s)|_{V_2^*}^2 ds\,. \tag{4.13}$$

Recalling that $w_{1m} \to w_1$ in H, $w_{0m} \to w_0$ in V (and thus $\{w_{1m}\}, \{w_{0m}\}$ are H and V bounded) and fixing ϵ such that $2(k_2 - \epsilon) = \delta > 0$, we conclude that the estimate (4.13) can be replaced by (for m sufficiently large)

$$|\dot{w}_m(t)|_H^2 + k_1|w_m(t)|_V^2 + \delta \int_0^t |\dot{w}_m(s)|_{V_2}^2 ds$$
$$\leq (M + 1) + 2\lambda_0 \int_0^t |\dot{w}_m(s)|_H^2 ds \tag{4.14}$$

where

$$M \equiv |w_1|_H^2 + c_1|w_0|_V^2 + \int_0^t \frac{1}{2\epsilon}|f(s)|_{V_2^*}^2 ds.$$

By ignoring the second and third terms in (4.14) and using Gronwall's in-equality, it follows that $\{\dot{w}_m\}$ is bounded in $C((0,T), H)$. Then using (4.14) again and knowing that $\{\dot{w}_m\}$ is bounded this manner, we may conclude that $\{w_m\}$ is $C((0,T), V)$ bounded. It also follows from (4.14) and the fact that $\{\dot{w}_m\}$ is $C((0,T), H)$ bounded that $\{\ddot{w}_m\}$ is $L^2((0,T), V_2)$ bounded. Hence we may find a subsequence $\{w_{m_k}\}$ and limit functions $w \in L^2((0,T), V)$ and $\hat{w} \in L^2((0,T), V_2)$ such that

$$w_{m_k} \to w \text{ weakly in } L^2((0,T), V)$$
$$\dot{w}_{m_k} \to \hat{w} \text{ weakly in } L^2((0,T), V_2). \tag{4.15}$$

We note, however, that for $t \in [0,T)$,

$$w_{m_k}(t) = w_{m_k}(0) + \int_0^t \dot{w}_{m_k}(s)ds \tag{4.16}$$

in the V (and hence V_2 and H) sense. Moreover, $w_{m_k}(0) = w_{0m_k} \to w_0$ in the V and hence V_2 sense whereas for each t, $\int_0^t \dot{w}_{m_k}(s)ds \to \int_0^t \hat{w}(s)ds$ weakly in V_2. Hence, taking the limit in the weak V_2 sense in (4.16) we obtain

$$w(t) = w_0 + \int_0^t \hat{w}(s)ds \quad \text{for } t \in [0,T)$$

in the V_2 sense. Thus, $\dot{w}(t)$ exists almost everywhere in the V_2 sense with $\dot{w} = \hat{w} \in L^2((0,T), V_2)$ and $w(0) = w_0$. It remains to argue that w is indeed a solution to (4.10).

To this end, we return to (4.11) and let $\psi \in C^1[0,T]$, with $\psi(T) = 0$, be arbitrarily chosen. Put $\psi_j(t) \equiv \psi(t)\xi_j$ where the $\{\xi_j\}$ are the same as in (4.11). Multiplying (4.11) by $\psi(t)$ and fixing $j < m$, we have, upon integration,

$$\int_0^T \{\langle \ddot{w}_m(t), \psi_j(t)\rangle_H + \sigma_2(\dot{w}_m(t), \psi_j(t)) + \sigma_1(w_m(t), \psi_j(t))\} dt$$
$$= \int_0^T \langle f(t), \psi_j(t)\rangle_{V_2^*, V_2} dt. \tag{4.17}$$

Integrating by parts in the first term, employing the convergences of (4.15), the inclusions $\sigma_2(\cdot, \psi_j(t)) \in V_2^*$, $\sigma_1(\cdot, \psi_j(t)) \in V^*$ for each t, and taking subsequential limits as $m = m_k \to \infty$ in this equation, we obtain

$$\int_0^T \{\langle -\dot{w}(t), \dot{\psi}_j(t)\rangle_H + \sigma_2(\dot{w}(t), \psi_j(t)) + \sigma_1(w(t), \psi_j(t))\} dt$$
$$= \int_0^T \langle f(t), \psi_j(t)\rangle_{V_2^*, V_2} dt + \langle w_1, \psi_j(0)\rangle_H \tag{4.18}$$

for each j. Recalling that $\psi_j(t) = \psi(t)\xi_j$ and further restricting ψ so that $\psi \in C_0^\infty(0,T)$, we obtain from (4.18)

$$\int_0^T \dot\psi(t)\langle -\dot w(t), \xi_j\rangle_H dt$$

$$+ \int_0^T \psi(t)\left\{\sigma_2(\dot w(t),\xi_j) + \sigma_1(w(t),\xi_j) - \langle f(t),\xi_j\rangle_{V_2^*,V_2}\right\} dt = 0$$

for each ξ_j. This implies for each ξ_j

$$\frac{d}{dt}\langle \dot w(t),\xi_j\rangle_H + \sigma_2(\dot w(t),\xi_j) + \sigma_1(w(t),\xi_j) = \langle f(t),\xi_j\rangle_{V_2^*,V_2}. \qquad (4.19)$$

Since $\{\xi_j\}$ is total in V we thus have that $\ddot w \in L_2((0,T),V^*)$ and for all $\phi \in V$

$$\langle \ddot w(t),\phi\rangle_{V^*,V} + \sigma_2(\dot w(t),\phi) + \sigma_1(w(t),\phi) = \langle f(t),\phi\rangle_{V_2^*,V_2}$$

which is the equation in (4.10).

We already have $w(0) = w_0$ and to argue $\dot w(0) = w_1$, we return to (4.18) which holds for all $\psi_j(t) = \psi(t)\xi_j$, $\psi \in C^1[0,T]$, $\psi(T) = 0$. Integrating by parts in the first term in (4.18) and using (4.19), we obtain

$$\langle -\dot w(t),\psi_j(t)\rangle_H\Big|_{t=0}^{t=T} = \langle w_1,\psi_j(0)\rangle_H$$

or

$$\langle \dot w(0),\xi_j\rangle_H\,\psi(0) = \langle w_1,\xi_j\rangle_H\,\psi(0) \quad \text{for all } j.$$

From this it follows that $\dot w(0) = w_1$ and thus w is indeed a solution to the initial value problem (4.10).

Continuous dependence of these solutions on the data (w_0, w_1, f) follows readily from some of the estimates used above in establishing existence. Returning to (4.13), we see that this estimate implies

$$|\dot w_m(t)|_H^2 \le K_m + 2\lambda_0 \int_0^t |\dot w_m(s)|_H^2 ds$$

where $K_m \equiv |w_{1m}|_H^2 + c_1|w_{0m}|_V^2 + \frac{1}{2\epsilon}|f|_{L^2((0,T),V_2^*)}$. Thus, using Gronwall's inequality we obtain

$$|\dot w_m(t)|_H^2 \le K_m e^{2\lambda_0 t}.$$

However, use of this inequality in (4.13) yields

$$k_1|w_m(t)|_V^2 + \delta \int_0^t |\dot w_m(s)|_{V_2}^2 ds \le K_m + K_m\, 2\lambda_0 \int_0^t e^{2\lambda_0 s} ds$$

$$\le K_m C_1$$

where C_1 is a constant. Integrating over $(0,T)$ we obtain

$$k_1|w_m|_{L^2((0,T),V)}^2 + \delta T|\dot w_m|_{L^2((0,T),V_2)}^2 \le K_m C_2$$

where again, C_2 is a constant. Recalling that $w_{1m} \to w_1$ in H, $w_{0m} \to w_0$ in V and using weak lower semicontinuity of norms, along with the weak convergences in (4.15), we may take limits in this last inequality to obtain

$$k_1 |w|^2_{L^2((0,T),V)} + \delta T |\dot{w}|^2_{L^2((0,T),V_2)}$$
$$\leq \left\{ |w_1|^2_H + c_1 |w_0|^2_V + \frac{1}{2\epsilon} |f|^2_{L^2((0,T),V_2^*)} \right\} C_2. \tag{4.20}$$

Since the mapping $(w_0, w_1, f) \to (w, \dot{w})$ is linear, this estimate yields the desired continuity statement of the theorem.

Finally, we turn to uniqueness of solutions of (4.10), observing that (4.20) shows that solutions constructed through the above limiting procedure are, of course, unique. However, we must argue that *any* two solutions corresponding to the same data w_0, w_1, f are the same. It suffices to argue that *only* the trivial solution $w \equiv 0$ of (4.10) can result from data $w_0 = 0$, $w_1 = 0$, $f = 0$. Again we follow standard arguments (for example, see Lions [137, pages 272-278]). Let w be a solution of (4.10) corresponding to $w_0 = 0$, $w_1 = 0$, $f = 0$ and define for s fixed but arbitrary in $(0, T)$

$$\psi(t) = \begin{cases} -\int_t^s w(\xi)d\xi & t < s \\ 0 & t \geq s \end{cases}$$

so that $\psi(T) = 0$. It is readily argued that $\psi(t) \in V$ for each t and hence we may choose $\phi = \psi(t)$ in (4.10) to obtain

$$\langle \ddot{w}(t), \psi(t) \rangle_{V^*,V} + \sigma_2(\dot{w}(t), \psi(t)) + \sigma_1(w(t), \psi(t)) = 0. \tag{4.21}$$

Since $\dot{\psi}(t) = w(t)$ for almost every $t < s$, we have

$$\int_0^s \left\{ \langle \ddot{w}(t), \psi(t) \rangle_{V^*,V} + \langle \dot{w}(t), w(t) \rangle_{V^*,V} \right\} dt$$

$$= \int_0^s \frac{d}{dt} \langle \dot{w}(t), \psi(t) \rangle_{V^*,V} dt$$

$$= \langle \dot{w}(t), \psi(t) \rangle_{V^*,V} \Big|_{t=0}^{t=s} = 0 .$$

Integrating (4.21) and using this last identity, we obtain

$$\int_0^s \left\{ \langle \dot{w}(t), w(t) \rangle_{V^*,V} - \sigma_2(\dot{w}(t), \psi(t)) - \sigma_1(w(t), \psi(t)) \right\} dt = 0$$

which can be equivalently written as

$$\int_0^s \frac{d}{dt} \left\{ |w(t)|^2_H - \sigma_1(\psi(t), \psi(t)) \right\} dt = 2 \, \mathrm{Re} \int_0^s \sigma_2(\dot{w}(t), \psi(t)) dt.$$

Since $\psi(s) = 0$ and $w(0) = 0$, this yields

$$|w(s)|^2_H + \sigma_1(\psi(0), \psi(0)) = 2 \, \mathrm{Re} \int_0^s \sigma_2(\dot{w}(t), \psi(t)) dt. \tag{4.22}$$

Moreover, since $\frac{d}{dt}\sigma_2(w(t), \psi(t)) = \sigma_2(\dot{w}(t), \psi(t)) + \sigma_2(w(t), \dot{w}(t))$, it follows that

$$\int_0^s \sigma_2(\dot{w}(t), \psi(t))dt = -\int_0^s \sigma_2(w(t), w(t))dt + \sigma_2(w(t), \psi(t)) \mid_{t=0}^{t=s}$$

$$= -\int_0^s \sigma_2(w(t), w(t))dt$$

because $\psi(s) = 0$, $w(0) = 0$. Using this in (4.22), we have

$$|w(s)|_H^2 + \sigma_1(\psi(0), \psi(0)) = 2\,\mathrm{Re}\int_0^s -\sigma_2(w(t), w(t))dt.$$

Recalling (H3) and (H5), we thus obtain

$$|w(s)|_H^2 + k_1|\psi(0)|_V^2 \le 2\lambda_0 \int_0^s |w(t)|_H^2 dt$$

or

$$|w(s)|_H^2 \le 2\lambda_0 \int_0^s |w(t)|_H^2 dt,$$

for s arbitrary in $(0, T)$. Once again, we can employ Gronwall's inequality and conclude that $w(t) = 0$ for $t \in (0, T)$.   ∎

Remark 4.2 One can actually strengthen the result of Theorem 4.1 to conclude that $w \in C((0, T), V)$ and $\dot{w} \in C((0, T), H)$ – (compare with [137, page 273], [138, Chapter 3]). To see this, we first consider the situation where $(w(0), \dot{w}(0)) = (w_0, w_1) = (0, 0)$. For $\epsilon > 0$ define

$$w_\epsilon(t) = \frac{1}{\epsilon}\int_{-\epsilon}^0 w(t + s)ds\,, \qquad f_\epsilon(t) = \frac{1}{\epsilon}\int_{-\epsilon}^0 f(t + s)ds$$

where we take $w(s) = w(0) = 0$, $f(s) = 0$ for $s < 0$. From this definition we see that $w_\epsilon \in C^1((0, T), V) \cap H^2((0, T), H)$. From (4.10) we have

$$\langle \ddot{w}_\epsilon(t), \phi \rangle + \sigma_2(\dot{w}_\epsilon(t), \phi) + \sigma_1(w_\epsilon(t), \phi) = \langle f_\epsilon(t), \phi \rangle \qquad (4.23)$$

for all $\phi \in V$ since

$$\frac{1}{\epsilon}\int_{-\epsilon}^0 \ddot{w}(t + s)ds = \ddot{w}_\epsilon(t)\,,$$

$$\frac{1}{\epsilon}\int_{-\epsilon}^0 \dot{w}(t + s)ds = \dot{w}_\epsilon(t)\,.$$

Let Z be the Banach space defined by

$$Z = \left\{ z \in C((0, T), V) \mid \dot{z} \in C((0, T), H) \cap L^2((0, T), V_2) \right\}$$

equipped with norm

$$|z| = \sup_{t \in [0,T]} \{|z(t)|_V + |\dot{z}(t)|_H\} + \left(\int_0^T |\dot{z}(s)|^2_{V_2} ds\right)^{1/2}.$$

Then it is readily seen that $\{w_\epsilon\}_{\epsilon>0}$ is Cauchy in Z. Indeed, for $\epsilon, \eta > 0$ we have from (4.23)

$$\langle \ddot{w}_\epsilon(t) - \ddot{w}_\eta(t), \phi \rangle + \sigma_2(\dot{w}_\epsilon(t) - \dot{w}_\eta(t), \phi) + \sigma_1(w_\epsilon(t) - w_\eta(t), \phi)$$
$$= \langle f_\epsilon(t) - f_\eta(t), \phi \rangle$$

for all $\phi \in V$. Putting $\phi = \dot{w}_\epsilon(t) - \dot{w}_\eta(t)$ and arguing as in (4.13), we obtain

$$|\dot{w}_\epsilon(t) - \dot{w}_\eta(t)|^2_H + k_1 |w_\epsilon(t) - w_\eta(t)|^2_V + \delta \int_0^t |\dot{w}_\epsilon(s) - \dot{w}_\eta(s)|^2_{V_2} ds$$

$$\leq 2\lambda_0 \int_0^t |\dot{w}_\epsilon(s) - \dot{w}_\eta(s)|^2_H ds + \frac{1}{2\epsilon} \int_0^t |f_\epsilon(s) - f_\eta(s)|^2_{V_2^*} ds .$$

Since $\{f_\epsilon\}_{\epsilon>0}$ is Cauchy in $L^2((0,T), V_2^*)$, our claim follows from this inequality combined with Gronwall's inequality. Moreover, since Z is complete, the limit w of $\{w_\epsilon\}$ as $\epsilon \to 0^+$ is in Z.

The case when $w \in Z$ for nontrivial $(w_0, w_1) \in V \times H$ and $f \equiv 0$ follows from Theorems 4.7 and 4.13 presented below. Moreover, the arguments above yield the stronger continuous dependence result which we state as a corollary.

Corollary 4.3 *The solutions of (4.10) guaranteed by Theorem 4.1 satisfy*

$$|\dot{w}(t)|^2_H + k_1 |w(t)|^2_V + \delta \int_0^t |\dot{w}(s)|^2_{V_2} ds$$

$$\leq |w_1|^2_H + k_1 |w_0|^2_V + 2\lambda_0 \int_0^t |\dot{w}(s)|^2_H ds + \frac{1}{2\epsilon} \int_0^t |f(s)|^2_{V_2^*} ds$$

for $(w_0, w_1) \in V \times H$ and $f \in L^2((0,T), V_2^*)$. Hence the map $(w_0, w_1, f) \to w$ is continuous from $V \times H \times L^2((0,T), V_2^*)$ to Z.

Remark 4.4 In the event that $k_2 = 0$ in (H5) is the best lower bound available, the arguments and results above are valid with the following modifications. One requires $f \in L^2((0,T), H)$ and replaces the V_2^* and V_2 norms in the estimates (H4) and (4.12) by the H norm ($\epsilon = \frac{1}{2}$ can then be chosen in (4.13)). One does not obtain $L^2((0,T), V_2)$ boundedness for $\{\dot{w}_m\}$, rather only $C((0,T), H)$ boundedness. Thus the convergence of $\dot{w}_{m_k}$ in (4.15) is weak in $L^2((0,T), H)$. One then obtains only that $\dot{w} \in L^2((0,T), H)$ in the statement of the theorem. Essentially, one replaces V_2 by H in the statement and hypotheses of the theorem and one obtains the same results as in the case of no damping ($\sigma_2 = 0$) given in [137, 185].

Remark 4.5 Since the arguments establishing Theorem 4.1 are constructive and involve *a priori* estimates, they can also be used rather efficiently to provide convergence arguments for certain approximation schemes which can be formulated in a semi-discrete Galerkin framework. We outline briefly the ideas.

Let H^N be a sequence of finite dimensional subspaces of H satisfying:

(D1N) The spaces H^N satisfy $H^N \subset V$ and if $P^N : H \to H^N$ is the orthogonal projection of H onto H^N, we have for each $\psi \in V$, $|\psi - P^N\psi|_V \to 0$ as $N \to \infty$.

(D2N) The spaces H^N satisfy the monotonicity condition $H^N \subset H^{N+1}$.

The approximation condition (D1N) is a rather standard one and is satisfied by many finite element, spectral and modal approximation families. The monotonicity condition (D2N) is not quite as widespread in approximation theory, but can be satisfied by construction in certain linear finite element families, in certain spectral approximations, and in the modal expansions employed in many applications. For example, let $V^K = span\{\psi_1^K, \ldots, \psi_K^K\} \subset L^2(0, \ell)$ where the ψ_j^K are the usual piecewise linear spline elements corresponding to discretizations of $(0, \ell)$ with uniform mesh $\Delta x = \ell/K$. Then $H^N = V^{2^N}$ satisfies conditions (D1N) and (D2N) with $V = H^1(0, \ell)$.

We define the Galerkin approximations $\{w^N\}$ to solutions of (4.10) by

$$\langle \ddot{w}^N(t), \phi \rangle + \sigma_2(\dot{w}^N(t), \phi) + \sigma_1(w^N(t), \phi) = \langle f(t), \phi \rangle \text{ for all } \phi \in H^N$$

$$(4.24)$$

$$w^N(0) = P^N w_0, \quad \dot{w}^N(0) = P^N w_1.$$

We then have the following convergence results.

Theorem 4.6 *Under (D1N), (D2N) and the assumption that $V_2 \hookrightarrow H$ is a compact embedding, we have $w^N \to w$ in $C((0, T), H)$ where w^N is the solution of (4.24) and w is the unique solution of (4.10).*

Proof: We only sketch the arguments here. First, we consider (4.24) with the choice of test function $\phi = \dot{w}^N(t)$. Using *a priori* type estimates similar to (4.14) we find that $\{\dot{w}^N\}$ is bounded in $C((0, T), H)$ while $\{w^N\}$ is bounded in $C((0, T), V)$. Moreover, $\{\dot{w}^N\}$ is also bounded in $L^2((0, T), V_2)$. Using this and the estimate

$$\left| w^N(t + \Delta) - w^N(t) \right|_{V_2} = \left| \int_t^{t+\Delta} \dot{w}^N(\tau) d\tau \right|_{V_2} \leq |\dot{w}^N|_{L^2((0,T),V_2)} |\Delta|,$$

we obtain that $\{w^N\}$ is equicontinuous and bounded in $C((0, T), V_2)$. Since the embedding $V_2 \hookrightarrow H$ is compact, we may invoke the Arzela-Ascoli theorem to extract a subsequence (which we again denote by $\{w^N\}$) converging to some element $\tilde{w}$ in $C((0, T), H)$. Again, using the boundedness of $\{\dot{w}^N\}$ in

$L^2((0,T), V_2)$ and $\{w^N\}$ in $C((0,T), H)$, we may take subsequences, again denoted by $\{w^N\}$, and argue (as in the proof of Theorem 4.1) that w^N converges
weakly to $\tilde{w}$ in $L^2((0,T), V)$ and $\dot{\tilde{w}}^N$ converges weakly to $\dot{\tilde{w}}$ in $L^2((0,T), V_2)$.

It remains to argue that $\tilde{w}$ is indeed a solution of (4.10). This is done in the
same manner as in the arguments for Theorem 4.1. For the limit arguments,
one chooses (in the analogue of (4.17)) a test function $\psi_j(t) = \psi(t)\phi_{N_j}$ where
ϕ_{N_j} is fixed in H^{N_j} and ψ is arbitrary with $\psi \in C^1(0,T)$ and $\psi(T) = 0$. Then,
considering the integrated form of (4.24) with $\phi = \psi_j(t)$ and $N \geq N_j$ (for
example, (4.17) with the w_m replaced by w^N), we note that by (D2N) we have
$\psi_j(t) \in H^N$ for all $N \geq N_j$ (it is here and only here that we use the monotone
property (D2N) of the H^N). Passing to the limit as $N \to \infty$ and arguing
exactly as in the proof of Theorem 4.1, we are able to conclude that indeed $\tilde{w}$
is a solution of (4.10). Thus, we can show that for some subsequence of the
original sequence $\{w^N\}$ defined via (4.24), we have convergence in $C((0,T), H)$
to a solution of (4.10). But solutions of (4.10) are unique and hence we may
conclude that any subsequence of the original Galerkin sequence $\{w^N\}$ has,
in turn, a subsequence converging to the unique solution of (4.10). Hence,
the original Galerkin sequence $\{w^N\}$ defined in (4.24) must converge to the
solution of (4.10).

■

As we shall see in the next chapter, the results given in Theorem 4.6 can
be established under considerably weaker assumptions; however, the proofs
then are no longer essentially the simple constructive arguments used above to
obtain existence of solutions. Specifically, one can omit the assumption that
$V_2 \hookrightarrow H$ is compact (this allows weaker damping models to be included in the
abstract framework) as well as omit the monotonicity condition (D2N). In that
case, a condition similar to (D1N) concerning convergence in V_2 must be added:
for $\psi \in V_2$, $|\psi - P^N\psi|_{V_2} \to 0$ as $N \to \infty$. If one additionally assumes that the
sesquilinear forms σ_1, σ_2 depend continuously on the parameters $q \in Q$, then
one can in actuality give a parameter-dependent version of the convergence
of Theorem 4.6 that is precisely what is needed to develop approximation
theorems for inverse problems governed by systems (4.10) and (4.24). Details
will be deferred until the next chapter.

4.3   Semigroup Formulation

In this section we turn to a semigroup formulation for solutions of (4.9), or
equivalently (4.10), under the same assumptions (H1)–(H6) of the previous
section. As a first step we rewrite (4.9) in first order form in the variables
$z = (w, \dot{w})^T$ on the state space $\mathcal{H} = V \times H$. We next establish that the
resulting operator generates a C_0-semigroup on $\mathcal{H}$.

We begin by defining the space $\mathcal{V} = V \times V$ and noting that under the
Gelfand triple formulation of the previous section, we also have $\mathcal{V} \hookrightarrow \mathcal{H} \hookrightarrow \mathcal{V}^*$

where $\mathcal{V}^* = V \times V^*$. We next define a sesquilinear form $\sigma : \mathcal{V} \times \mathcal{V} \to \mathbb{C}$ by

$$
\begin{aligned}
\sigma(\Phi, \Psi) &= \sigma\left((\phi_1, \phi_2), (\psi_1, \psi_2)\right) \\
&= -\langle \phi_2, \psi_1 \rangle_V + \sigma_1(\phi_1, \psi_2) + \sigma_2(\phi_2, \psi_2)
\end{aligned}
\tag{4.25}
$$

for $\Phi = (\phi_1, \phi_2)$, $\Psi = (\psi_1, \psi_2) \in \mathcal{V}$. For $z(t) = (w(t), \dot{w}(t))$, (4.10) can be rewritten as

$$
\begin{aligned}
\langle \dot{z}(t), \chi \rangle_{\mathcal{V}^*, \mathcal{V}} + \sigma(z(t), \chi) &= \langle F(t), \chi \rangle_{\mathcal{V}^*, \mathcal{V}} \quad \text{for } \chi \in \mathcal{V} \\
z(0) &= z_0 = (w_0, w_1)
\end{aligned}
\tag{4.26}
$$

where $F(t) = (0, f(t))$. This is formally equivalent to the strong form of the equation given by (we won't distinguish between a vector and its transpose in this section)

$$
\begin{aligned}
\dot{z}(t) &= \mathcal{A}z(t) + F(t) \\
z(0) &= z_0 = (w_0, w_1)
\end{aligned}
\tag{4.27}
$$

where $\mathcal{A}$ is given by

$$
\mathrm{dom}\mathcal{A} = \{ \chi = (\phi, \psi) \in \mathcal{H} \mid \psi \in V \text{ and } A_1\phi + A_2\psi \in H \}
\tag{4.28}
$$

and $\mathcal{A}\chi = (\psi, -A_1\phi - A_2\psi)$, or (in matrix operator form)

$$
\mathcal{A} = \begin{bmatrix} 0 & I \\ -A_1 & -A_2 \end{bmatrix}.
\tag{4.29}
$$

We note that $\mathcal{A}$ is the negative of the restriction to $\mathrm{dom}\mathcal{A}$ of the operator $\tilde{\mathcal{A}} \in \mathcal{L}(\mathcal{V}, \mathcal{V}^*)$ defined by $\sigma(\Phi, \Psi) = \langle \tilde{\mathcal{A}}\Phi, \Psi \rangle_{\mathcal{V}^*, \mathcal{V}}$ so that $\sigma(\Phi, \Psi) = \langle -\mathcal{A}\Phi, \Psi \rangle_{\mathcal{H}}$ for $\Phi \in \mathrm{dom}\mathcal{A}, \Psi \in \mathcal{V}$. Under (H1)–(H3), it is readily observed that (recall that $\sigma_1(\phi, \phi)$ is real)

$$
k_1 |\phi|_V^2 \leq \sigma_1(\phi, \phi) \leq c_1 |\phi|_V^2
$$

so that σ_1 and the V inner product are equivalent. We may then define V_1 as the set V taken with σ_1 as inner product, i.e., $\langle \phi, \psi \rangle_{V_1} = \sigma_1(\phi, \psi)$, thus obtaining a space that is setwise equal and topologically equivalent to V. We shall argue that $\mathcal{A}$ generates a C_0-semigroup in $\mathcal{H}_1 = V_1 \times H$ and hence also in $\mathcal{H}$. In the space $\mathcal{H}_1$, the operator $\mathcal{A}$ is associated with a $\mathcal{V}_1 = V_1 \times V_1$ sesquilinear form $\sigma^{(1)}$ given by

$$
\sigma^{(1)}(\Phi, \Psi) = \langle \mathcal{A}\Phi, \Psi \rangle_{\mathcal{H}_1}, \quad \Phi \in \mathrm{dom}\mathcal{A}, \ \Psi \in \mathcal{V}_1.
$$

To argue that $\mathcal{A}$ is an infinitesimal generator, we employ the Lumer-Phillips theorem [152, page 14]. In the Hilbert space $\mathcal{H}_1$ it suffices to argue that $\mathcal{A} - \lambda$ is dissipative and that the range $\mathcal{R}(\lambda - \mathcal{A})$ of $\lambda - \mathcal{A}$ is all of $\mathcal{H}_1$ for some $\lambda > 0$.

Dissipativeness follows immediately from the definition of $\mathcal{A}$, (H1) and (H5) since

$$
\begin{aligned}
\mathrm{Re}\, \langle \mathcal{A}\Phi, \Phi \rangle_{\mathcal{H}_1} &= \mathrm{Re}\, \{ \sigma_1(\phi_2, \phi_1) - \sigma_1(\phi_1, \phi_2) - \sigma_2(\phi_2, \phi_2) \} \\
&= \mathrm{Re}\, \{ \overline{\sigma_1(\phi_2, \phi_1)} - \sigma_1(\phi_1, \phi_2) - \sigma_2(\phi_2, \phi_2) \} \\
&\leq \lambda_0 |\phi_2|_H^2 - k_2 |\phi_2|_{V_2}^2 \\
&\leq \lambda_0 |\Phi|_{\mathcal{H}_1}^2
\end{aligned}
\tag{4.30}
$$

where $\lambda_0 \geq 0$ is the constant of (H5). Thus Re $\langle (\mathcal{A} - \lambda_0)\Phi, \Phi \rangle \leq 0$ and it remains only to argue the range condition.

We wish to argue that for some $\lambda > 0$, the range of $\lambda - \mathcal{A}$ is $\mathcal{H}_1$. Thus given $\xi = (\eta, \zeta) \in \mathcal{H}_1$, we wish to establish solvability of $(\lambda - \mathcal{A})\chi = \xi$ for $\chi = (\phi, \psi) \in \text{dom}\mathcal{A}$. This equation, however, is equivalent to

$$\lambda\phi - \psi = \eta$$
$$\lambda\psi + A_1\phi + A_2\psi = \zeta. \tag{4.31}$$

If we formally solve the first equation for $\psi = \lambda\phi - \eta$ and substitute this into the second equation, we obtain

$$\lambda^2\phi + A_1\phi + \lambda A_2\phi = \zeta + \lambda\eta + A_2\eta \tag{4.32}$$

which must be solved for $\phi \in V_1$ (and then ψ defined by $\psi = \lambda\phi - \eta$ will also be in V_1, with $A_1\phi + A_2\psi$ in H). These formal calculations suggest that we define for $\lambda > 0$ the sesquilinear form on $V \times V$ given by

$$\sigma_\lambda(\phi, \psi) = \lambda^2 \langle \phi, \psi \rangle_H + \sigma_1(\phi, \psi) + \lambda\sigma_2(\phi, \psi).$$

Since σ_1, σ_2 satisfy (H3), (H5) we have

$$
\begin{aligned}
\text{Re } \sigma_\lambda(\phi, \phi) &= \lambda^2|\phi|_H^2 + \text{Re } \sigma_1(\phi, \phi) + \lambda\text{Re } \sigma_2(\phi, \phi) \\
&\geq \lambda^2|\phi|_H^2 + k_1|\phi|_V^2 + \lambda\left(k_2|\phi|_{V_2}^2 - \lambda_0|\phi|_H^2 \right) \\
&= \lambda(\lambda - \lambda_0)|\phi|_H^2 + k_1|\phi|_V^2 + \lambda k_2|\phi|_{V_2}^2 \\
&> k_1|\phi|_V^2
\end{aligned}
$$

for $\lambda \geq \lambda_0 \geq 0$. Hence σ_λ is V-elliptic for $\lambda \geq \lambda_0$ and thus (4.32) is solvable for $\phi \in V$. It follows, as noted above, that (4.31) is thus solvable for $(\phi, \psi) \in \text{dom}\mathcal{A}$ and hence $\mathcal{R}(\lambda - \mathcal{A}) = \mathcal{H}_1$. Thus for $\lambda \geq \lambda_0$, $\mathcal{A} - \lambda$ generates a contraction C_0-semigroup on $\mathcal{H}_1$ and hence we have established the following.

Theorem 4.7   *Under hypotheses (H1)–(H5) on σ_1, σ_2, the operator $\mathcal{A}$ defined in (4.28), (4.29) generates a C_0-semigroup $\mathcal{T}(t)$ on $\mathcal{H} = V \times H$ which satisfies $|\mathcal{T}(t)|_{\mathcal{H}_1} \leq e^{\lambda t}$ for any $\lambda \geq \lambda_0$.*

As usual, we can use this semigroup to define mild solutions for (4.27). For $z_0 = (w_0, w_1) \in \mathcal{H}$ and $F = (0, f)$ with $f \in L^2((0, T), H)$ the "variation-of-parameters" representation

$$z(t) = \mathcal{T}(t)z_0 + \int_0^t \mathcal{T}(t - s)F(s)ds \tag{4.33}$$

defines a mild solution for (4.27) which for $z_0 \in \text{dom}\mathcal{A}$ and $f \in C^1((0, T), H)$ is the unique strong solution [172, page 64] to (4.26) and hence to (4.10). Thus this solution must be the same as the weak or variational solution guaranteed by Theorem 4.1 of Section 4.2 whenever $z_0 = (w_0, w_1)$ is in dom$\mathcal{A}$ and $f \in C^1((0,T), H)$.

We wish, of course, to extend the formula (4.33) and the concept of mild solution to allow for $f \in L^2((0,T), V^*)$ and to establish equivalence to weak solutions in this case.

We consider first the case in which σ_2 is V-elliptic and V-continuous. That is, we take $V_2 = V$ in hypotheses (H4) and (H5) which we now denote (H4'), (H5'). That is,

(H4') The form σ_2 satisfies (H4) with V_2 replaced by V.

(H5') The form σ_2 satisfies (H5) with V_2 replaced by V.

Under these assumptions, consider the operator $\tilde{A}$ (which is an extension of $\mathcal{A}$ from dom$\mathcal{A}$ to $\mathcal{V}_1 = V_1 \times V_1$) given by

$$\sigma^{(1)}(\Phi, \Psi) \equiv -\sigma_1(\phi_2, \psi_1) + \sigma_1(\phi_1, \psi_2) + \sigma_2(\phi_2, \psi_2)$$

$$= \langle \tilde{A}\Phi, \Psi \rangle_{\mathcal{V}_1^*, \mathcal{V}_1}.$$

Note that $\sigma^{(1)}$ is just σ of (4.25) where we use the σ_1 inner product with V, i.e., V_1. The arguments in (4.30) yield immediately that

$$\text{Re } \sigma(\Phi, \Phi) \geq k_2 |\phi_2|_V^2 - \lambda_0 |\phi_2|_H^2$$

$$\geq k_2 \left(|\phi_1|_V^2 + |\phi_2|_V^2 \right) - k_2 |\phi_1|_V^2 - \lambda_0 |\phi_2|_H^2$$

$$\geq \tilde{k}_2 |\Phi|_{\mathcal{V}_1}^2 - \tilde{\lambda}_0 |\Phi|_{\mathcal{H}_1}^2$$

where $\tilde{k}_2$ and $\tilde{\lambda}_0$ are positive constants. Thus $\sigma(\cdot, \cdot) + \tilde{\lambda}_0 \langle \cdot, \cdot \rangle_{\mathcal{H}_1}$ is a $\mathcal{V}_1$ elliptic sesquilinear form. That σ is $\mathcal{V}_1$ continuous also follows readily from (H2) and (H4'). Thus we have [167, page 99], [172, page 76] the following results.

Theorem 4.8 *Under hypotheses (H1)–(H3), (H4'), (H5') on σ_1, σ_2, the operator $\tilde{A}$ with domain $\mathcal{V}_1$ is an extension of $\mathcal{A}$ and generates an analytic semigroup $\tilde{T}(t)$ on $\mathcal{H}_1$ and $\mathcal{V}_1^*$ (and hence V^*). We have $\tilde{T}(t) = T(t)$ on $\mathcal{H}$ and (4.33) can be extended to define mild solutions for (4.27) whenever $z_0 \in V^*$ and $F = (0, f)$ with $f \in L^2((0,T), V^*)$.*

For the general case of (H4), (H5) for σ_2 with $V_2 \neq V$, the concept of mild solution (i.e., a variation of parameters representation) is somewhat more delicate. We still desire, however, to have a concept of mild solution whenever $f \in L^2((0,T), V_2^*)$. To this end we must extend the semigroup $T(t)$ on $\mathcal{H}$ of Theorem 4.7 to a larger space $\mathcal{W}$ where $\mathcal{H} \subset \mathcal{W}$ as well as $\{0\} \times V_2^* \subset \mathcal{W}$. This space $\mathcal{W}$ will be chosen as $\mathcal{W} = \mathcal{Y}^*$ where $\mathcal{Y} = [\text{dom}\mathcal{A}^*]$ will be carefully defined below and, hence, before carrying out this extension, it is useful to characterize in some detail the adjoint $\mathcal{A}^*$ of the operator $\mathcal{A}$ of (4.28), (4.29). The form of the adjoint $\mathcal{A}^*$ of $\mathcal{A}$ in the space $\mathcal{H}_1 = V_1 \times H$ is characterized in the next lemma.

Lemma 4.9 *The adjoint $\mathcal{A}^*$ of $\mathcal{A}$ in $\mathcal{H}_1$ is given by*

$$\operatorname{dom}\mathcal{A}^* = \{\Psi = (\psi_1, \psi_2) \in \mathcal{H} \mid \psi_2 \in V, \ A_1^*\psi_1 - A_2^*\psi_2 \in H\} \tag{4.34}$$

and

$$\mathcal{A}^*\Psi = (-\psi_2, \ A_1^*\psi_1 - A_2^*\psi_2) \tag{4.35}$$

or, in matrix operator form,

$$\mathcal{A}^* = \begin{bmatrix} 0 & -I \\ A_1^* & -A_2^* \end{bmatrix}.$$

Proof: From the definition of the adjoint, we may characterize $\operatorname{dom}\mathcal{A}^*$ and $\mathcal{A}^*$ by

$$\operatorname{dom}\mathcal{A}^* = \{\Psi = (\psi_1, \psi_2) \in \mathcal{H} \mid \Phi \to (\mathcal{A}\Phi)(\Psi) \text{ is continuous on } \operatorname{dom}\mathcal{A}\}$$

and

$$\langle \mathcal{A}\Phi, \Psi \rangle_{\mathcal{H}_1} = \langle \Phi, \mathcal{A}^*\Psi \rangle_{\mathcal{H}_1}$$

for all $\Phi = (\phi_1, \phi_2) \in \operatorname{dom}\mathcal{A}$ and $\Psi = (\psi_1, \psi_2) \in \operatorname{dom}\mathcal{A}^*$. From the definition of $\mathcal{A}^*$, we see that $\Psi \in \operatorname{dom}\mathcal{A}^*$ if and only if there exists $\Gamma = (\gamma_1, \gamma_2) \in \mathcal{H}_1 = V_1 \times H$ such that

$$\langle \mathcal{A}\Phi, \Psi \rangle_{\mathcal{H}_1} = \langle \Phi, \Gamma \rangle_{\mathcal{H}_1}$$

for all $\Phi \in \operatorname{dom}\mathcal{A}$. Thus we have the defining condition

$$\langle \phi_2, \psi_1 \rangle_{V_1} + \langle -A_1\phi_1 - A_2\phi_2, \psi_2 \rangle_H = \langle \phi_1, \gamma_1 \rangle_{V_1} + \langle \phi_2, \gamma_2 \rangle_H \tag{4.36}$$

for all $(\phi_1, \phi_2) \in \operatorname{dom}\mathcal{A}$, $(\psi_1, \psi_2) \in \operatorname{dom}\mathcal{A}^*$. Observing that (4.36) must hold for $(\phi_1, \phi_2) \in \mathcal{H}_1$ with $\phi_2 = 0$, $A_1\phi_1 \in H$, we find

$$\langle -A_1\phi_1, \psi_2 \rangle_H = \langle \phi_1, \gamma_1 \rangle_{V_1} = \langle A_1\phi_1, \gamma_1 \rangle_H$$

for all $\phi_1 \in V$ with $A_1\phi_1 \in H$, or

$$\langle A_1\phi_1, -\psi_2 - \gamma_1 \rangle_H = 0.$$

Since the set of all such ϕ_1 is dense in H and since A_1 is V elliptic, this implies that $\gamma_1 = -\psi_2$. Since $\gamma_1 \in V$, we find that $\psi_2 \in V$.

Using $\gamma_1 = -\psi_2$ in (4.36), we obtain

$$\langle A_1\phi_2, \psi_1 \rangle_H + \langle -A_2\phi_2, \psi_2 \rangle_H = \langle \phi_2, \gamma_2 \rangle_H$$

for all $(\phi_1, \phi_2) \in \operatorname{dom}\mathcal{A}$. This establishes that the map $\phi_2 \to \langle A_1\phi_2, \psi_1 \rangle_H + \langle -A_2\phi_2, \psi_2 \rangle_H$ is continuous in H for $(\phi_1, \phi_2) \in \operatorname{dom}\mathcal{A}$, which implies that $A_1^*\psi_1 - A_2^*\psi_2 \in H$ and that

$$\langle \phi_2, A_1^*\phi_1 - A_2^*\psi_2 \rangle_H = \langle \phi_2, \gamma_2 \rangle_H$$

for all $\phi_2 \in V$. It follows that $\gamma_2 = A_1^*\psi_1 - A_2^*\psi_2$ and hence $\mathrm{dom}\mathcal{A}^*$ and $\mathcal{A}^*$ are as stated in the lemma.

∎

We note that since $\psi_2 \in V$ for any $\Psi = (\psi_1, \psi_2)$ in $\mathrm{dom}\mathcal{A}^*$, we have $\mathrm{dom}\mathcal{A}^* \subset V \times V = \mathcal{V}$. We next define an inner product (equivalent to a graph norm for $\mathcal{A}^*$) under which $\mathrm{dom}\mathcal{A}^*$ will be a Hilbert space $\mathcal{Y}$ contained in, but not densely embedded in $\mathcal{V}$.

For our extension formulation, we use "extrapolation space" techniques that are similar to those used by Weissler [183], DaPrato and Grisvard [76], Haraux [103], and numerous other investigators in recent years. The form of the results used here follows the ideas of Haraux in [103]. We summarize the necessary material, referring the reader to [22, 103] for more specific details.

Let $\lambda > \lambda_0$ be fixed so that $\lambda - \mathcal{A}^*$ is invertible in $\mathcal{H}_1$. We define an inner product on $\mathrm{dom}\mathcal{A}^*$ by

$$\langle \Phi, \Psi \rangle_{\mathcal{Y}} \equiv \langle (\lambda - \mathcal{A}^*)\Phi, (\lambda - \mathcal{A}^*)\Psi \rangle_{\mathcal{H}_1} \tag{4.37}$$

with corresponding norm $|\Phi|_{\mathcal{Y}}^2 = \langle \Phi, \Phi \rangle_{\mathcal{Y}}$. We then define $\mathcal{Y} \equiv [\mathrm{dom}\mathcal{A}^*]$ as the space consisting of $\mathrm{dom}\mathcal{A}^*$ taken with this inner product and norm. It can be shown (e.g., see Lemma 3.1 of [22]) that this norm is equivalent to the graph norm of $\mathcal{A}^*$, that is, for some constants C_1, C_2 we have

$$C_1|\Phi|_{\mathcal{Y}} \le |\Phi|_{\mathcal{H}_1} + |\mathcal{A}^*\Phi|_{\mathcal{H}_1} \le C_2|\Phi|_{\mathcal{Y}} \tag{4.38}$$

and thus the embedding $\mathcal{Y} \hookrightarrow \mathcal{H}_1$ is continuous. Since $\mathcal{A}^*$ (which also generates a C_0-semigroup on $\mathcal{H}_1$) is a closed operator, we have that $\mathcal{Y}$ is closed and hence $\mathcal{Y} = [\mathrm{dom}\mathcal{A}^*]$ is a Hilbert space which is dense in $\mathcal{H}_1$. Moreover $\mathcal{Y} \hookrightarrow \mathcal{H}_1 \cong \mathcal{H}_1^* \hookrightarrow \mathcal{Y}^*$ forms a Gelfand triple, where $\mathcal{Y}^*$ is the conjugate dual of $\mathcal{Y}$. We then define $\mathcal{W} = \mathcal{Y}^*$ so that $\mathcal{Y} \hookrightarrow \mathcal{H}_1 \hookrightarrow \mathcal{W}$. It should be noted that while $\mathcal{Y} \subset \mathcal{V}$, we do not have $\mathcal{Y} \hookrightarrow \mathcal{V}$ and hence do not obtain $\mathcal{V}^* \hookrightarrow \mathcal{Y}^* = \mathcal{W}$. If we had $\mathcal{A}^*$ satisfying a $\mathcal{V}$ ellipticity condition, then the embedding would be dense and continuous; however, that is not the situation in the general case. As we shall see, however, we can argue that $\mathcal{V}^* = V \times V^* \subset \mathcal{W}$ and $|\cdot|_{\mathcal{W}} \le C_3|\cdot|_{\mathcal{V}^*}$, which suffices for our purposes. It is not difficult to argue (see [22]) that $\mathcal{W}$ is isomorphic to the completion of $\mathcal{H}_1$ in the norm $|w|_{\mathcal{W}} \equiv |(\lambda - \mathcal{A})^{-1}w|_{\mathcal{H}_1}$.

We next extend the operator $\mathcal{A}$ defined on $\mathrm{dom}\mathcal{A} \subset \mathcal{H}_1$ to all of $\mathcal{H}_1$. Such an extension $\hat{\mathcal{A}}$ is the generator of a semigroup $\hat{\mathcal{T}}(t)$ on $\mathcal{W}$ that is an extension of $\mathcal{T}(t)$. Such an extension is achieved through a sesquilinear form $\hat{\sigma}$ defined on $\mathcal{H}_1 \times \mathcal{Y}$ by

$$\hat{\sigma}(\Phi, \Psi) \equiv \langle \Phi, \mathcal{A}^*\Psi \rangle_{\mathcal{H}_1}, \qquad \Phi \in \mathcal{H}_1, \quad \Psi \in \mathcal{Y}. \tag{4.39}$$

From (4.38) we have immediately that

$$|\hat{\sigma}(\Phi, \Psi)| \le |\Phi|_{\mathcal{H}_1}|\mathcal{A}^*\Psi|_{\mathcal{H}_1} \le C_2|\Phi|_{\mathcal{H}_1}|\Psi|_{\mathcal{Y}}$$

which implies that the mapping $\Psi \to \hat{\sigma}(\Phi, \Psi)$ is in $\mathcal{Y}^* = \mathcal{W}$ for each $\Phi \in \mathcal{H}_1$. Hence there exists $\hat{\mathcal{A}} : \mathcal{H}_1 \to \mathcal{W}$ such that $(\hat{\mathcal{A}}\Phi)(\Psi) = \hat{\sigma}(\Phi, \Psi)$, or equivalently

$$(\hat{\mathcal{A}}\Phi)(\Psi) = \langle \Phi, \mathcal{A}^*\Psi \rangle_{\mathcal{H}_1} \tag{4.40}$$

for $\Phi \in \mathcal{H}_1$, $\Psi \in \mathcal{Y}$. Then we have for $\Phi \in \text{dom}\mathcal{A}$, $\Psi \in \mathcal{Y}$

$$(\hat{\mathcal{A}}\Phi)(\Psi) = \langle \Phi, \mathcal{A}^*\Psi \rangle_{\mathcal{H}_1} = \langle \mathcal{A}\Phi, \Psi \rangle_{\mathcal{H}_1} = (\mathcal{A}\Phi)(\Psi)$$

so that $\hat{\mathcal{A}}$ is indeed an extension of $\mathcal{A}$ from $\text{dom}\mathcal{A}$ to $\mathcal{H}_1$.

In [22], generic arguments are given (see Lemma 3.2, Theorem 3.1 of that paper) to show that for operators $\mathcal{A}, \mathcal{A}^*$ and $\hat{\mathcal{A}}$ as defined above, we have that $\hat{\mathcal{A}} - \lambda$ is dissipative in $\mathcal{W}$, with $\text{dom}\hat{\mathcal{A}} = \mathcal{H}_1$ and that $\mathcal{R}(\lambda - \hat{\mathcal{A}}) = \mathcal{W}$. Moreover, $\hat{\mathcal{A}}$ is the infinitesimal generator of a C_0-semigroup $\hat{T}(t)$ on $\mathcal{W}$ that is an extension of $T(t)$ from $\mathcal{H}_1$ to $\mathcal{W}$. Hence we have an extension of the representation (4.33) given by

$$\hat{z}(t) = \hat{T}(t)z_0 + \int_0^t \hat{T}(t-s)F(s)ds \tag{4.41}$$

which is valid for $z_0 \in \mathcal{W}$ and $F \in L^2((0,T),\mathcal{W})$. In general the space $\mathcal{W}$ is difficult to characterize precisely. As noted earlier, it suffices for our purposes to argue that $\mathcal{V}^* = V \times V^*$ is contained in $\mathcal{W}$ and $|\cdot|_{\mathcal{W}} \leq C_3|\cdot|_{\mathcal{V}^*}$.

Lemma 4.10 *Under (H1)–(H5), we have $\mathcal{V}^* = V \times V^* \subset \mathcal{W} = \mathcal{Y}^* = [\text{dom}\mathcal{A}^*]^*$ and for some constant $C_3 > 0$*

$$|\Phi|_{\mathcal{W}} \leq C_3|\Phi|_{\mathcal{V}^*} \text{ for all } \Phi \in \mathcal{V}^*. \tag{4.42}$$

Proof: First note that $\mathcal{Y} \subset \mathcal{V}$ and from (4.38), (4.35) and (H3) we have for $\Psi = (\psi_1, \psi_2) \in \mathcal{Y}$

$$C_2|\Psi|_{\mathcal{Y}} \geq |\Psi|_{\mathcal{H}_1} + |\mathcal{A}^*\Psi|_{\mathcal{H}_1} \geq |\psi_1|_{V_1} + |\psi_2|_{V_1}$$

$$\geq \sqrt{k_1}\left(|\psi_1|_V + |\psi_2|_V\right) \geq \tilde{k}|\Psi|_{\mathcal{V}}.$$

Hence, for some constant $C_3 > 0$, we have for all $\Psi \in \mathcal{Y}$,

$$|\Psi|_{\mathcal{V}} \leq C_3|\Psi|_{\mathcal{Y}}. \tag{4.43}$$

Now suppose $\Lambda \in \mathcal{V}^* = V \times V^*$. Then for all $\Psi \in \mathcal{V}$ we have $|\Lambda(\Psi)| \leq |\Lambda|_{\mathcal{V}^*}|\Psi|_{\mathcal{V}}$. In particular, if $\Psi \in \mathcal{Y} \subset \mathcal{V}$ we then find using (4.43) that

$$|\Lambda(\Psi)| \leq C_3|\Lambda|_{\mathcal{V}^*}|\Psi|_{\mathcal{Y}}$$

and thus $\Lambda \in \mathcal{Y}^* = \mathcal{W}$ with $|\Lambda|_{\mathcal{W}} \leq C_3|\Lambda|_{\mathcal{V}^*}$. That is, $\mathcal{V}^* \subset \mathcal{W}$ with (4.42) holding.   ∎

Remark 4.11 Under (H4$'$), (H5$'$), we have that dom$\mathcal{A}^*$ is dense in $\mathcal{V}$ (since $\mathcal{A}^*$ generates an analytic semigroup on $\mathcal{V}$ [172, page 28]) and hence $\mathcal{Y} \hookrightarrow \mathcal{V} \hookrightarrow \mathcal{H} \hookrightarrow \mathcal{V}^* \hookrightarrow \mathcal{W}$.

As a consequence of this lemma and the considerations preceding it, we have the following extension results.

Theorem 4.12 *Under hypotheses (H1)–(H5) on σ_1, σ_2, the operator $\hat{\mathcal{A}}$ with domain $\mathcal{H}_1$ defined in (4.40) is an extension of $\mathcal{A}$ and generates a C_0-semigroup $\hat{T}(t)$ on $\mathcal{W} = \mathcal{Y}^* = [\text{dom}\mathcal{A}^*]^*$ with $\hat{T}(t) = T(t)$ on $\mathcal{H}$. Moreover, $\mathcal{V}^* \subset \mathcal{W}$ and (4.33) can be extended so that (4.41) defines a mild solution $\hat{z}$ for (4.27) whenever $z_0 \in \mathcal{W}$ and $F = (0, f)$ with $f \in L^2((0, T), \mathcal{V}^*)$.*

4.4 Equivalence of Solution Formulations

Clearly it is of interest to determine conditions on problem data, that is $z_0 = (w_0, w_1)$, f, σ_1, σ_2, such that one has equivalence between solutions of (4.9) given in terms of a semigroup formulation such as those in Section 4.3 and solutions obtained from a variational formulation as given in Section 4.2. We first consider the case for $f \in L^2((0, T), \mathcal{H})$ and $z_0 = (w_0, w_1) \in \mathcal{H}$ under the general conditions (H1)–(H5) on σ_1, σ_2.

Let $z_{sg} = z(z_0, f)$ denote the semigroup solution to (4.27) given by (4.33) and guaranteed by Theorem 4.7, and let $z_{var} = (w(z_0, f), \dot{w}(z_0, f))$ denote the weak or variational solution guaranteed by Theorem 4.1 corresponding to initial data $z_0 = (w_0, w_1)$ and nonhomogeneous term f. The following theorem summarizes the correspondence between solutions.

Theorem 4.13 *Let (H1)–(H5) hold and suppose $z_0 = (w_0, w_1) \in \mathcal{H} = V \times H$ and $f \in L^2((0, T), \mathcal{H})$. Then $z_{sg}(z_0, f) = z_{var}(z_0, f)$.*

Proof: First suppose that $z_0 \in \text{dom}\mathcal{A} \subset \mathcal{V}$ which is dense in $\mathcal{H}$ and that $f \in C^1((0, T), \mathcal{H})$ which is dense in $L^2((0, T), \mathcal{H})$. By standard results (e.g., see [152, Corollary 2.11] or [26, Theorem 1.11]), in this case the mild solution $z_{sg} = z(z_0, f)$ is the unique strong solution to (4.27), or equivalently (4.26) (which is the same as (4.10)) and, by the uniqueness statement in Theorem 4.1, must agree with $z_{var}(z_0, f)$ of Theorem 4.1. Hence $z_{sg} = z_{var}$ in this case. Recall that $(z_0, f) \rightarrow z_{var}(z_0, f)$ is continuous from $\mathcal{H} \times L^2((0, T), \mathcal{H})$ to $L^2((0, T), \mathcal{H})$ while it is readily seen from (4.33) that $(z_0, f) \rightarrow z_{sg}(z_0, f)$ is continuous in the same sense. Thus, the continuous functions z_{sg} and z_{var} agree on the dense set $(\text{dom}\mathcal{A}) \times C^1((0, T), \mathcal{H})$ and the desired result follows immediately.

∎

We next consider the special case where σ_2 is V-elliptic and V-continuous (i.e., (H4'), (H5') hold). As before, let z_{var} denote the solution guaranteed by Theorem 4.1. Let z_{sg} denote the mild solution given by (4.33) with the semigroup now defined on $\mathcal{V}^*$ as guaranteed in Theorem 4.8. We then have the following result.

Theorem 4.14   *Under hypotheses (H1)–(H3), (H4'), (H5'), we have that* $z_{sg}(z_0, f) = z_{var}(z_0, f)$ *for* $z_0 \in \mathcal{H} = V \times H$ *and* $f \in L^2((0, T), \mathcal{V}^*)$.

Proof: From Theorem 4.13, the solutions agree for initial data $z_0 \in \mathcal{H}$ and force $f \in L^2((0, T), H)$, where $L^2((0, T), H)$ is dense in $L^2((0, T), \mathcal{V}^*)$. Moreover $(z_0, f) \to z_{sg}(z_0, f)$ is obviously continuous from $\mathcal{H} \times L^2((0, T), \mathcal{V}^*)$ to $L^2((0, T), \mathcal{V}^*)$ and $(z_0, f) \to z_{var}(z_0, f)$ is continuous from $\mathcal{H} \times L^2((0, T), \mathcal{V}^*)$ to $L^2((0, T), V) \times L^2((0, T), V)$ and hence to $L^2((0, T), \mathcal{V}^*)$, $\mathcal{V}^* = V \times V^*$. Again the result follows immediately. ∎

Finally we turn to the general case to establish general equivalence of mild solutions and variational or weak solutions to (4.27). Let $\hat{z} = \hat{z}(z_0, f)$ denote the mild solution given by (4.41) for $z_0 \in \mathcal{H} = V \times H$ and $F = (0, f)$ with $f \in L^2((0, T), V_2^*)$.

Theorem 4.15   *Under hypotheses (H1)–(H5), we have* $\hat{z}(z_0, f) = z_{var}(z_0, f)$ *for* $z_0 \in \mathcal{H} = V \times H$ *and* $f \in L^2((0, T), V_2^*)$.

Proof: From Theorem 4.13 we have the desired equivalence for $z_0 \in \mathcal{H}$ and $f \in L^2((0, T), H)$ where $L^2((0, T), H)$ is dense in $L^2((0, T), V_2^*)$. We recall from Theorem 4.1 that $(z_0, f) \to z_{var}(z_0, f)$ is continuous from $\mathcal{H} \times L^2((0, T), V_2^*)$ to $L^2((0, T), \mathcal{V}_2)$ and hence to $L^2((0, T), \mathcal{W})$ since $\mathcal{V}_2 \hookrightarrow \mathcal{H} \hookrightarrow \mathcal{W}$, where $\mathcal{V}_2 = V \times V_2$. From (4.41) it is obvious that $(z_0, f) \to \hat{z}(z_0, f)$ is continuous from $\mathcal{H} \times L^2((0, T), \mathcal{W})$ to $L^2((0, T), \mathcal{W})$ and hence from $\mathcal{H} \times L^2((0, T), \{0\} \times V_2^*)$ to $L^2((0, T), \mathcal{W})$ since (4.42) implies $|\Phi|_{\mathcal{W}} \leq \tilde{C}_3 |\Phi|_{V_2^*}$ for some $\tilde{C}_3$ because $V_2^* \hookrightarrow V^*$. The desired result once again follows from equivalence on the dense subset $\{(z_0, f) \in \mathcal{H} \times L^2((0, T), H)\}$ of $\mathcal{H} \times L^2((0, T), V_2^*)$ and the continuity statements. ∎

We point out that $z_{sg}(z_0, f)$ of Theorem 4.14 is the same as $\hat{z}(z_0, f)$ for $f \in L^2((0, T), V_2^*)$ (and both equal z_{var}) under (H1)–(H3), (H4'), (H5'). Indeed, in this case the analytic semigroups $\tilde{T}(t)$ of Theorem 4.8 and $\hat{T}(t)$ of Theorem 4.12 are the same on $\mathcal{V}^*$ and thus $\hat{T}(t)$ is an extension of $\tilde{T}(t)$ from $\mathcal{V}^*$ to $\mathcal{W}$.

4.5   Examples

We return to the motivating examples of Section 4.1 to briefly indicate how a wide variety of structural vibration examples might be treated in the context of the framework of this chapter. To illustrate, we consider the cantilevered Euler-Bernoulli beam with a piezoceramic patch pair as described in Section 4.1 and treat only the transverse vibrations; that is, equation (4.8) along with the cantilever boundary conditions. Defining $H = L^2(0, \ell)$ with the weighted inner product $\langle \phi, \psi \rangle_H = \int_0^\ell \tilde{\rho} \phi \psi \, dy = \langle \tilde{\rho} \phi, \psi \rangle_{L^2}$, the space

$$V = H_L^2(0, \ell) = \left\{ \phi \in H^2(0, \ell) \mid \phi(0) = \phi'(0) = 0 \right\},$$

and the sesquilinear form

$$\sigma_1(\phi, \psi) = \langle \widetilde{EI} \phi'', \psi'' \rangle_{L^2}, \tag{4.44}$$

with $\widetilde{EI}$ given in (4.3), we readily see that (H1)–(H3) are satisfied. For the *Kelvin-Voigt* damping of this model that we formulated in Section 4.1, we define

$$V_2 = V = H_L^2(0, \ell) \quad \text{and} \quad \sigma_2(\phi, \psi) = \langle \widetilde{c_D I} \phi'', \psi'' \rangle_{L^2}, \tag{4.45}$$

with $\widetilde{c_D I}$ given in (4.3). Then (H4)–(H5) is easily verified. Finally (see (4.4)) for $f(t, y) \equiv \mathcal{K}^B \chi_{pe}''(y)[V_1(t) - V_2(t)]$, we see that for $V_1(\cdot) - V_2(\cdot) \in L^2(0, T)$ we have $f \in L^2((0, T), V^*)$. This follows since the derivatives χ_{pe}'' (in the sense of distributions) are in $V^* = (H_L^2(0, \ell))^*$. That is, for $\phi \in V = H_L^2(0, \ell)$,

$$(\chi_{pe}'')(\phi) = \langle \chi_{pe}, \phi'' \rangle_H \leq |\chi_{pe}|_H \, |\phi''|_H$$

$$\leq |\tilde{\rho}|_{L^\infty} \sqrt{y_2 - y_1} \, |\phi|_V \, .$$

Hence (H6) is satisfied for this example and the results of Theorems 4.1, 4.8 and 4.14 are applicable.

   We remark that if $\widetilde{EI}$ and $\widetilde{c_D I}$ are constant, the operator $\mathcal{A}$ of (4.28)-(4.29) reduces to the usual infinitesimal generator widely discussed in the literature, that is, the *closure* of the operator $\mathcal{A}$ given by (4.29) on $\text{dom}\mathcal{A} = (H^4(0, \ell) \cap H_L^2(0, \ell)) \times (H^4(0, \ell) \cap H_L^2(0, \ell))$.

   As we mentioned earlier, a wide variety of damping models may be used with examples involving beams, plates, and shells. We list several in the context of transverse vibrations of the cantilevered Euler-Bernoulli beam of Section 4.1. The spaces H and V and the sesquilinear form σ_1 remain as above in all these examples.

   Viscous damping. This damping model, also called air damping, involves a velocity proportional damping term with damping coefficient $\gamma \in L^\infty(0, \ell)$. We choose $V_2 = H$ and

$$\sigma_2(\phi, \psi) = \langle \gamma \phi, \psi \rangle_{L^2}.$$

For systems modeled with internal Kelvin-Voigt damping and external viscous air damping, the appropriate sesquilinear form is

$$\sigma_2(\phi, \psi) = \langle \widetilde{c_D I} \phi'', \psi'' \rangle_{L^2} + \langle \gamma \phi, \psi \rangle_{L^2}. \tag{4.46}$$

Spatial hysteresis damping. This model, explained in some detail in [162], has been shown to be appropriate for composite material beams where graphite fibers are embedded in an epoxy matrix [48]. The damping sesquilinear form here can be given in terms of a compact operator G on $L^2(0, \ell)$ defined by

$$(G\phi)(y) = \int_0^\ell \gamma(y, s)\phi(s)\, ds$$

where γ is a symmetric, non-negative kernel in $L^\infty((0, \ell) \times (0, \ell))$. The space V_2 is chosen as $H^1(0, \ell)$ and

$$\sigma_2(\phi, \psi) = \langle (\nu I - G)\phi', \, \psi' \rangle_{L^2}$$

where $\nu(y) = \int_0^\ell b(y, s)\, ds$.

For a fixed-end Euler-Bernoulli beam which is modeled with the boundary conditions $w(t, 0) = w'(t, 0) = w(t, \ell) = w'(t, \ell) = 0$, all of the above damping models may also be used. In this case, we again choose $H = L^2(0, \ell)$ with $\tilde{\rho}$-weighted inner product, but $V = H_0^2(0, \ell) = \{\phi \in H^2(0, \ell) \mid \phi, \phi' \text{ vanish at } y = 0, \ell\}$ so that $V^* = H^{-2}(0, \ell)$. The sesquilinear form σ_1 is as before and the damping forms as defined above, except we replace $H_L^2(0, \ell)$ by $H_0^2(0, \ell)$ throughout. In addition to Kelvin-Voigt, viscous, and spatial hysteresis damping models for this example, one can also consider so-called "structural" (square-root or $A^{1/2}$) damping. In this case, we choose $V = H_0^2(0, \ell)$, $V_2 = H_0^1(0, \ell) = \{\phi \in H^1(0, \ell) \mid \phi(0) = \phi(\ell) = 0\}$ and

$$\sigma_2(\phi, \psi) = \langle c\phi', \, \psi' \rangle_{L^2}$$

where $c \in L^\infty(0, \ell)$.

As we have noted, many other examples can be treated in the abstract framework developed in this chapter including Timoshenko beams [47, 180], 2-D plates and grid structures [8, 29, 30, 158], shells [39], and structural acoustic models [12, 39]. We shall discuss some of these in subsequent chapters of this monograph.

4.6   Extensions to Nonlinear Systems

The ideas and techniques developed in Sections 4.2 and 4.3 above can be combined with monotonicity methods to treat certain nonlinear systems that arise in recent smart material studies [28]. In particular, the techniques and results of Sections 4.2 and 4.3 were used in [13] to treat systems of the form

$$\ddot{w}(t) + A_2\dot{w}(t) + A_1 w(t) + \mathcal{N}^* g(\mathcal{N} w(t)) = f(t) \qquad (4.47)$$

in V^* where $\mathcal{N} \in \mathcal{L}(V_2, H)$ with $\mathcal{N}(V)$ dense in H, and g is a nonlinear mapping of gradient type satisfying certain "monotonicity" type conditions. Here the operator $\mathcal{N}$ is typically a differentiation type operator (e.g., $\mathcal{N} = \frac{\partial}{\partial y}$ in examples of type (4.7), $\mathcal{N} = \frac{\partial^2}{\partial y^2}$ in examples such as (4.8)) that arises along with the function g in a nonlinear constitutive law relating internal stress and strain (e.g., a nonlinear "Hooke's" law). In [13] the ideas of this chapter are combined with monotonicity ideas to obtain well-posedness and representation results for solutions of (4.47).

Remark 4.16 The well-posedness literature on second-order partial differential equations (PDE) in both variational and semigroup formulations is rather extensive, dating back to the early work of Lions (see [137, 138] and the references therein to other earlier efforts) in the late 1960's to the more recent treatments by Showalter [167], Tanabe [172] and Wloka [185]. While results are scattered, most of the results presented in this chapter can be found in one form or another in earlier publications. Our well-posedness treatment for systems with general damping sesquilinear forms were first given in [20] and these results cannot, to our knowledge, be found in this form elsewhere. Most authors consider undamped [137, 185] second-order hyperbolic systems or assume damping sufficiently strong so as to make the system essentially parabolic in nature. In recent years, owing in part to increased computational capabilities, serious interest in and efforts on use of PDE models in complex structures (for parameter estimation, system identification, and control) have substantially increased. All experimental structures and data for them exhibit damping of some form and thus, in model fitting and control design, it is important to include damping in mathematical models which are supposed to describe these structures.

To our knowledge, Theorems 4.1 and 4.13, which yield equivalence for $f \in L^2((0, T), H)$, appeared first in [20] and give results for general (including weak but nontrivially) damped systems with input or inhomogeneous term bounded in the state space.

Theorems 4.8 and 4.14, the case for strong damping (σ_2 is V elliptic and V continuous) can be obtained directly from Tanabe's results by rewriting the second-order system as a first order system with $\mathcal{V}$ elliptic form (as we did in Section 4.3 in obtaining Theorem 4.8) and then appealing to Proposition 5.5.1, page 153 of [172].

The results of Section 4.3 on applying Haraux's techniques for the treatment (involving (4.41), Lemma 4.10 and Theorem 4.12) of weakly damped second-order systems along with the equivalence results of Theorem 4.15 first appeared in [20]. These results play a primary role when developing models involving piezoceramic actuators as discussed in this monograph.

Chapter 5

Estimation of Parameters and Inverse Problems

In this chapter we shall discuss techniques for inverse problems that arise in many smart material applications involving simulations, stabilization and feedback control. The problems on which we focus concern estimation of parameters or model-based system identification. We shall return to the specific example of a cantilevered beam with piezoceramic patches of Section 4.1 to motivate and illustrate our presentation. We discuss concepts in the context of this example as we formulate generic estimation problems and state results for general abstract parameter dependent systems in the form (4.9).

5.1 Formulation of the Parameter Estimation Problem

To motivate our discussions, we shall use the cantilevered beam example with transverse and longitudinal vibrations described by equations (4.7), (4.8), (4.3) with input terms given in (4.4). To this model, we also add air damping as described in (4.46). If we are to use such models for simulations, control or even simply modal analysis, we need values for the parameters $\mathcal{K}^A, \mathcal{K}^B$ in (4.4), $\tilde{\rho}, \widetilde{Eb}, \widetilde{c_D b}, \widetilde{EI}, \widetilde{c_D I}$ in (4.7),(4.8), and γ of (4.46). These parameters can be nominally given (for example, see (4.3),(4.4)) in terms of basic geometric quantities (for example, $h, b, h_{pe}, y_1, y_2, a_3$) and physical or material characteristic parameters such as $\rho, \rho_{pe}, E, E_{pe}, c_D, c_{Dpe}, d_{31}, \gamma$. Some of these can often be measured directly (for example, h, b, h_{pe}, et cetera) while others might be obtained from manufacturers specifications (so-called "book values" given in engineering tables) such as ρ, E for aluminum and ρ_{pe}, E_{pe} for lead zirconate titinate (PZT). However, some of these (for example, the damping coefficients c_D, c_{Dpe}, γ) cannot be measured or obtained in this way. Moreover, "book values" given by manufacturers are only average values which can vary substantially between samples. More importantly, when such products are combined as components in a composite material configuration (the standard situation in many smart material structures), the effective values of such parameters are usually found to be quite different as we shall see in later experimental results in this chapter. In response to these difficulties, one can attempt to estimate these "unknown" parameters through dynamic experiments with the physical structure.

In the generic abstract parameter estimation problem, we consider a dynamic model of the form (4.9) where the operators A_1, A_2 and possibly the input f depend on some unknown (i.e., to be estimated) parameters q with values in an admissible family Q of parameters. That is, we have

$$\ddot{w}(t) + A_2(q)\dot{w}(t) + A_1(q)w(t) = f(t,q) \quad \text{in } V^*$$
$$w(0) = w_0, \qquad \dot{w}(0) = w_1, \tag{5.1}$$

or, equivalently, in terms of parameter dependent sesquilinear forms (see(4.10))

$$\langle \ddot{w}(t), \phi \rangle + \sigma_2(q)(\dot{w}(t), \phi) + \sigma_1(q)(w(t), \phi) = \langle f(t,q), \phi \rangle$$
$$w(0) = w_0, \qquad \dot{w}(0) = w_1 \tag{5.2}$$

for all $\phi \in V$. As in (4.10), $\langle \cdot, \cdot \rangle$ denotes the duality product $\langle \cdot, \cdot \rangle_{V^*,V}$. In some problems the initial data w_0, w_1 may also depend on parameters to be estimated, i.e., $w_0 = w_0(q)$, $w_1 = w_1(q)$, but we shall not discuss such problems here. The ideas and techniques we present can be used to treat such problems; we refer readers to [26] for discussion of general estimation problems where not only the initial data but even the spaces V and H depend on unknown parameters.

For the problem at hand, it is assumed that the parameter-dependent sesquilinear forms $\sigma_1(q), \sigma_2(q)$ of (5.2) satisfy the continuity and ellipticity conditions (H2)-(H5) of Chapter 4 uniformly in $q \in Q$; that is, the constants $c_1, k_1, c_2, \lambda_0, k_2$ of (H2)-(H5) can be found independently of $q \in Q$.

As we have noted, one must estimate the parameters q from dynamic observations of the system (5.1) or (5.2). This raises the important consideration as to what will be measured in the dynamic experiments from which we obtain our observations. To discuss these measurements in a specific setting, we consider just the transverse vibrations (for example, equation (4.8) with the second expression in (4.4)) of our beam example of Chapter 4. Measurements, of course, depend on the sensors available. Any truly smart material structure contains both sensors and actuators which may or may not rely on the same device. In usual mechanical experiments, there are several popular measurement devices, some of which could possibly be used in a smart material configuration. If one uses an accelerometer placed at the point $\bar{y} \in (0, \ell)$ along the beam, then one obtains observations $\ddot{w}(t, \bar{y})$ of beam *acceleration*. A laser vibrometer will yield data of *velocity* $\dot{w}(t, \bar{y})$ while proximity probes including displacement solenoids produce measurements of *displacement* $w(t, \bar{y})$.

In the case of a beam (or structure) with piezoceramic patches, the patches may be used as sensors as well as actuators. In this case one obtains observations of voltages which are proportional to the *accumulated strain*, i.e., the observations are given by

$$\mathcal{K}^S \int_{y_1}^{y_2} \frac{\partial^2 w}{\partial y^2}(t,y)dy = \mathcal{K}^S \left\{ \frac{\partial w}{\partial y}(t,y_2) - \frac{\partial w}{\partial y}(t,y_1) \right\} \tag{5.3}$$

where $\mathcal{K}^S$ is a sensor constant which is a piezoceramic material and geometry related quantity [80]. Like the actuator constants $\mathcal{K}^A$ and $\mathcal{K}^B$, the patch constant $\mathcal{K}^S$ is in general included among the parameters q to be estimated.

Whatever the measuring devices, the resulting observations can be used in a general least squares output formulation of the parameter estimation problem. In such cases, the problems are stated in terms of finding parameters which give the best fit of the parameter-dependent solutions of the partial differential equations to dynamic system response data collected with various excitations.

In the beam example, the parameters to be estimated include the composite mass density $\tilde{\rho}(y)$, stiffness coefficient $\widetilde{EI}(y)$, Kelvin-Voigt damping parameter $\widetilde{c_D I}(y)$, viscous air damping coefficient γ and piezoelectric material parameters $\mathcal{K}^B$, $\mathcal{K}^S$. In this case, we let the collection of unknown parameters be denoted by $q = (\tilde{\rho}(y), \widetilde{EI}(y), \widetilde{c_D I}(y), \gamma, \mathcal{K}^B, \mathcal{K}^S)$. Details regarding the estimation of these parameters for an experimental beam are given in Section 5.4. Similar experimental results for a plate are summarized in Section 5.5.

The general least squares parameter estimation problem can be formulated as follows. For a given discrete set of measured observations $z = \{z_i\}_{i=1}^{N_t}$ corresponding to model observations $z_{ob}(t_i)$ at times t_i as obtained in most practical cases, we consider the problem of minimizing over $q \in Q$ the least squares output functional

$$J(q,z) = \left| \tilde{C}_2 \left\{ \tilde{C}_1 \{w(t_i,\cdot;q)\} - \{z_i\} \right\} \right|^2, \tag{5.4}$$

where $\{w(t_i,\cdot;q)\}$ are the parameter dependent solutions of (5.1) or (5.2) evaluated at each time t_i, $i = 1, 2, \cdots, N_t$ and $|\cdot|$ is an appropriately chosen Euclidean norm. Here the operators $\tilde{C}_1$ and $\tilde{C}_2$ are observation operators that depend on the type of observed or measured data that is available. The operator $\tilde{C}_1$ may have several forms depending on the type of sensors being used. When the collected data z_i consists of time domain displacement, velocity, or acceleration values at a point $\bar{y}$ on the beam as discussed above, the functional takes the form

$$J_\nu(q,z) = \sum_{i=1}^{N_t} \left| \frac{\partial^\nu w}{\partial t^\nu}(t_i,\bar{y};q) - z_i \right|^2, \tag{5.5}$$

for $\nu = 0, 1, 2$, respectively. In this case the operator $\tilde{C}_1$ involves differentiation (either $\nu = 0, 1$ or 2 times, respectively) with respect to time followed by pointwise evaluation in t and y. When the piezoelectric is used as a sensor, the functional to be minimized is

$$J_{pe}(q,z) = \sum_{i=1}^{N_t} \left| \mathcal{K}^S \left\{ \frac{\partial w}{\partial y}(t_i,y_2;q) - \frac{\partial w}{\partial y}(t_i,y_1;q) \right\} - z_i \right|^2, \tag{5.6}$$

for the piezoelectric elements being located on the beam between y_1 and y_2. The data $\{z_i\}$ then consists of measured voltages across the piezoelectric elements.

While the most commonly encountered least squares problems involve time domain data, it is often advantageous to fit the data in a frequency domain setting. To treat these alternate possibilities, we introduce a second observation operator $\tilde{C}_2$. This operator $\tilde{C}_2$ may be the identity (corresponding to time domain identification procedures as in (5.5) and (5.6)) or may be related to the Fourier transform (corresponding to identification in the frequency domain). If the identification is carried out in the frequency domain and the operator $\tilde{C}_2$ is a Fourier transform related operator, then an appropriate cost functional is

$$\hat{J}(q,z) = \sum_{\ell=1}^{N_f} \left(\epsilon_1 \left| f_{k_\ell^w}(q) - f_{k_\ell^z} \right|^2 \right.$$

$$\left. + \epsilon_2 \sum_{j=-n_\ell}^{N_\ell} \{ |W(k_\ell^w + j; q)| - |Z(k_\ell^z + j)| \}^2 \right). \tag{5.7}$$

Here $W(k;q)$ and $Z(k)$ are the Fourier series coefficients of $\tilde{C}_1\{w(t_i, \bar{y}; q)\}$ and $\{z_i\}$ respectively, and $f_{k_\ell^w}$ and $f_{k_\ell^z}$ are the $(k_\ell^w)^{th}$ vibration frequency of the solution $W(k;q)$ and the $(k_\ell^z)^{th}$ frequency of the observation data $Z(k)$. Moreover, ϵ_1, ϵ_2 are weighting constants, and n_ℓ, N_ℓ are certain lower and upper limits associated with the width (or the support) of the ℓth spike. In formulating (5.7), we have assumed that there are a finite and distinct number N_f of nontrivial "spikes", i.e. vibration frequencies or "significant modes", among the $Z(k)$ and the number of nontrivial spikes of the solution $W(k;q)$ is the same as N_f.

More precisely, if the measurements are taken with fixed sampling time, i.e., $\Delta t = t_i - t_{i-1} = t_{i+1} - t_i$ for all i and with a total of N_t samples at the fixed space position $y = \bar{y}$ (as is often the case in experiments), then the Fourier series coefficients for $0 \le i \le N_t$ are given by

$$\tilde{C}_2\left\{ \tilde{C}_1\{w(t_i, \bar{y}; q)\} \right\}_k = W(k;q) = \frac{1}{N_t} \sum_{i=0}^{N_t-1} \tilde{C}_1\{w(t_i, \bar{y}; q)\} e^{-jk(2\pi/N_t)i}$$

$$\tilde{C}_2\{z(t_i)\}_k = Z(k) = \frac{1}{N_t} \sum_{i=0}^{N_t-1} z(t_i) e^{-jk(2\pi/N_t)i} \tag{5.8}$$

where $t_i = i\Delta t$ and $k = 0, 1, \ldots, N_t - 1$. With Δt as the sampling time, the kth value of the frequency corresponding to the kth coefficient is given by

$$f_k = \frac{1}{\Delta t \cdot N_t} k \tag{5.9}$$

and the corresponding magnitudes are given by $|W(k;q)|$ and $|Z(k)|$.

Having assumed that both $Z(k)$ and $W(k;q)$ have the same number $N_f <$ N_t of significant modes, we may re-index coefficients of these spikes among the $Z(k)$ and $W(k;q)$, denoting these indices by k_ℓ^z, k_ℓ^w for $\ell = 1, \ldots, N_f$ with $0 \leq k_\ell^z$, $k_\ell^w \leq N_t - 1$. The frequency domain cost function can then be appropriately expressed by (5.7) with the weights ϵ_1, ϵ_2 chosen to balance frequency fit with the fit in width and magnitude of the spike. Thus the first part of the cost function (5.7) is related to the frequencies and the second part is related to the magnitude and the width of each spike. The limits n_ℓ and N_ℓ are the last i and j respectively, for which the following conditions are satisfied:

$$|Z(k_\ell^z - i)| \geq 20\%|Z(k_\ell^z)|$$

for $i = 1, 2, \ldots, n_\ell$, and

$$|Z(k_\ell^z + j)| \geq 20\%|Z(k_\ell^z)|$$

for $j = 1, 2, \ldots, N_\ell$. The motivation behind this choice, which is related to the width of the spike, is that in traditional modal analysis, the width at approximately 30% of the peak value of the spike is used to estimate the damping ratio for the ℓth mode. Hence taking a conservative approach and using the width at 20% of the peak value should guarantee the inclusion of substantial damping information in the Fourier coefficients.

5.2 Approximation and Convergence

The minimization in our general abstract parameter estimation problems involves an infinite dimensional state space and possibly an infinite dimensional admissible parameter set (of functions). To obtain computationally tractable methods, we thus consider Galerkin type approximations in the context of the variational formulation (5.2). Let H^N be a sequence of finite dimensional subspaces of H, and Q^M be a sequence of finite dimensional sets approximating the parameter set Q. We denote by P^N the orthogonal projections of H onto H^N. Then a family of approximating estimation problems with finite dimensional state spaces and parameter sets can be formulated by seeking $q \in Q^M$ which minimizes

$$J^N(q, z) = \left| \tilde{C}_2 \left\{ \tilde{C}_1 \{ w^N(t_i, \cdot; q) \} - \{ z_i \} \right\} \right|^2, \qquad (5.10)$$

where $w^N(t; q) \in H^N$ is the solution to the finite dimensional approximation of (5.2) given by

$$\langle \ddot{w}^N(t), \phi \rangle + \sigma_2(q)(\dot{w}^N(t), \phi) + \sigma_1(q)(w^N(t), \phi) = \langle f(t, q), \phi \rangle$$

$$w^N(0) = P^N w_0, \quad \dot{w}^N(0) = P^N w_1, \qquad (5.11)$$

for $\phi \in H^N$.

For the parameter sets Q and Q^M, and state spaces H^N, we make the following hypotheses.

(A1M) The sets Q and Q^M lie in a metric space $\tilde{Q}$ with metric d. It is assumed that Q and Q^M are compact in this metric and there is a mapping $i^M :$ $Q \to Q^M$ so that $Q^M = i^M(Q)$. Furthermore, for each $q \in Q, i^M(q) \to q$ in $\tilde{Q}$ with the convergence uniform in $q \in Q$.

(A2N) The finite dimensional subspaces H^N satisfy $H^N \subset V$ as well as the approximation properties of the next two statements.

(A3N) For each $\psi \in V$, $|\psi - P^N \psi|_V \to 0$ as $N \to \infty$.

(A4N) For each $\psi \in V_2$, $|\psi - P^N \psi|_{V_2} \to 0$ as $N \to \infty$.

The reader is referred to Chapter 4 for a complete discussion motivating the space V_2.

We also need some regularity with respect to the parameters q in the parameter dependent sesquilinear forms σ_1, σ_2. In addition to (uniform in Q) ellipticity and continuity conditions (H2)-(H5), the sesquilinear forms $\sigma_1 = \sigma_1(q)$ and $\sigma_2 = \sigma_2(q)$ are assumed to be defined on Q and satisfy the continuity-with-respect-to-parameter conditions

(H7) $|\sigma_1(q)(\phi, \psi) - \sigma_1(\tilde{q})(\phi, \psi)| \leq \gamma_1 d(q, \tilde{q}) |\psi|_V \, |\psi|_V$, for $\phi, \psi \in V$

(H8) $|\sigma_2(q)(\xi, \eta) - \sigma_2(\tilde{q})(\xi, \eta)| \leq \gamma_2 d(q, \tilde{q}) |\xi|_{V_2} \, |\eta|_{V_2}$, for $\xi, \eta \in V_2$

for $q, \tilde{q} \in Q$ where the constants γ_1, γ_2 depend only on Q.

Solving the approximate estimation problems involving (5.10), (5.11), we obtain a sequence of parameter estimates $\{\bar{q}^{N,M}\}$. It is of paramount importance to establish conditions under which $\{\bar{q}^{N,M}\}$ (or some subsequence) converges to a solution for the original infinite dimensional estimation problem involving (5.2),(5.4). Toward this goal we have the following results.

Theorem 5.1 *To obtain convergence of at least a subsequence of $\{\bar{q}^{N,M}\}$ to a solution $\bar{q}$ of minimizing (5.4) subject to (5.2), it suffices, under assumption (A1M), to argue that for arbitrary sequences $\{q^{N,M}\}$ in Q with $q^{N,M} \to q$ in Q, we have*

$$\tilde{C}_2 \tilde{C}_1 w^N(t; q^{N,M}) \to \tilde{C}_2 \tilde{C}_1 w(t; q). \tag{5.12}$$

Proof: Under the assumptions (A1M), let $\{\bar{q}^{N,M}\}$ be solutions minimizing (5.10) subject to the finite dimensional system (5.11) and let $\hat{q}^{N,M} \in Q$ be such that $i^M(\hat{q}^{N,M}) = \bar{q}^{N,M}$. From the compactness of Q, we may select subsequences, again denoted by $\{\hat{q}^{N,M}\}$ and $\{\bar{q}^{N,M}\}$, so that $\hat{q}^{N,M} \to \bar{q} \in Q$ and $\bar{q}^{N,M} \to \bar{q}$ (the latter follows the last statement of (A1M)). The optimality of $\{\bar{q}^{N,M}\}$ guarantees that for every $q \in Q$

$$J^N(\bar{q}^{N,M}, z) \leq J^N(i^M(q), z). \tag{5.13}$$

Using (5.12), the last statement of (A1M) and taking the limit as $N, M \to \infty$ in the inequality (5.13), we obtain $J(\bar{q}, z) \leq J(q, z)$ for every $q \in Q$, or that $\bar{q}$ is a solution of the problem for (5.2),(5.4). We note that under the uniqueness assumptions on the problems, one can actually guarantee convergence of the entire sequence $\{\bar{q}^{N,M}\}$ in place of subsequential convergence to solutions.

∎

We note that the essential aspects in the arguments given above involve compactness assumptions on the sets Q^M and Q. Such compactness ideas play a fundamental role in other theoretical and computational aspects of these problems. For example, one can formulate distinct concepts of *problem stability* and *method stability* involving some type of continuous dependence of solutions on the observations z, and use conditions similar to those of (5.12) and (A1M), with compactness again playing a critical role, to guarantee stability. We illustrate with a simple form of *method stability* (other stronger forms are also amenable to this approach).

We might say that an *approximation method*, such as that formulated above involving Q^M, H^N and (5.10), is *stable* if

$$dist(\tilde{q}^{N,M}(z^k), \tilde{q}(z^*)) \to 0$$

as $N, M, k \to \infty$ for any $z^k \to z^*$ (in this case in the appropriate Euclidean space), where $\tilde{q}(z)$ denotes the set of all solutions of the problem for (5.4) and $\tilde{q}^{N,M}(z)$ denotes the set of all solutions of the problem for (5.10). Here "dist" represents the usual distance set function. Under (5.12) and (A1M), one can use arguments very similar to those sketched above to establish that one has this method stability. If the sets Q^M are not defined through a mapping i^M as supposed above, one can still obtain this method stability if one replaces the last statement of (A1M) by the assumptions:

(i) If $\{q^M\}$ is *any* sequence with $q^M \in Q^M$, then there exist q^* in Q and subsequence $\{q^{M_k}\}$ with $q^{M_k} \to q^*$ in the $\tilde{Q}$ topology.

(ii) For *any* $q \in Q$, there exists a sequence $\{q^M\}$ with $q^M \in Q^M$ such that $q^M \to q$ in $\tilde{Q}$.

Similar ideas may be employed to discuss the question of *problem stability* for the problem of minimizing (5.4) over Q (i.e., the original problem) and again compactness of the admissible parameter set plays a critical role. For a sample of discussions of other questions related to problem stability, we refer the reader to [63, 69].

Compactness of parameter sets also plays an important role in computational considerations. In certain problems, the formulation outlined above (involving $Q^M = i^M(Q)$) results in a computational framework wherein the Q^M and Q all lie in some uniform set possessing compactness properties. The compactness criteria can then be reduced to uniform constraints on the derivatives of the admissible parameter functions. There are numerical examples (for example, see [14]) which show that imposition of these constraints is necessary (and sufficient) for convergence of the resulting algorithms. (This offers a possible explanation for some of the numerical failures of such methods reported in the engineering literature, for example, see [186].)

Thus we have that compactness of admissible parameter sets plays a fundamental role in a number of aspects, both theoretical and computational, in parameter estimation problems. This compactness may be assumed (and imposed) explicitly as we have outlined here, or it may be included implicitly in the problem formulation through Tikhonov regularization as discussed for example by Kravaris and Seinfeld [121] and recently by many others. In the regularization approach one restricts consideration to a subset Q_1 of parameters which has compact embedding in Q and modifies the least-squares criterion to include a term which insures that minimizing sequences will be Q_1 bounded and hence compact in the original parameter set Q.

After this short digression on general inverse problem concepts, we return to the condition (5.12). To demonstrate that this condition can be readily established in many problems of interest to us here, we first give the following general convergence results.

Theorem 5.2   *Suppose that H^N satisfies (A2N),(A3N),(A4N) and assume that the sesquilinear forms $\sigma_1(q)$ and $\sigma_2(q)$ satisfy (H7),(H8), as well as (H1)-(H5) of Chapter 4 (uniformly in $q \in Q$). Furthermore, assume that*

$$q \to f(\,\cdot\,; q) \text{ is continuous from } Q \text{ to } L^2((0,T), V_2^*) . \qquad (5.14)$$

Let q^N be arbitrary in Q such that $q^N \to q$ in Q. Then as $N \to \infty$, we have

$$
\begin{aligned}
w^N(t; q^N) &\to w(t; q) &&\text{in } V \text{ norm for each } t > 0 \\
\dot{w}^N(t; q^N) &\to \dot{w}(t; q) &&\text{in } V_2 \text{ norm for almost every } t > 0,
\end{aligned}
$$

where $w^N, \dot{w}^N$ are the solutions to (5.11) and $w, \dot{w}$ are the solutions to (5.2).

Proof: In Chapter 4 it was proven that the solution of (5.2) satisfies $w(t) \in V$ for each t, $\dot{w}(t) \in V_2$ for almost every $t > 0$. Since

$$|w^N(t; q^N) - w(t; q)|_V \le |w^N(t; q^N) - P^N w(t; q)|_V + |P^N w(t; q) - w(t; q)|_V,$$

and (A3N) implies that second term on the right side converges to 0 as $N \to \infty$, it suffices for the first convergence statement to show that

$$|w^N(t; q^N) - P^N w(t; q)|_V \to 0 \quad \text{as } N \to \infty.$$

Similarly, we note that this same inequality with w^N, w replaced by $\dot{w}^N, \dot{w}$ and the V-norm replaced by the V_2-norm along with (A4N) permits us to claim that the convergence

$$|\dot{w}^N(t; q^N) - P^N \dot{w}(t; q)|_{V_2} \to 0 \quad \text{as } N \to \infty,$$

is sufficient to establish the second convergence statement of the theorem. We shall, in fact, establish the convergence of $\dot{w}^N - P^N\dot{w}$ in the stronger V norm.

Let $w^N = w^N(t; q^N)$, $w = w(t; q)$, and $\Delta^N \equiv w^N(t; q^N) - P^N w(t; q)$. Then

$$\dot{\Delta}^N = \dot{w}^N - \frac{d}{dt}P^N w = \dot{w}^N - P^N\dot{w}$$

and

$$\ddot{\Delta}^N = \ddot{w}^N - \frac{d^2}{dt^2}P^N w$$

since $\dot{w} \in L^2((0,T), V_2)$, $\ddot{w} \in L^2((0,T), V^*)$. From (5.2) and (5.11), we have for $\psi \in H^N$

$$
\begin{aligned}
\langle \ddot{\Delta}^N, \psi \rangle_{V^*,V} &= \langle \ddot{w}^N - \ddot{w} + \ddot{w} - \frac{d^2}{dt^2}P^N w, \psi \rangle_{V^*,V} \\
&= \langle f(q^N), \psi \rangle_{V_2^*,V_2} - \sigma_2(q^N)(\dot{w}^N, \psi) - \sigma_1(q^N)(w^N, \psi) \\
&\quad -\langle f(q), \psi \rangle_{V_2^*,V_2} + \sigma_2(q)(\dot{w}, \psi) + \sigma_1(q)(w, \psi) \\
&\quad +\langle \ddot{w} - \frac{d^2}{dt^2}P^N w, \psi \rangle_{V^*,V} .
\end{aligned}
$$

This can be written as

$$
\begin{aligned}
\langle \ddot{\Delta}^N, \psi \rangle_{V^*,V} &+ \sigma_1(q^N)(\Delta^N, \psi) \\
&= \langle \ddot{w} - \frac{d^2}{dt^2}P^N w, \psi \rangle_{V^*,V} - \langle f(q) - f(q^N), \psi \rangle_{V_2^*,V_2} \\
&\quad +\sigma_2(q^N)(\dot{w} - P^N\dot{w}, \psi) + \sigma_2(q)(\dot{w}, \psi) - \sigma_2(q^N)(\dot{w}, \psi) \qquad (5.15) \\
&\quad +\sigma_1(q^N)(w - P^N w, \psi) + \sigma_1(q)(w, \psi) - \sigma_1(q^N)(w, \psi) \\
&\quad -\sigma_2(q^N)(\dot{\Delta}, \psi) .
\end{aligned}
$$

Choosing $\dot{\Delta}^N$ as the test function ψ in (5.15) and employing the equality $\langle \ddot{\Delta}^N, \dot{\Delta}^N \rangle_{V^*,V} = \frac{1}{2}\frac{d}{dt}|\dot{\Delta}^N|_V^2$ (this follows using definitions of the duality mapping – see [22] and the hypothesis (A2N)), we have using the symmetry of σ_1

$$
\begin{aligned}
\frac{1}{2}\frac{d}{dt}\{|\dot{\Delta}^N|_V^2 &+ \sigma_1(q^N)(\Delta^N, \Delta^N)\} \\
&= Re\left\{ \langle \ddot{w} - \frac{d^2}{dt^2}P^N w, \dot{\Delta}^N \rangle_{V^*,V} - \langle f(q) - f(q^N), \dot{\Delta}^N \rangle_{V_2^*,V_2} \right. \\
&\quad +\sigma_2(q^N)(\dot{w} - P^N\dot{w}, \dot{\Delta}^N) + \sigma_2(q)(\dot{w}, \dot{\Delta}^N) - \sigma_2(q^N)(\dot{w}, \dot{\Delta}^N) \qquad (5.16) \\
&\quad +\sigma_1(q^N)(w - P^N w, \dot{\Delta}^N) + \sigma_1(q)(w, \dot{\Delta}^N) - \sigma_1(q^N)(w, \dot{\Delta}^N) \\
&\quad \left. -\sigma_2(q^N)(\dot{\Delta}^N, \dot{\Delta}^N) \right\} .
\end{aligned}
$$

Integrating the terms in (5.16) from 0 to t, applying (H2)-(H5),(H7),(H8), and the initial conditions

$$\Delta^N(0) = w^N(0) - P^N w(0) = w^N(0) - P^N w_0 = 0$$

$$\dot{\Delta}^N(0) = \dot{w}^N(0) - P^N \dot{w}(0) = \dot{w}^N(0) - P^N w_1 = 0 \, ,$$

we obtain

$$|\dot{\Delta}^N|_V^2 + k_1 |\Delta^N|_V^2 \le \nu_1 \delta^N(t) + \nu_2 \int_0^t \{|\dot{\Delta}^N|_V^2 + k_1 |\Delta^N|_V^2\} \, ds, \qquad (5.17)$$

where ν_1, ν_2 and k_1 are positive constants not dependent on N, and

$$\delta^N = \int_0^t \left\{ Re \, \langle \ddot{w} - \frac{d^2}{dt^2} P^N w, \dot{\Delta}^N \rangle_{V^*,V} + |f(q) - f(q^N)|_{V_2^*}^2 + c_2^2 |\dot{w} - P^N \dot{w}|_{V_2}^2 \right.$$

$$\left. + \gamma_2^2 d^2(q, q^N) |\dot{w}|_{V_2}^2 + c_1^2 |w - P^N w|_V^2 + \gamma_1^2 d^2(q, q^N) |w|_V^2 \right\} ds \, .$$

We claim that for $q^N \to q \in Q$, the term δ^N converges to 0 as $N \to \infty$. To establish this, we first note that $\langle \ddot{w} - \frac{d^2}{dt^2} P^N w, \dot{\Delta}^N \rangle_{V^*,V} \equiv 0$. Indeed, for any $\psi \in H^N$, we have

$$\langle \ddot{w} - \frac{d^2}{dt^2} P^N w, \psi \rangle_{V^*,V} = \langle \frac{d^2}{dt^2}(w - P^N w), \psi \rangle_{V^*,V}$$

$$= \frac{d^2}{dt^2} \langle w - P^N w, \psi \rangle_{V^*,V} \qquad (5.18)$$

$$= \frac{d^2}{dt^2} \langle w - P^N w, \psi \rangle_H \, .$$

However, it follows that $\langle w - P^N w, \psi \rangle_H \equiv 0$ since $(w - P^N w)$ is orthogonal to elements in H^N. We note that the last equality in (5.18) follows from the fact that the duality pairing $\langle \cdot, \cdot \rangle_{V^*,V}$ is the extension by continuity of the inner product $\langle \cdot, \cdot \rangle_H$ from $H \times V$ to $V^* \times V$ and hence for $h \in V$, $\langle g, h \rangle_{V^*,V} = \langle g, h \rangle_H$ whenever $g \in H = H^*$. The remainder of the terms in δ^N approach zero as $N \to \infty$ due to (A3N)-(A4N), the continuity condition (5.14), and $q^N \to q$.

Applying Gronwall's inequality to (5.17), we obtain for each $t > 0$

$$|\dot{\Delta}^N(t)|_V^2 + k_1 |\Delta^N(t)|_V^2 \to 0$$

as $N \to \infty$, and hence the convergence statement of the theorem. ∎

As we have indicated, the results of Theorem 5.2 can be used to establish (subsequential) convergence of solutions $\bar{q}^{N,M}$ of the approximation problems in many cases. In particular, these results along with those of Theorem 5.1 can be used to establish convergence of solutions of minimizing J_ν^N, $\nu = 0, 1$, or J_{pe}^N

(i.e., (5.5) or (5.6) with w replaced by w^N of (5.11)) over Q^M to solutions $\bar{q}$ of the original (infinite dimensional) problems of minimizing J_ν or J_{pe} of (5.5) or (5.6) and (5.2) over Q. Indeed, under (A1M)-(A4N) one can establish results in the more general sense of the ideas of "method stability" discussed briefly above.

The results of Theorems 5.1 and 5.2 can also be used to treat estimation problems in the frequency domain format associated with the least squares performance criterion $\hat{J}(q, z)$ of (5.7). We take $\tilde{C}_1$ as the identity (differentiation zero times) followed by pointwise evaluation in t and y. Let $W^N(k; q)$ be the Fourier coefficients associated with w^N of (5.11); i.e., for $k = 0, 1, \ldots, N_t - 1$,

$$\tilde{C}_2\left\{\tilde{C}_1\{w^N(t_i, \bar{y}; q)\}\right\}_k = W^N(k; q) = \frac{1}{N_t} \sum_{i=0}^{N_t - 1} w^N(t_i, \bar{y}; q) e^{-jk(2\pi/N_t)i} \quad (5.19)$$

where we have tacitly assumed that the number of "significant modes" present in the approximate solution is the same as N_f, the number present in w and the data z. (This is easy to guarantee by choosing N sufficiently large - for example, $N \geq 2N_f$). The desired convergence results for the $\bar{q}^{N,M}$ follows readily from the following convergence result for solutions in the frequency domain and the observation that the convergence of (5.20) guarantees the required convergence in the cost functional $\hat{J}(q, z)$ of (5.7).

Theorem 5.3 *Suppose $\{q^N\} \subset Q$ with $q^N \to \bar{q}$ as $N \to \infty$. Let $W^N(k; q^N)$ denote the Fourier series coefficients of the unique solution $w^N(t; q^N)$ to the initial value problem (5.11) corresponding to q^N and let $W(k; \bar{q})$ denote the Fourier series coefficients of the unique solution $w(t; \bar{q})$ to the initial value problem (5.2) corresponding to $\bar{q}$. If $w^N(t; q^N) \to w(t; \bar{q})$ in V norm, and pointwise evaluation is continuous in the V norm, then*

$$\sum_{\ell=1}^{N_f} (|f_{k_\ell^{w^N}}(q^N) - f_{k_\ell^w}(\bar{q})|^2 + \{|W^N(k_\ell^{w^N}; q^N)| - |W(k_\ell^w; \bar{q})|\}^2) \to 0 \quad (5.20)$$

as $N \to \infty$.

Proof: Since pointwise evaluation is continuous in the V topology, we have that $w^N(t; q^N)$ converging to $w(t; \bar{q})$ in V norm implies that $w^N(t, \bar{y}; q^N)$ converges to $w(t, \bar{y}; q)$ for each t. From (5.8) and (5.19), it follows that $W^N(k; q)$ converges to $W(k; \bar{q})$ for each k. Then we have $k_\ell^{w^N} \to k_\ell^w$ as $N \to \infty$ since the significant coefficients of $W^N(k; q^N)$ converge to those of $W(k; \bar{q})$ both in magnitude and phase due to the Fourier series coefficients convergence for each k. Hence, $f_{k_\ell^{w^N}} \to f_{k_\ell^w}$ and $|W^N(k_\ell^{w^N}; q^N)| \to |W(k_\ell^w; \bar{q})|$ as $N \to \infty$ and the result follows.

■

The case of minimizing J_ν^N and J_ν for $\nu = 2$ (i.e., accelerometer data) is slightly more delicate. In the case that $V_2 = V$ (strong damping such as the case of Kelvin-Voigt damping in the beam example) one can use ideas from analytic semigroup theory (see similar examples in [30]) to establish an analogue of Theorem 5.2 involving convergence of $\ddot{w}^N(t; q^N)$ to $\ddot{w}(t; q)$, thereby obtaining a desired convergence of solutions for the corresponding minimization problems. In the case that $V_2 \neq V$, we do not have a general theory for convergence in the case of accelerometer data.

We return to the example of Section 4.5 with σ_1, σ_2 defined by (4.44) and (4.46), respectively. This involves the transverse vibrations of an Euler-Bernoulli beam, and we argue that all the assumptions for Theorem 5.2 can be satisfied. First, we note that the sesquilinear forms defined by (4.44) and (4.46) satisfy (H7) and (H8) while f defined by

$$f(t, y) = \mathcal{K}^B \chi''_{pe}(y)[V_1(t) - V_2(t)], \tag{5.21}$$

(see (4.4) and Section 4.5) satisfies the required continuity condition (5.14) if we choose $H = L^2(0, \ell), V = V_2 = H_L^2(0, \ell)$ and $q = (\tilde{\rho}, \widetilde{EI}, \widetilde{c_D I}, \gamma, \mathcal{K}^B, \mathcal{K}^S)$ in Q, a compact subset of $\tilde{Q} = [L^\infty(0, \ell)]^3 \times \mathbb{R}^3$. Moreover, $w^N(t; q^N) \to w(t; q)$ in this V norm implies $w^N(t; q^N)' \to w(t; q)'$ in the H^1 norm. Hence we have the pointwise convergence required if J_{pe} of (5.6) is used as the least squares criterion. If we choose cubic splines for the basis of the approximation scheme and the parameter set Q as a uniformly bounded collection of piecewise constant functions, then we can ensure that (A1M)-(A4N) are satisfied (verification will be given below) and the desired convergence of Theorem 5.2 will be guaranteed.

With the chosen Q and $\tilde{Q}$, i^M is taken as the identity so that (A1M) is met. To argue (A2N)-(A4N), we construct H^N as follows. On the spatial interval $\Omega = [0, \ell]$, corresponding to the equidistant partition $\{\frac{i\ell}{N}\}_{i=0}^N$ consider the family given by

$$S_{3,B}^N(0, \ell) = \{p \in C^2(0, \ell) \mid p \text{ is a cubic polynomial on each}$$

$$\text{subinterval } [i\ell/N, (i+1)\ell/N], \quad 0 \le i \le N\}$$

and define H^N as the span of the basis set for $S_{3,B}^N(0, \ell)$ (see [156]) with the basis set modified to satisfy the essential boundary conditions $\phi(0) = \phi'(0) = 0$. It follows immediately that $H^N \subset V$.

We conclude this section with a theorem which guarantees the hypothesis (A3N) (and also (A4N) in this case since $V = V_2$).

Theorem 5.4   *Let P^N denote the orthogonal projection of the state space $L^2(0, \ell)$ onto $S_{3,B}^N(0, \ell)$ and let $\psi \in H^2(0, \ell)$. Then there exist constants α_1 and α_2 independent of N and ψ such that*

$$|\psi - P^N \psi|_{L^2} \le \alpha_1 N^{-2} |\psi'|_{L^2}$$

$$|(\psi - P^N \psi)'|_{L^2} \le \alpha_2 N^{-1} |\psi'|_{L^2} .$$

Furthermore, $|(\psi - P^N \psi)''|_{L^2} \to 0$ as $N \to \infty$.

Proof: The proof employs arguments very similar to those in the proof of Theorem A.5.4 in [26]. For completeness, we give a brief outline here.

From [166], for $\psi \in H^2(0, \ell)$ we have the following estimates (we denote the L^2-norm by $|\cdot|$)

$$|\psi - I_Q^N \psi| \leq 4\mu_1 N^{-2} |\psi''|, \tag{5.22}$$

$$|(\psi - I_Q^N \psi)'| \leq 4\mu_1 N^{-1} |\psi''|, \tag{5.23}$$

$$|(I_Q^N \psi)'| \leq 4\mu_1 |\psi''| \tag{5.24}$$

where I_Q^N is the quasi-interpolation operator (e.g., see Theorem A.5.2 of [26] for details) corresponding to N^{-1}, and μ_1 is a fixed constant independent of N and ψ. The first estimate of the theorem follows (with $\alpha_1 = \alpha_2 = 4\mu_1$) from (5.22) and the minimizing property of the projection P^N, $|\psi - P^N \psi| \leq |\psi - I_Q^N \psi|$. The second estimate follows from (5.22), (5.23) along with use of the Schmidt inequality.

From Theorem A.1.2 of [26], the Schmidt inequality and (5.24), we have for $\psi \in H^2(0, \ell)$

$$|(P^N \psi)''| \leq K |\psi''|$$

with $K = 4\mu_1(2\tilde{C} + 1)$, where $\tilde{C}$ is independent of h and N and can be calculated explicitly. Then from density arguments (e.g., take $g \in C^4$) and the inequality

$$\begin{aligned} |(\psi - P^N \psi)''| &\leq |(\psi - g)''| + |(g - P^N g)''| + |(P^N(g - \psi))''| \\ &\leq (1 + K)|(\psi - g)''| + |(g - P^N g)''|, \end{aligned}$$

the convergence statement of the theorem follows from the standard estimates of Theorem A.4.2 of [26] and use of the Schmidt inequality again (see [26, page 79]).

∎

5.3 Experimental and Numerical Results

Methods based on the approximation ideas given in the previous section have been extensively tested computationally and used with experimental data from a number of different smart material structures. We summarize here some of the findings for beams and plates with bonded piezoceramic patches.

The experiments summarized in the next section were performed with a thin aluminum beam with a pair of surface-mounted piezoceramic patches. The thin beam models described in Sections 2.2.6 and 3.3.2 were used to model the beam dynamics with the techniques described in previous sections used to estimate the material and patch strain parameters. The results from three experiments are summarized to illustrate the consistency of estimates across experiments with a variety of actuators and sensors.

The estimation of parameters for a thin circular plate, modeled by equations in Sections 2.2.5 and 3.2.2, is summarized in Section 5.5. In addition to consistency across experiments, these results illustrate the distributed nature of the PDE-based models and passive damping effects that can result when the circuit involving the patches is shunted or closed.

In the computations used to estimate parameters for both the beam and plate, minimization of the respective penalty functionals was accomplished using a Levenberg-Marquardt algorithm. This algorithm can be formulated for the functional (5.10) by defining $g : Q^M \rightarrow \mathbb{R}^{N_t}$, where N_t again denotes the number of temporal observations, as

$$g_i = \tilde{C}_2 \left\{ \tilde{C}_1 \{ w^N(t_i, \cdot; q) \} - \{ z_i \} \right\},$$

and considering an iteration of the form $q^{(k+1)} = q^{(k)} + p^{(k)}$. With $g^{(k)}$ and $J^{(k)}$ denoting the evaluation of g and its Jacobian with respect to $q^{(k)}$, the search vector $p^{(k)}$ is determined by solving the constrained subproblem

$$\min_{p \in Q^M} \left| J^{(k)} p + g^{(k)} \right|_2^2$$

$$\text{subject to } |p|_2 \leq \delta^{(k)} .$$

The radius $\delta^{(k)}$ is chosen in a manner which ensures adequate descent.

While further details regarding this algorithm are found in [77, pages 227-228] and [96, pages 133-137], we do point out that due to the enforced constraint, compactness is automatically imposed by the algorithm. As discussed in Section 5.2, such compactness is a necessary requirement in the minimization procedure and must be considered when solving such parameter estimation problems. Other techniques for ensuring such compactness include explicit constraints on the parameters, compactification through regularization (e.g., Tikhonov) or employment of other optimization routines which lead to constrained subproblems (e.g., the trust region methods that evolved from the Levenberg-Marquardt method).

5.4   Experimental Results: Beams

In early tests of the previously described theory and computational procedures, a series of experiments were carried out at the Mechanical Systems Laboratory, then located at the State University of New York at Buffalo. A cantilevered 2024-T4 aluminum beam of length $\ell = 46.73\,cm$, width $b = 2.03\,cm$ and thickness $h = .16\,cm$ with surface-mounted G-1195 PZT piezoceramic patches was used as a test structure. The patches had length $\ell_{pe} = 6.37\,cm$, width $2.03\,cm$, thickness $h_{pe} = .0254\,cm$ and, as depicted in Figure 4.1 of Chapter 4, were located between $y_1 = 2.54\,cm$ and $y_2 = 8.89\,cm$ on opposite sides of the beam. The patches were bonded with an epoxy adhesive with a 0.64 cm wide, 0.0076 cm thick piece of copper foil glued between each patch and the beam to act as a conductive medium.

As summarized in Column 3 of Table 5.1, the given handbook values of the Young's modulus and mass density for the beam and patches were $E = 7.3 \times 10^{10}\, N/m^2$, $\rho = 2.766 \times 10^3\, kg/m^3$ and $E_{pe} = 6.3 \times 10^{10}\, N/m^2$, $\rho_{pe} = 7.6 \times 10^3\, kg/m^3$, respectively. While these provided starting values for the optimization routines, they were not adequate for obtaining accurate model fits to experimental data due to material nonhomogeneities, adhesive layers, et cetera. Instead, with $a_3 \equiv (h/2 + h_{pe})^3 - (h/2)^3$, the piecewise constant expressions

$$\tilde{\rho}(y) = \rho h b + 2\rho_{pe} h_{pe} b \chi_{pe}(y)$$

$$\widetilde{EI}(y) = \frac{1}{12} h^3 bE + \frac{2}{3} a_3 b E_{pe} \chi_{pe}(y) \,,$$

(5.25)

derived in Chapter 3 (see (3.21) and (4.3)), were used to model the density and stiffness of the structure (Figure 5.1 depicts the general form of these functions). Again, the characteristic function

$$\chi_{pe}(y) = \begin{cases} 1 & y_1 < y < y_2 \\ 0 & \text{otherwise} \end{cases}$$

(5.26)

was used to isolate those regions of the structure that are covered by patches. Similarly, the Kelvin-Voigt parameter $\widetilde{c_D I}$ was taken to be piecewise constant to account for patch contributions to the structural damping. The air damping coefficient γ which multiplies the velocity term was assumed to be constant since it is unaffected by nonhomogeneities in the beam. The final parameters to be estimated were the patch constants $\mathcal{K}^B$ and $\mathcal{K}^S$ which relate the strains and voltages generated during actuation and sensing, respectively.

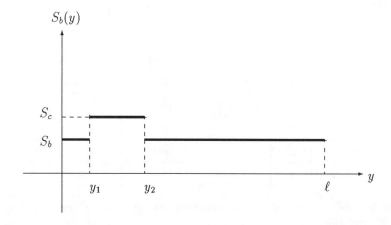

Figure 5.1: Form of the functional parameters $\tilde{\rho}(y), \widetilde{EI}(y), \widetilde{c_D I}(y)$ where S_b indicates the value for the basic structure (the beam) and S_c the value for the composite (beam and piezoceramic) structure.

For the choice of parameters $q = (\tilde{\rho}(y), \widetilde{EI}(y), \widetilde{c_D I}(y), \gamma, \mathcal{K}^B, \mathcal{K}^S)$, the parameter space $\tilde{Q}$ was taken to be $[L^\infty(0, \ell)]^3 \times \mathbb{R}^3$, and the parameter set Q was chosen to be a collection of uniformly bounded piecewise constant functions each having jump discontinuities at most at y_1 and y_2. It can then be verified that for these choices with the mapping i^M taken as the identity, the hypothesis (A1M) is satisfied.

The results presented here were obtained with $N = 10$ in the cubic spline approximations described in the previous section since this resolved the system dynamics in the sense that eigenvalues of the approximate systems did not change significantly with larger N.

To test the smart material methodology for estimating parameters in the PDE-based model, several experiments were conducted with a variety of sensing and excitation devices. In the remainder of this section, results from the following experiments are summarized:

Experiment B1: Impulse hammer excitation and piezoceramic patch sensing

Experiment B2: Piezoceramic patch excitation and accelerometer sensing

Experiment B3: Piezoceramic patch excitation and sensing.

In each experiment, time response data and input signals from the experimental beam were obtained using a Tektronix Analyzer (Model 2600) with a sample rate of 256 Hz for 16 seconds. In response to various forms of excitation, two modes at 6.6 Hz and 38.4 Hz were observed in the experiments. The parameters estimated in each case are summarized in Table 5.1.

Beam Experiment		Given	Exp. B1	Exp. B2	Exp. B3
Actuator/Sensor			Hammer/PZT	PZT/Acce.	PZT/PZT
E	E (beam)	7.300+10	7.062+10	7.068+10	7.062+10
(N/m^2)	E_{pe} (PZT)	6.300+10	4.772+10	4.449+10	4.772+10
ρ	ρ (beam)	2.766+3	2.943+3	2.943+3	2.943+3
(kg/m^3)	ρ_{pe} (PZT)	7.600+3	16.405+3	15.823+3	15.944+3
c_D	c_D (beam)	—	0.896+6	1.061+6	1.040+6
(Ns/m^2)	c_{Dpe} (PZT)	—	0.396+6	0.396+6	0.396+6
γ (Ns/m^2)		—	1.299-2	1.2996-2	1.301-2

Table 5.1: Given and estimated structural parameters for beam experiments B1, B2 and B3.

Experiment B1: Impulse hammer excitation, piezoceramic patch sensing

In the first experiment reported here, the beam was excited by an impulse force delivered by a hammer impact to the beam $2.54\,cm$ from the clamped end. This impact was modeled by a forcing function of the form $f(t, y) = \tilde{f}(t)\delta(y - 2.54)$ where $\tilde{f}(t)$ is triangular in shape. The voltages generated by the patches in response to the beam vibrations were collected as data. Parameters estimated using the finite dimensional system analogue of the penalty functional (5.6) are summarized in Column 4 of Table 5.1 while model responses obtained with these estimated parameters are compared with the observed trajectory and frequency response in Figures 5.2 and 5.3.

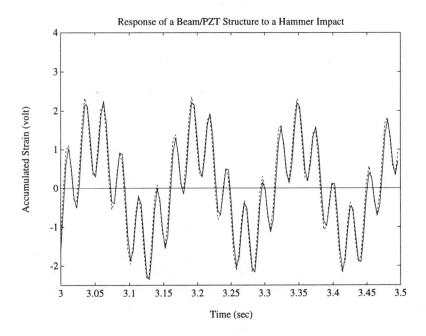

Figure 5.2: Time history of experimental beam and model response on the time interval $[3, 3.5]$ seconds for Experiment B1, —— (experimental data), - - - (model response).

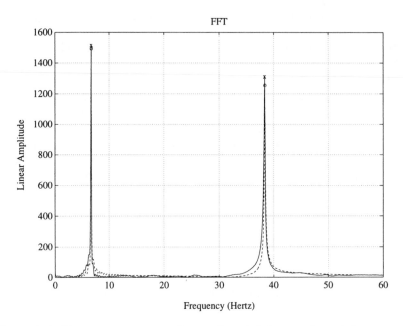

Figure 5.3: Frequency content of experimental data and model response for Experiment B1, x ——— (experimental data), o – – – (model response).

Experiment B2: Piezoceramic patch excitation and accelerometer sensing

The input to the beam in the second experiment was provided by a narrow triangular shaped voltage (approximating an impulse) applied to the patches. This voltage was applied out-of-phase to the patches so as to produce pure bending moments. Hence the input function was given by (5.21) or

$$f(t, y) = \mathcal{K}^B \chi''_{pe}(y)[V_1(t) - V_2(t)] \, .$$

An accelerometer weighing 0.5 grams located at $\bar{y} = 2.14\,cm$ was used as a sensor. The penalty functional (5.5), with w replaced by w^N, was then used when estimating the coefficients which are summarized in Column 5 of Table 5.1.

Experiment B3: Piezoceramic patch excitation and sensing

To test the capability of piezoelectric materials in smart structures, we also designed and performed experiments on the same beam in which the piezoceramic patches were used as actuator and sensor in self-identification of material parameters. A PC relay device was used as a switch to allow the dual use of the piezoceramics in this manner. In this third type of experiment, the structure was excited by an input of sinusoidal voltage, out-of-phase, to the PZT patches for a time period of $[0, t_0)$ and was left to free vibrate for $t \geq t_0$.

The parameters estimated in this manner are summarized in Column 6 of Table 5.1 while the circuit setup for this experiment is depicted in Figure 5.4. The full experimental setup is similar to that described in Section 6.3.1 and details can be found in that section.

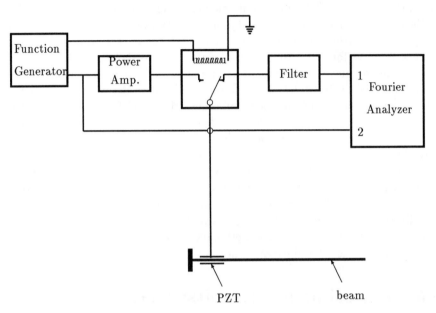

Figure 5.4: Circuit for the experiments using piezoceramic patches as actuators/sensors.

It can be observed that identification results for $\rho(y), y_1 < y < y_2$, in all three examples, yield values that are substantially larger than the handbook values. One explanation for this is that equation (5.25) is simply a superposition of linear mass densities of the beam and the patches which ignores the mass of the bonding glue and the conducting copper foil.

In all our parameter estimation procedures, we adopted a so-called hybrid method (mentioned previously), wherein we started with the finite dimensional version of the cost function (5.7) (the cost function corresponding to frequency domain data) and then switched to (5.5) or (5.6) (time domain data). Our experiences on these and numerous other inverse problems reveal that parameter identification in time domain is more sensitive and accurate if the initial guesses of parameters are close to the optimal ones. On the other hand, parameter identification in the frequency domain will yield quick and rough estimates with the resulting parameters in a neighborhood of the optimal values.

From the above reported results, we can make several conclusions. First, we have observed that for the beam responses under consideration here, the use of the piezoceramic as a sensor is as reliable as the accelerometer and even better in the sense that one doesn't have to worry about the effects of concentrated weight and the wire connected to the accelerometer.

An equally important discovery is that standard "finite element" methods for approximating models such as (5.1), (5.2) are often inadequate. Our findings, in addition to difficulties experienced by others [80], suggest that a straightforward use of standard finite element packages will not be successful for the most part in quantitative studies unless extreme care is taken to account for spatially varying coefficients in the model. In the parameter identification process, we have found it essential to model the micro-structure related to the patch/beam geometry and elastic characteristics. Models with constant (in y) mass density $\tilde{\rho}$, stiffness $\widetilde{EI}$, and damping $\widetilde{c_D I}$ fail to describe experimental data in an adequate fashion (one cannot match model frequencies with the data in experiments where multiple mode excitation is present). Our efforts with variable geometry and variable elasticity parameters promise to alleviate these difficulties. They also emphasize the importance of correct physical models in identification and feedback control design. This, among other findings including the results for plates in the next section, strongly supports our own belief that partial differential equation techniques play a very important role in modeling for the emerging technology of adaptive or smart material structures. Our conclusions are based on the methodology developed here which presents theory, computational and experimental tests, all of which are consistent.

5.5   Experimental Results: Plates

To further illustrate the consistency and accuracy of distributed structural models, we summarize here experimental results for a clamped circular aluminum plate with a single centered circular patch (see Figure 5.5). The dimensions of the plate and piezoceramic patches are summarized in Table 5.2. We note that the patch has a radius that is $\frac{1}{12}$ that of the plate and a thickness that is approximately $\frac{1}{7}$ of the plate thickness; hence it is quite small in relation to the plate. Table 5.2 also contains "handbook" values of the density, Young's modulus and Poisson ratio for the plate. We reiterate that while these values provide a starting point in the parameter estimation routine, they usually cannot be used in the final system model with any accuracy due to nonuniformities in the plate or boundary conditions, variations in materials, and the contributions due to the presence of the patches (this fact is illustrated in the examples).

The plate was excited so as to produce transverse vibrations; thus we used the model (2.63), (2.64) of Chapter 2, Section 2.2.5 with moments and patch contributions as in Chapter 3, Section 3.2.2 (see (3.14)–(3.16)). We ignored contributions from the bonding layers and used Fourier-Galerkin approximations as described in Chapter 8 below (see Section 8.3.1) for our calculations in the approximate estimation problems for (5.10) and (5.11).

To provide a basis for comparison between measured experimental natural frequencies and the analytic frequencies for a plate of this size to which no patches are bonded, analytic values were calculated using the plate dimensions and "handbook" parameter values summarized in Table 5.2. These calculated values are compiled in Table 5.3. In this latter table, m refers to the Fourier number and n denotes the order of the root to the Bessel functions which comprise the analytic solutions. Hence the analytic frequencies of the first four axisymmetric modes are $61.9, 241.2, 540.5$ and 959.5 hertz corresponding to $m = 0$, $n = 0, 1, 2$ and 3, respectively. The Fourier number m can also be interpreted as the number of nodal diameters while n is the number of nodal circles, not including the boundary. As will be seen in the examples, the experimental frequencies are, in many cases, significantly lower than the corresponding analytic values. This is due to variations in material properties, potential energy loss in boundary conditions, and the presence of material damping in the physical structure.

	Plate Properties	Patch Properties
Radius	$a = .2286\,m$	$rad = .01905\,m$
Thickness	$h = .00127\,m$	$h_{pe} = .0001778\,m$
Young's Modulus	$E = 7.1 \times 10^{10}\,N/m^2$	$E_{pe} = 6.3 \times 10^{10}\,N/m^2$
Density	$\rho = 2700\,kg/m^3$	$\rho_{pe} = 7600\,kg/m^3$
Poisson ratio	$\nu = .33$	$\nu_{pe} = .31$
Strain Coefficient		$d_{31} = 190 \times 10^{-12}\,m/V$

Table 5.2: Dimensions and "handbook" characteristics of the plate and PZT piezoceramic patch.

n	$m = 0$	$m = 1$	$m = 2$	$m = 3$	$m = 4$	$m = 5$	$m = 6$
0	61.9	129.0	211.6	309.6	422.6	550.4	692.8
1	241.2	368.9	513.0	673.3	849.8		
2	540.5	728.3	932.9				
3	959.5						

Table 5.3: Plate frequencies calculated using "handbook" dimensions and parameters under the assumption of fixed boundary conditions.

Time domain data was collected using accelerometers located at the points $A_\ell = (2'', 0)$, $A_c = (0'', 0)$ and $A_r = (2'', \pi)$ as depicted in Figure 5.5 (all plate points in this section are expressed in polar coordinates). This orientation of accelerometers permitted the collection of both axisymmetric and nonaxisymmetric data with the locations chosen to avoid low-order nodal lines and

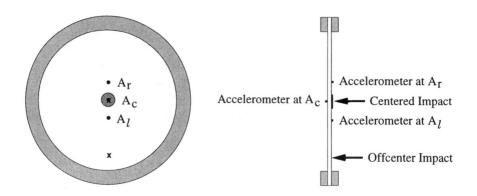

Figure 5.5: Clamped circular plate with a single centered piezoceramic patch. Accelerometers located at $A_\ell = (2'', 0)$, $A_c = (0'', 0)$ and $A_r = (2'', \pi)$. Centered hammer impact at $(0'', 0)$, offcenter hammer impact at $(7.27'', 0)$.

circles. In all cases, data was obtained at a 12 KHz sample rate so as to resolve any high frequency responses.

The experimental results reported here can be summarized as follows:

Experiment P1: Axisymmetric excitation with large hammer - Open circuit

Experiment P2: Axisymmetric excitation with large hammer - Closed circuit

Experiment P3: Axisymmetric excitation with small hammer

Experiment P4: Axisymmetric excitation through a voltage spike to patch

Experiment P5: Nonaxisymmetric excitation with small hammer.

Note that both axisymmetric and general nonaxisymmetric responses are considered with input provided by various impact hammers as well as from the patches themselves. The damping effects due to the electric circuit containing the patch are also investigated. In each experiment, the goal is the estimation of the various physical parameters and the results from all experiments discussed here are summarized in Table 5.4. The consistency and/or variability of these estimates is discussed in the following presentation.

Experiment P1: Axisymmetric excitation with large hammer – Open circuit

In the first set of experiments, the plate was excited through an impulse delivered by a large impact hammer having a plastic tip (the force transducer on the hammer delivered 50 mV/lb). The impact was delivered to the center of the plate and data was collected from accelerometers located at the points

		Book	Ex. P1	Ex. P2	Ex. P3	Ex. P4	Ex. P5
$\rho \cdot thick.$	Plate	3.429	3.107	3.123	3.114	3.170	3.165
(kg/m^2)	Plate + Pzt		3.131	3.230	2.993	3.216	3.179
D	Plate	13.601	11.310	11.270	11.205	11.151	11.361
(Nm)	Plate + Pzt		11.381	11.302	11.674	11.506	11.783
c_D	Plate		1.161-4	1.443-4	9.358-6	2.816-5	2.598-5
$(Nm\,s)$	Plate + Pzt		1.290-4	2.031-4	9.392-6	3.211-5	2.693-5
ν	Plate	.33	.331	.331	.330	.326	.330
	Plate + Pzt		.326	.325	.327	.325	.328
$\gamma\,(Ns/m^3)$			11.57	17.02	58.57	58.97	45.71
$\mathcal{K}^B\,(N/V)$			.006074				

Table 5.4: "Handbook" and estimated parameter values obtained in plate experiments P1 - P5.

A_ℓ, A_c and A_r (see Figure 5.5). The excitation of the structure in this manner provided a primarily axisymmetric response with the purely axisymmetric component being measured by the centered accelerometer. Data obtained from off-center accelerometers indicated that while slight nonaxisymmetric vibrations were present, their effect was minimal.

During the collection of this data, the circuit involving the piezoceramic patch was left open to minimize piezoelectric effects due to the bending patch (with a closed circuit, the voltage produced when the patch vibrates is fed back to the patch which in turn produces a bending moment; the damping and stiffening effects which occur in this case are investigated in the next experiment).

Minimization of the function J_ν^N for $\nu = 2$ (i.e., (5.5) with w replaced by w^N and evaluation at $(\bar{r}, \bar{\theta})$ instead of $\bar{y}$) was performed. That is,

$$J_2^N(q, z) = \sum_{i=1}^{N_t} \left| \frac{\partial^2 w^N}{\partial t^2}(t_i, \bar{r}, \bar{\theta}; q) - z_i \right|^2 \qquad (5.27)$$

was used. In this experiment, minimization of (5.27) was performed using the time history of the acceleration obtained from the *centered* accelerometer (at $A_c = (0'', 0)$). For this experiment, fixed-edge boundary conditions were assumed. The estimated parameters ρ, γ, D, ν and c_D ($\mathcal{K}^B$ was not estimated in this case since there is no patch input) are recorded in Table 5.4 while model-based results obtained with these values are plotted against experimental results in Figures 5.6 and 5.7. We reiterate that in these plots, both the data and calculated model response were obtained at the center point of the plate.

As indicated by the frequency results in Figure 5.7, four axisymmetric modes, having frequencies of 59.3, 227.8, 516.4 and 917.7 Hz, were excited in this experiment. The results in both figures demonstrate that the parameter estimates in Table 5.4 lead to a very close matching of the first two frequencies. The overdamping of the higher frequency modes is characteristic of the Kelvin-Voigt damping model and this leads to the very slight variation seen in the time history when comparing the experimental data and model response.

To demonstrate the distributed nature of the model, the parameters obtained using data from the centered accelerometer, as summarized in Table 5.4, were used to calculate the model response at the offcenter point $A_r = (2'', \pi)$. The results are plotted along with the experimental data at that point in Figures 5.8 and 5.9. From the frequency results in Figure 5.9, it can be seen that the primary response at that point is in the first two axisymmetric modes. While the model response in the first mode is slightly larger than the corresponding experimental result, the agreement is very close in light of the fact that experimental data from this accelerometer was *not* used when determining the physical parameters. Similar results were found at the point $A_\ell = (2'', 0)$, thus demonstrating our ability to identify parameters using data from one accelerometer that produces a reasonable distributed model.

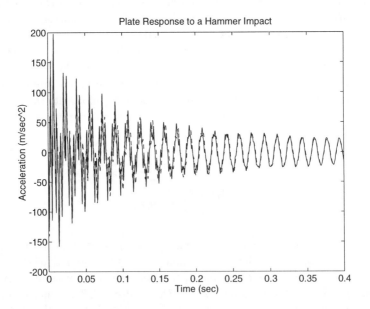

Figure 5.6: Time history of Experiment P1 data measured at $A_c = (0'', 0)$ and model response with estimated parameters, ——— (experimental data), $- - -$ (model response).

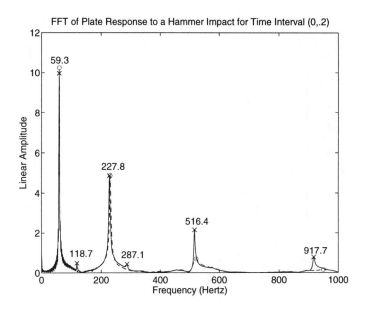

Figure 5.7: Frequency content of Experiment P1 data measured at $A_c = (0'', 0)$ and thin plate model with estimated parameters, x —— (experimental data), o − − − (model response).

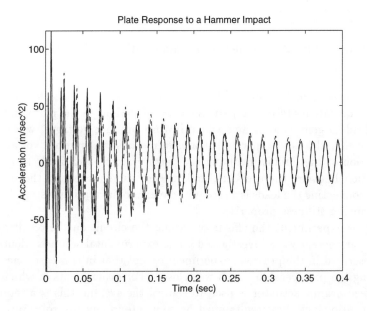

Figure 5.8: Time history of Experiment P1 data measured at $A_r = (2'', \pi)$ and model response with estimated parameters, —— (experimental data), − − − (model response).

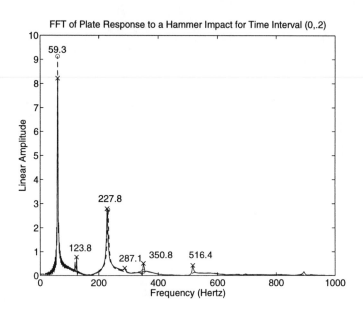

Figure 5.9: Frequency content of Experiment P1 data measured at $A_r = (2'', \pi)$ and thin plate model with estimated parameters, x ——— (experimental data), o – – – (model response).

Experiment P2:   Axisymmetric excitation with large hammer – Closed circuit

As discussed in Chapter 1, the piezoelectric effect is manifested in two ways in the patches (see (1.1)). In one case, vibrations in the plate and hence patch lead to generated strains which in turn produce voltages, whereas the converse effect leads to generated strains in response to an input voltage. The completion of the circuit involving the piezoceramic patch leads to a strong interaction between these effects, and indeed, the shunting of the patch by simply connecting the leads is a recognized means of increasing system damping and changing stiffness properties [99].

In this experiment, the effects of closing the circuit on the resulting estimated parameters were investigated. The experimental setup is identical to that described in the previous experiment except that in this case, the circuit involving the piezoceramic patch was closed. For experiments in which input to a piezoceramic actuator is used to control the system, this is a more realistic scenario since the circuits must be complete in any control setup. As in Experiment P1, an impact hammer hit to the plate center was used to obtain an axisymmetric response and data was obtained from accelerometers located at the points A_c, A_r and A_ℓ depicted in Figure 5.5.

To obtain model responses for this case, three sets of parameters, as summarized in Table 5.5, were used. The first set of parameters was obtained by minimizing the least squares functional (5.27) using data from the centered accelerometer. These parameters can be compared with those in the second set which were obtained in Experiment P1. To obtain the third set, the analytic values for the density, flexural rigidity and Poisson ratio for the plate were used throughout the structure while γ and c_D from the first data set were used as damping values. The use of the third data set simulates the results that are obtained if one simply uses "handbook" values for the density, flexural rigidity and Poisson ratio.

As demonstrated by the time history and corresponding frequency plots in Figure 5.10a, results comparable to those obtained in Experiment P1 can be obtained when the physical parameters are obtained using fit-to-data techniques. By comparing the parameters obtained here with those of the first experiment, however, one sees some variation due to the circuit effects on the piezoceramic patches. The most marked difference is an increase in damping which results when the system is closed. Since the damping effects due to the circuit are not included in the model, the optimization routine increased the viscous damping coefficient γ and Kelvin-Voigt parameter c_D. As noted in the plots of Figure 5.10a, this compensation for the damping leads to a good model fit to the data even though the mechanism for the unmodeled circuit damping differs from the internal and viscous damping included in the model. While some difference in density and stiffness also occur, these effects are less pronounced due to the small size of the patch in relation to the plate.

The experimental data and model response obtained with parameters from Experiment P1 (open circuit) are plotted in Figure 5.10b. As noted in these plots, the model response is significantly underdamped since the effects of damping due to the closed circuit were not considered in Experiment P1. Moreover, a slight shift in frequency due to changes in ρ, D and ν can also be noted. This illustrates some of the variations which result from changing the configuration of the electric circuit and highlights the fact that identification procedures should be performed in the setting in which applications or control are to be considered.

Finally, the experimental data and model response obtained with the third set of parameters (analytic values of ρ, D and ν) are plotted in Figure 5.10c. As noted in both time domain and frequency plots, the frequency of the model response is much too large in this case, namely due to the fact that the analytic value of the flexural rigidity is approximately 17% larger than the estimated values. This illustrates the fact that even those parameters for which "handbook" values exist must be estimated through parameter identification techniques in order to guarantee an accurate model.

(a) Plate Data and Model Response with Experiment P2 Parameters

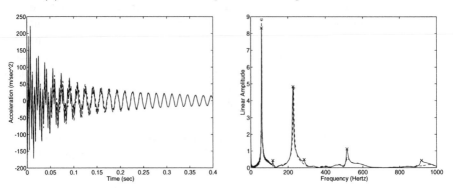

(b) Plate Data and Model Response with Experiment P1 Parameters

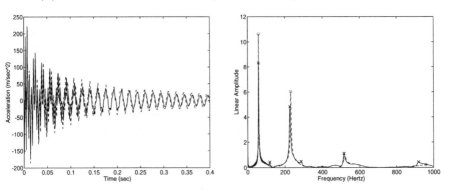

(c) Plate Data and Model Response with Analytic Parameters

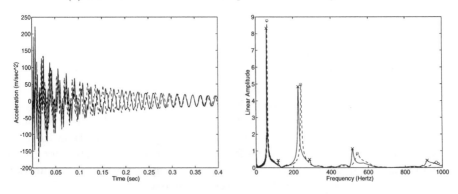

Figure 5.10: Experiment P2 data measured at $A_c = (0'', 0)$ and model response with (a) estimated parameters, (b) parameters from Experiment P1, and (c) analytic parameters; x ——— (experimental data), o – – – (model response).

		Estimated Parameters	Experiment P1 Parameters	Handbook Parameters
$\rho \cdot thickness$	Plate	3.123	3.107	3.429
(kg/m^2)	Plate + Pzt	3.230	3.131	3.429
D	Plate	11.270	11.310	13.601
$(N \cdot m)$	Plate + Pzt	11.302	11.381	13.601
c_D	Plate	1.443-4	1.161-4	1.161-4
$(N \cdot m \cdot sec)$	Plate + Pzt	2.031-4	1.290-4	1.290-4
ν	Plate	.331	.331	.330
	Plate + Pzt	.325	.326	.330
$\gamma \, (sec \cdot N/m)$		17.021	11.569	11.569

Table 5.5: Handbook and experimental values of the physical parameters. The first column of estimated parameters was obtained using Experiment P2 data.

Experiment P3:   Axisymmetric excitation with small hammer

In the previous two experiments, the plate was excited through impacts from a large hammer having a soft tip. This resulted in the excitation of four axisymmetric modes having frequencies ranging from approximately 60 Hz to 920 Hz. To investigate the suitability of the model when a wider range of frequencies are excited, a small impact hammer (with a force transducer delivering 100 mV/lb) having a metal tip was also used with the results being reported as Experiment P3.

As in the previous experiments, a centered hit was used to evoke an axisymmetric response with data being obtained from accelerometers located at A_c, A_ℓ and A_r. The leads to the piezoceramic patches were left disconnected, thus minimizing the damping effects due to the circuit and patch. The minimization of the functional (5.27) was performed with data from the centered accelerometer and the resulting estimated parameters are summarized in Table 5.4. The model response and experimental data from the centered accelerometer are plotted in Figure 5.11 and 5.12. As indicated by the frequency plots in Figure 5.12, six modes were accurately matched with these estimated parameter values. The expected overdamping, due to the Kelvin-Voigt model, of the high frequency 2814 and 3661 Hz modes is apparent.

In comparing the parameter estimates of Experiments P1 and P2 in Table 5.4, it can be seen that while little change occurs in ρ, D and ν, there is some variation in the viscous damping constant γ and the internal Kelvin-Voigt parameter c_D. This is due to the different frequency responses in the two experiments and again reflects some limitations in the damping model. When the large hammer was used to excite the plate, the primary response was in the lower frequency modes and the parameters obtained from the minimization of (5.27) yielded a model which matched the lower frequencies but overdamped

the higher frequencies having less energy. The use of the small hammer with a
metal head resulted in data in which the primary response was in the 918 Hz
mode with very little energy in the 60 Hz mode. This shift in the excited
frequencies generally leads to a reduction in the estimated values of c_D and an
increase in γ (see also Experiments P4 and P5 below).

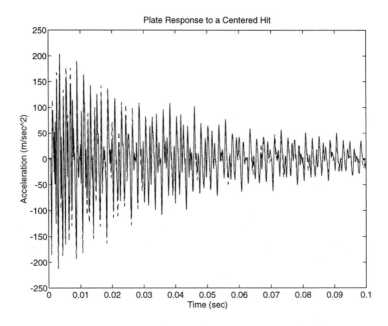

Figure 5.11: Time history of Experiment P3 data measured at $A_c = (0'', 0)$
and model response with estimated parameters, —— (experimental data),
– – – (model response).

Experiment P4:   Axisymmetric excitation – Voltage spike to patch

A second means of exciting the plate is through a voltage spike to the piezoceramic patch and results obtained in that manner are reported here. Because
the active patch was centered on the plate, this yielded an axisymmetric response and data from the centered accelerometer was used when minimizing
the functional (5.27).

The estimated physical parameters ρ, D, ν, c_D and γ as well as the patch
input parameter $\mathcal{K}^B$ are summarized in Table 5.4 and the resulting model
response is plotted along with the experimental data from the centered accelerometer in Figures 5.13 and 5.14. As indicated by the time and frequency
plots in the figures, the plate response obtained in this manner is quite similar
to that obtained by exciting the plate with the small metal-tipped hammer.

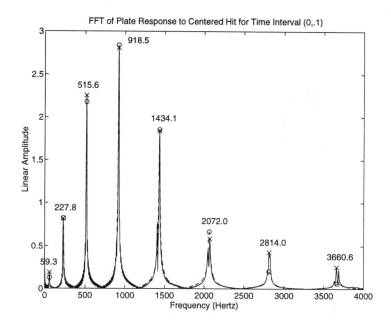

Figure 5.12: Frequency content of Experiment P3 data measured at $A_c = (0'', 0)$ and thin plate model with estimated parameters, x ——— (experimental data), o − − − (model response).

In comparing the estimated parameters from Experiments P3 and P4, it is noted that there is very little variation in either ρ, D, ν or c_D, γ, in spite of the differing mechanisms for exciting the system.

The estimated value .006074 for the patch parameter $\mathcal{K}^B$ is seen to be approximately 48% of the value .0126 predicted by the model

$$\mathcal{K}^B = \frac{1}{2} \frac{E_{pe}}{1 - \nu_{pe}} d_{31}(h + h_{pe} + 2h_{b\ell}) \qquad (5.28)$$

(see (3.16) for the analogous expression for two patches) with the values of $E_{pe}, \nu_{pe}, h, h_{pe}$ and d_{31} specified in Table 5.4 and $h_{b\ell}$ taken to be 0. Some of this variation can be attributed to patch material values which differ slightly from those summarized in Table 5.4. While differences occur between the "handbook" values of the Young's modulus and Poisson ratio and the "true" parameters for the experimental patch, perhaps the biggest source of variation occurs in the values for the strain constant d_{31}. The value reported in the product literature (and given in Table 5.4) was obtained through static tests while the estimated value is obtained in a dynamic setting which tends to decrease the realized values of strain parameters such as d_{31} [83, 151, 154]. Hence while the analytic values given by the model (3.16) or (5.28) can be used as starting values in the optimization routine, they will not in general yield an accurate model response due to physical variations in the material patch properties.

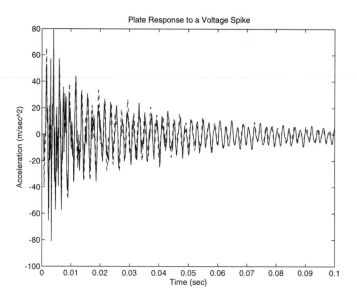

Figure 5.13: Time history of the Experiment P4 data measured at $A_c = (0'', 0)$ and model response with estimated parameters, ——— (experimental data), – – – (model response).

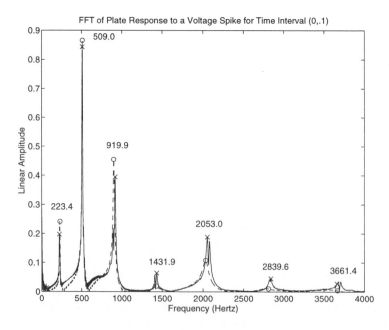

Figure 5.14: Frequency content of Experiment P4 data measured at $A_c = (0'', 0)$ and thin plate model with estimated parameters, x ——— (experimental data), o – – – (model response).

Experiment P5:   Nonaxisymmetric excitation with small hammer

In this experiment, a nonaxisymmetric response was obtained through a small hammer impact at the point $(7.27'', 0)$ (see Figure 5.5). The leads to the piezoceramic patch were disconnected in this experiment to minimize damping effects due to the circuit and piezoelectric properties of the patch. Data was again measured via the three accelerometers located at $A_r = (2'', 0), A_c = (0'', 0)$ and $A_\ell = (2'', \pi)$. Optimization was performed using the data from the accelerometer located at $A_r = (2'', 0)$ and the estimated parameters values are summarized in Table 5.4.

Time and frequency plots of the experimental data from the right (A_r), centered (A_c) and left (A_ℓ) accelerometers as well as corresponding model responses are given in Figure 5.15a, b and c, respectively. The observed experimental frequencies as well as the calculated model frequencies at the three accelerometers are tabulated in Table 5.6.

From the frequency plots in Figure 5.15a, it can be seen that the model very accurately matches the $(n, m) = (0,0), (0,2), (0,3), (1,1), (1,2), (2,0), (2,1)$ and $(0,4)$ modes while significantly underdamping the $(1,0)$ and $(0,1)$ modes (see Table 5.3 to compare the observed frequencies with the corresponding modes). As expected, the higher-order modes are overdamped as is typical with the Kelvin-Voigt damping mechanism.

Similar results are observed in the plots in Figure 5.15c which depict the acceleration data and model response at A_ℓ (recall that the data from the right accelerometer was used for obtaining the parameters). In addition to the previously matched modes, this data contains a stronger response in the $(0,4)$ mode (408 Hz) which is accurately matched by the model. Although the $(1,0)$ and $(0,1)$ modes are still underdamped, the accurate matching of 9 modes demonstrates the distributed nature and accuracy of this model.

The underdamping of the $(1,0)$ mode is very evident in both the time domain and frequency plots of the data and model response at the centered accelerometer (Figure 5.15b). By comparing the relative degree of underdamping that is observed at A_c with that seen at A_r or A_ℓ, it can be seen that the results are comparable. However, the different distribution of energy in the axisymmetric and nonaxisymmetric modes leads to larger discrepancies between the model and experimental data measured at A_c than were observed at the noncentered points. The underdamping of the $(1,0)$ and $(0,1)$ modes again illustrates some of the limitations of the damping model being used in these investigations.

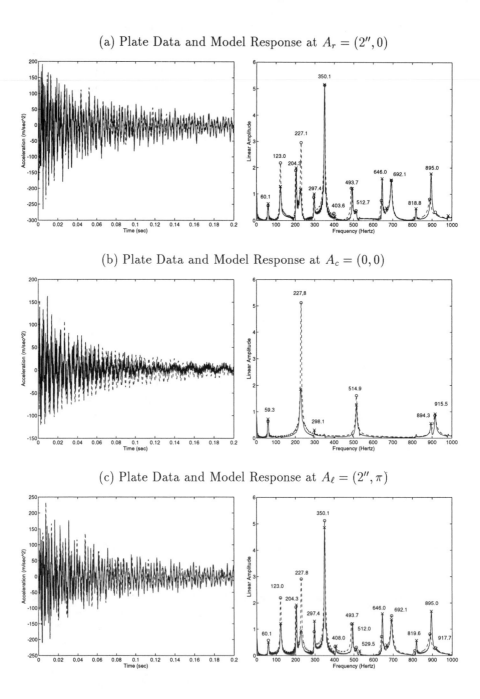

Figure 5.15: Experiment P5 data and model response at (a) $A_r = (2'', 0)$, (b) $A_c = (0, 0)$ and (c) $A_\ell = (2'', \pi)$; x ———— (experimental data), o – – – (model response).

Right Accel. (Measured)					Right Accel. (Model Response)				
60.1	123.0	204.3	297.4	403.6	59.3	123.0	202.1	295.2	399.2
227.1	350.1	493.7	646.0	818.8	229.2	350.8	488.5	642.3	
512.7	692.1	895.0			512.7	692.9	887.7		
					916.3				
Center Accel. (Measured)					Center Accel. (Model Response)				
59.3		298.1			59.3				
227.8					230.0				
514.9		894.3			514.9				
915.5					914.1				
Left Accel. (Measured)						Left Accel. (Model Response)				
60.1	123.0	204.3	297.4	408.0	529.5	59.3	122.3	201.4	295.9	403.6
227.8	350.1	493.7	646.0	819.6		229.9	350.8	488.5	641.6	
512.0	692.1	895.0				512.7	692.9	887.7		
917.7						914.8				

Table 5.6: Observed frequencies in Experiment P5 data and model response.

We have considered here issues associated with the estimation of parameters in a PDE-based model for a vibrating plate. The unknown parameters in the model included structural parameters (density, flexural rigidity, Poisson ratio and material and air damping) and patch parameters. The structural parameters were taken to be piecewise constant in order to account for the presence and differing material properties of the patches.

When designing and performing experiments, two issues were considered. The first concerned the ability of the PDE model to accurately and consistently describe the physics of the system under a variety of inputs and responses. Secondly, it is well-known that closing or shunting the circuit containing the piezoceramic patch provides additional damping, and this was investigated in the context of the PDE model.

With regards to the first issue, experiments were performed in which the plate was excited with a variety of inputs (including impact hammers and voltage spikes to the patches) which excited from four to fifteen frequencies ranging from 60 Hz to 4000 Hz. The matching of up to six axisymmetric and eight nonaxisymmetric frequencies illustrated that the thin plate model was appropriate and sufficiently accurate for the experimental plate under consideration. Moreover, the distributed nature of the PDE model means that it accurately describes the physics of the *entire* plate including points *not* used in the optimization process. As demonstrated by results reported in [42] and Chapter 8 below, the accuracy of the model, with parameters estimated in the manner discussed here, contributed to the good vibration attenuation attained when the model was incorporated in a PDE-based controller.

When comparing the parameters estimated in the various experiments, it was noted that the density, flexural rigidity and Poisson ratios were consistent across *all* experiments. There was some variation in the damping parameters depending on the frequency content of the data. In experiments with minimal low frequency excitation but substantial energy in the high frequencies, the Kelvin-Voigt damping coefficient c_D was smaller and air damping γ higher than in experiments in which the response was dominated by the primary mode. This indicates the necessity of estimating parameters with a response in the frequency range under consideration and illustrates a limitation in the damping model we have used here.

The damping which results when the circuit involving the piezoceramic patch is closed was investigated by performing a series of experiments with open and closed circuits. The estimated parameters and model responses for the two cases were then compared. As expected, the plate response with the closed circuit was more highly damped than that obtained with the open circuit, and the optimization routine compensated by increasing the material damping coefficients. While the damping provided by the circuit is not directly modeled by the Kelvin-Voigt or viscous damping terms (see (1.1)), it does produce an effect in the system which is phenomenologically similar to Kelvin-Voigt and viscous damping, and hence accurate model fits were obtained with the estimated parameters. We emphasize that if the applications of interest involve such a closed circuit, parameter estimation should be performed in this regime so as to account for the additional damping.

We reiterate that while the fixed-edge boundary conditions (2.64) adequately modeled the boundary dynamics for the setup under consideration, in many cases, energy loss through the boundary clamps will make the fixed-edge model inadequate. In such cases, an "almost fixed" boundary model of the type discussed in Chapter 2 (see (2.65)) may provide a more accurate description of edge physics. Experimental results pertaining to the use of that model for describing the plate dynamics when boundary clamps are loosened can be found in [5].

Chapter 6

Damage Detection in Smart Material Structures

It has been known for some time that damage such as cracks, corrosion, and delaminations in a structure produces changes in mass, stiffness, damping and other characteristics and material parameters in dynamic models for the structure. Our focus in this chapter is on the development of vibration response ideas for piezoceramic based smart material structures in which self-testing nondestructive evaluation (NDE) techniques may be employed.

A summary of some of the previous efforts in damage detection is presented in [15]. While the substantial literature on vibration-related damage detection is mostly based on modal methods, the utility of these methods has been highly debated. The most convincing conclusion is (e.g., [164]) that one should not use modal methods based on uniform undamped simple beams or plates as is often done in the engineering literature in addressing damage assessment methodologies. Cawley and Ray in [62] give a strong argument for including geometry of the damage in any diagnostic testing scheme. In Sato's investigation of free vibration of beams with abrupt changes of cross section [164], it is argued with rather forceful theoretical and experimental evidence that the simple beam theory (i.e., a constant parameter beam equation) may lead to incorrect results. Another important fact is that damage is a local phenomenon while modal information is a reflection of the global system properties [15, 182]. In [182], the author provides the evidence that the mode shapes do not change significantly as local mass and stiffness are modified.

In light of the above comments, a question of great interest is the following. Can one develop analytically sound, non-modal based self-excitation/self-sensing methods for detection and characterization (geometrical and quantitative) of damage in smart material structures? Here we address this question in the context of structures with surface-mounted or embedded piezoceramic elements.

For a piezoceramic smart material damage detection and characterization methodology, there are several distinct requirements. These include:

(a) One must be able to *estimate reliably* (repeatable across experiments) the *spatially variable material parameters* of a piezoceramic loaded structure. This must be done using piezoelectric materials actuation and sensing with accuracy comparable to that achievable with standard actuating devices (impulse hammers, solenoidal actuators) and sensing devices (accelerometers, strain gauges, laser vibrometers) in non-smart material testing schemes.

(b) One must be able to use the actuation and sensing properties of the piezoceramics to *excite the structure and analyze the response* (in a single experiment) for a reliable methodology that is the basis of self-excitation/self-sensing.

(c) One must be able to *detect and characterize damage via vibration self-excitation/self-sensing* that relies only on the input/output signals for the piezoceramics.

The first two requirements were studied analytically, numerically, and experimentally in Chapter 5. In those discussions, a strong case was made that requirements (a) and (b) can be satisfied for structures having surface-mounted or embedded piezoceramic actuators/sensors. In this chapter we address the last requirement in the context of a piezoceramic-loaded beam. This particular structure is sufficiently representative to make a compelling case for feasibility of the ideas proposed. As noted above, it is essential to model the micro-structure related to the local geometry and elastic characteristics. In this chapter, the partial differential equations describing the dynamics of a damaged beam with surface bonded piezoceramic patches are outlined. The damage detection problem is then formulated as an optimization problem and experimental verification with regard to requirement (c) is detailed.

6.1   Model for Damaged Structures

The test structure considered here is a cantilever beam with two identical piezoceramic patches attached on the opposite side of the beam as shown in Figure 6.1. This structure is the same as that depicted in Figure 4.1. For the convenience of the reader, we recall the specifics of the structure here. The beam of length ℓ is assumed to be homogeneous and satisfy the Euler-Bernoulli/Kelvin-Voigt hypotheses. Two piezoceramic patches are bonded to the beam (one on each side) in the region $y_1 < y < y_2$. For simplicity, the bonding layer material properties and geometry are ignored. The composite linear mass density, stiffness and Kelvin-Voigt damping coefficient are denoted by $\tilde{\rho}$, $\widehat{EI}$ and $\widetilde{c_D I}$, respectively, while γ is used to denote the viscous air damping coefficient.

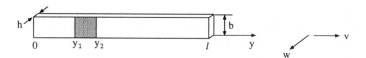

Figure 6.1: Cantilever beam with piezoceramic patches.

Our choice of structure is motivated by its simplicity and its representative nature. This configuration has also been well studied by conventional approaches such as finite element methods and provides a standard test model for comparison. The structure and model reveal the difficulties and possibilities inherent in developing damage models and methods for more complex structures such as those discussed in Chapters 2 and 3.

Without loss of generality, let us consider the beam described above under out-of-phase excitation, resulting in pure bending (transverse vibrations). The flat-beam equation (see (2.70)) and patch contribution equation (see (3.21)) lead to the dynamic system equation (see also (4.1)-(4.4))

$$\tilde{\rho}(y)\frac{\partial^2 w}{\partial t^2}(t,y) + \frac{\partial^2}{\partial y^2}\left(\widetilde{EI}(y)\frac{\partial^2 w}{\partial y^2}(t,y) + \widetilde{c_D I}(y)\frac{\partial^3 w}{\partial y^2 \partial t}(t,y)\right)$$
$$+ \gamma\frac{\partial w}{\partial t}(t,y) = -2\mathcal{K}^B \chi_{pe}''(y)\, u(t) \tag{6.1}$$

coupled with the boundary conditions for a cantilever beam

$$w(t,y)|_{y=0} = \left.\frac{\partial w}{\partial y}(t,y)\right|_{y=0} = 0,$$

$$\left.\left(\widetilde{EI}\frac{\partial^2 w}{\partial y^2}(t,y) + \widetilde{c_D I}\frac{\partial^3 w}{\partial y^2 \partial t}(t,y)\right)\right|_{y=\ell} = 0, \tag{6.2}$$

$$\left.\left(\frac{\partial}{\partial y}\left(\widetilde{EI}\frac{\partial^2 w}{\partial y^2}(t,y) + \widetilde{c_D I}\frac{\partial^3 w}{\partial y^2 \partial t}(t,y)\right)\right)\right|_{y=\ell} = 0,$$

and initial conditions

$$w(t,y)|_{t=0} = \phi(y),$$
$$\frac{\partial w}{\partial t}(t,y)|_{t=0} = \psi(y), \tag{6.3}$$

for the transverse displacement $w = w(t,y)$. In (6.1), $u(t)$ denotes the voltage to the patches, and χ_{pe} is the patch characteristic function.

In our investigation here of damage detection capabilities, we focus on corrosion types of damages characterized by holes. Since the beam model is based on the Euler-Bernoulli theory, we assume that the center of the damage coincides with the neutral axis (the axis parallel to the longest edges) of the beam and the damage is symmetric with respect to this center line. Furthermore, we assume that the damage, located between y_{d_1} and y_{d_2}, is characterized by shape functions. The shape functions represent a change in the geometry of the beam, resulting in the thickness and the width of the beam no longer being uniform. With the damage, the coefficients in the (6.1) are given by

$$\tilde{\rho}(y) = \rho h b + 2\rho_{pe} h_{pe} b \chi_{pe}(y) - \rho S_A(y) \chi_d(y)$$

$$\widetilde{EI}(y) = \frac{1}{12} h^3 b E + \frac{2}{3} a_3 b E_{pe} \chi_{pe}(y) - E S_I(y) \chi_d(y)$$

$$\widetilde{c_D I}(y) = \frac{1}{12} h^3 b c_D + \frac{2}{3} a_3 b c_{Dpe} \chi_{pe}(y) - c_D S_I(y) \chi_d(y)$$

$$\mathcal{K}^B = -\frac{1}{2}(h + h_{pe}) b E_{pe} d_{31}$$

(6.4)

where $a_3 = (h/2 + h_{pe})^3 - (h/2)^3$ as was derived in Chapter 3, and the characteristic function $\chi_d(y)$ is 1 for $y_{d_1} \leq y \leq y_{d_2}$, and zero elsewhere. The patch constant $\mathcal{K}^B$ and the first two terms in the expressions for $\tilde{\rho}(y)$, $\widetilde{EI}(y)$, and $\chi_{pe}(y)$ are the same as those for the undamaged beam (see (4.3) or (5.25)-(5.26)). The shape function $S_A(y)$ is the missing area of the cross section due to the damage whereas the function $\frac{1}{12} h^3 b - S_I(y)$ is the inertia of the cross section of the beam containing damages.

Instead of seeking solutions to the partial differential equations as formulated in strong form (6.1) with the boundary condition (6.2), we seek the solutions to its variational formulation

$$\int_0^{\ell} \left\{ \tilde{\rho} w_{tt} \phi + \left(\widetilde{EI} w'' + \widetilde{c_D I} w_t'' \right) \phi'' + \gamma w_t \phi \right\} dy = \left(\int_{y_1}^{y_2} -2\mathcal{K}^B \phi'' dy \right) u(t) \quad (6.5)$$

for $\phi \in V = H_L^2(0, \ell)$. The well-posedness and convergence of computational techniques have been established in previous chapters.

The parameters to be estimated include composite beam linear mass density $\tilde{\rho}(y)$, stiffness coefficient $\widetilde{EI}(y)$ as well as damping parameters $\widetilde{c_D I}(y)$, γ and piezoelectric material parameters $\mathcal{K}^B$, $\mathcal{K}^S$ (e.g., see (5.3)). This collection of unknown parameters is denoted by

$$q = (\tilde{\rho}(y), \widetilde{EI}(y), \widetilde{c_D I}(y), \gamma, \mathcal{K}^B, \mathcal{K}^S).$$

6.2   Damage Detection Technique

Damage detection is carried out by determining the shape functions $S_A(y)$ and $S_I(y)$ in (6.4) using observations of the system output response to excitations

$\{u_i\}$ to the patches. This estimation problem is formulated as an enhanced least squares fit to observations; that is, we seek $\bar{q} \in Q$ which minimizes

$$J(q) = \sum_{k=1}^{M_v} c_k J_{pe}(q, z^k; u_k) \tag{6.6}$$

where M_v denotes the number of experiments corresponding to different inputs u_k. The penalty functional $J_{pe}(q, z^k; u_k)$ is as given by (5.6) or

$$J_{pe}(q, z^k; u_k) = \sum_{i=1}^{N_t} \left| \mathcal{K}^S \left(\frac{\partial w}{\partial y}(t_i, y_2; q, u_k) - \frac{\partial w}{\partial y}(t_i, y_1; q, u_k) \right) - z_i^k \right|^2. \tag{6.7}$$

In the cost function, $\{z_i^k\}$ are measured voltages at time t_i across the piezoelectric elements due to bending, and $\{w(t_i, \cdot; q, u_k)\}$ are the parameter dependent weak solutions of (6.5) for a given input u_k and zero initial conditions, evaluated at each time t_i, $i = 1, 2, \ldots, N_t$. In (6.7), $|\cdot|$ is an appropriately chosen Euclidean norm. The set Q is some admissible parameter set which arises in parameterization of the desired shape functions.

In (6.6), the coefficients $\{c_k\}$ are chosen so that amplitudes of the weighted responses $\{c_k w(t, y; q, u_k)\}_{k=1}^{M_v}$ are of the same order. Furthermore, the system is excited in a manner so that the dominant mode in the response data corresponding to the input u_k is different from one excited by u_j for $j \neq k$.

For a unique solution of the damage detection problem, the standard least squares cost function (6.7) corresponding to a single excitation alone is not adequate. We illustrate this point with the following example in which the simulated piezoceramic responses corresponding to different damage locations are compared. One response data set is generated on a beam with a centered circular hole at $2.413\,cm$ away from the clamped end. The second is generated with a hole at $14.478\,cm$. In both cases, identical broadband input signals are applied. The time response data are plotted in Figure 6.2. The solid line in the figure corresponds to one damage location and dashed line corresponds to second damage location. It is obvious that vibration data and location of damages do not have a one-to-one relationship. Hence there may exist more than one set of damage parameters which yield the same value of cost function (6.7).

If one examines closely the very beginning of the response data, i.e., the enlargements in Figure 6.2, the second mode responses do not match while the first mode responses match quite well. This difference could be observed more clearly if the second mode were more substantially excited. When a bandwidth signal concentrated on the second mode is applied, the responses are quite different as the data depicted in Figure 6.3 clearly demonstrates.

This nonuniqueness was observed implicitly by Armon et al. [2]. The rank-ordering method in their efforts is based on the fact that the amount of phase shift in each mode differs from one type of damage (characterized by location and dimensions) to another, and the amount of shifts yield different order in response to different damages.

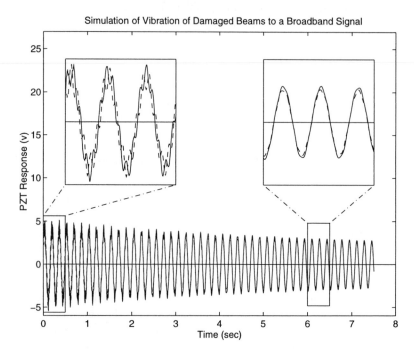

Figure 6.2: Simulated piezoceramic response to a broadband signal. The solid line is the response of the beam with damage located at $2.413\,cm$, and dashed line corresponds to the damage at $14.478\,cm$.

The above remarks raise the question as to whether it is possible to use a single response data set in which multiple modes are excited. Since the amount of change in the response due to the damage is very small, recorded data in a short time period would not provide sufficient information in the sense of displaying the difference caused by damages. On the other hand, the higher vibrational modes are damped out in the response over a longer time period. The example above demonstrates difficulties which can arise in damage detection by using only one vibration response data set. This motivates the use of the cost functional (6.6). The basic idea underlying this enhanced cost function is to use several excitations in time responses over a long period, thereby at the same time taking into account the information in several modes so that the cost function is sensitive to changes in any mode.

In the enhanced cost function, a time domain cost function is adopted. We demonstrate by using examples that frequency data (such as embodied in transfer functions) does not provide adequate information to detect damage. Two experimental data sets, one recorded from an undamaged beam and one from a damaged beam, are displayed in Figure 6.4. In the figure, part (a) is the frequency response, and part (b) is the time response. Clearly, with limited resolution of data acquisition equipment, the frequency response does

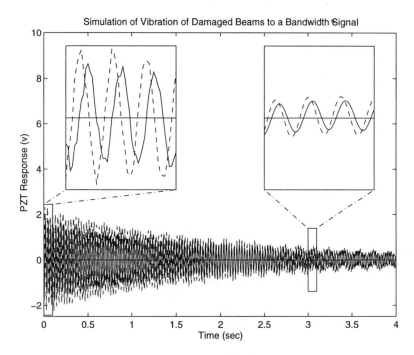

Figure 6.3: Simulated piezoceramic response to a bandwidth signal concentrated on the second mode. The solid line is the response of the beam with damage located at $2.413\,cm$ and dashed line corresponds to the damage at $14.478\,cm$.

not provide damage knowledge, while the time response data indicates a change in the structure. A detailed description of the experiment is presented in next section.

In seeking the weak solutions to the variational formulation (6.5), we consider Galerkin type approximations as outlined in Chapter 5. The beam displacement $w(t, y)$ is approximated by

$$w^N(t, y) = \sum_{i=1}^{N} w_i^N(t)\, \phi_i^N(y) \tag{6.8}$$

in an appropriate N^{th}-order finite dimensional space. The basis elements $\{\phi_i^N\}$ are chosen to be in V, i.e., standard elements with modifications such that the essential boundary conditions are satisfied. The generalized Fourier coefficients $\{w_i^N(t)\}$ represent the state relative to this basis. Replacing $w(t, y)$ and ϕ in the variational formulation (6.5) by (6.8) and $\{\phi_j^N\}_{j=1}^N$ respectively, we obtain the N-vector ordinary differential equation system

$$M^N\, \ddot{\vartheta}(t) + C^N\, \dot{\vartheta}(t) + K^N\, \vartheta(t) = F^N\, u(t) \tag{6.9}$$

where $\vartheta(t) = [w_1^N(t), w_2^N(t), \cdots, w_N^N(t)]^T$.

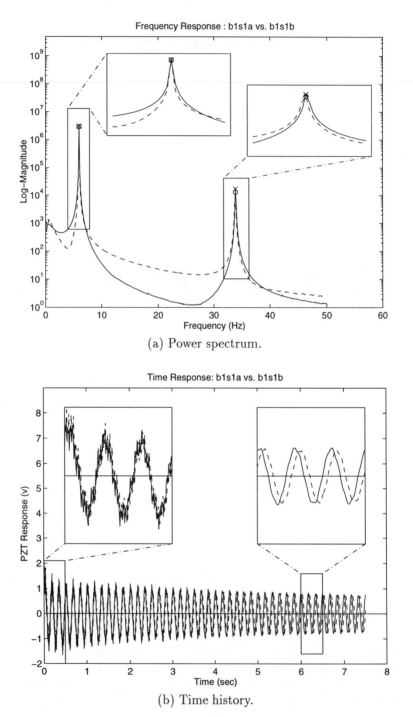

(a) Power spectrum.

(b) Time history.

Figure 6.4: Responses of an undamaged beam ($48.26 \times 2.032 \times 0.15875\,cm^3$) and the damaged beam (same beam with a $3.175\,mm$ hole at $2.54\,cm$) to a broadband signal.

The usual Galerkin coefficient matrices are derived by integrating each term in (6.5). This yields mass, damping, stiffness and forcing terms having the components

$$[M^N]_{i,j} = \int_0^\ell \tilde{\rho}(y)\, \phi_i^N(y)\, \phi_j^N(y)\, dy$$

$$[C^N]_{i,j} = \int_0^\ell \widetilde{c_D I}(y)\, \frac{d^2\phi_i^N(y)}{dy^2}\, \frac{d^2\phi_j^N(y)}{dy^2}\, dy + \int_0^\ell \gamma\, \phi_i^N(y)\, \phi_j^N(y)\, dy$$

$$[K^N]_{i,j} = \int_0^\ell \widetilde{EI}(y)\, \frac{d^2\phi_i^N(y)}{dy^2}\, \frac{d^2\phi_j^N(y)}{dy^2}\, dy$$

$$[F^N]_i = -2\mathcal{K}^B \int_0^\ell \frac{d^2\phi_i^N(y)}{dy^2}\, dy \ .$$

The dimension and location of the damage are unknown and must be estimated using the vibration data. The functions $S_A(y)$ and $S_I(y)$ which characterize the damage are elements to be chosen from an infinite class of functions. Rather than attempting to reconstruct $S_A(y)$ and $S_I(y)$, we considered appropriate parameterizations. One choice is to search for the projections of $S_A(y)$ and $S_I(y)$ onto the linear span of finite dimensional sets $\{\Phi_i\}_{i=1}^{M_A}$ and $\{\Psi_j\}_{j=1}^{M_I}$, respectively. In that case, S_A and S_I are parameterized by

$$S_A(y) = \sum_{i=1}^{M_A} \alpha_i\, \Phi_i(y) \qquad 0 < y_{d_1} \le y \le y_{d_2} < \ell \qquad (6.10)$$

$$S_I(y) = \sum_{j=1}^{M_I} \beta_j\, \Psi_j(y) \qquad 0 < y_{d_1} \le y \le y_{d_2} < \ell\ . \qquad (6.11)$$

A family of approximating estimation problems with finite dimensional state spaces and parameters is then formulated by seeking a vector parameter

$$q^M = (\alpha_1, \alpha_2, \cdots, \alpha_{M_A}, \beta_1, \beta_2, \cdots, \beta_{M_I}, y_{d_1}, y_{d_2})$$

which minimizes

$$J^N(q^M) = \sum_{k=1}^{M_v} c_k\, J_{pe}^N(q^M, z^k; u_k) \qquad (6.12)$$

where $J_{pe}^N(q^M, z^k; u_k)$ is given by (6.7) with $w(t, y)$ replaced by $w^N(t, y)$. The dimension of the parameter space is then $M = M_A + M_I + 2$. The coefficients $\{w_i(t)\}$ in (6.8) are the solutions to (6.9).

Based on our previous computational experiences (see, for example, Chapter 5 and the references given therein), we chose cubic spline basis elements modified to satisfy the essential boundary conditions at $x = 0$. For one choice of the parameter space, linear splines were selected as the basis elements Φ_i, Ψ_j in (6.10) and (6.11).

Solving the approximate minimization problems, we obtained a sequence of estimates $\{\bar{q}^{N,M}\}$. The sequence admits a convergent subsequence which converges to some $\bar{q} \in Q$ under the assumptions that Q is a compact set and the parameter functions $S_A(y)$ and $S_I(y)$ are bounded above. Furthermore the limit parameter $\bar{q}$ is a solution to the original infinite dimensional optimization problem. The relevant parameter estimate convergence and continuous dependence with respect to the observations follow directly from results in Chapter 5.

In preliminary investigations as to the possibility of detecting and geometrically characterizing damage such as holes, simulation studies were carried out. The initial findings reported in [46] suggested that such methods offer great promise.

6.3   Experimental and Numerical Results

To demonstrate the capability of damage detection with the algorithm outlined in the previous section, experiments with two different beams and three different types of damage were carried out. Each of them had a distinct damage location and size which allowed us to conduct a sensitivity study. The detailed experimental procedure is described in the following section and damage detection results are reported afterwards.

6.3.1   Test specimens and procedures

The test articles for our non-destructive damage detection studies were cantilever aluminum beams that were $48.26\,cm$ ($19''$) long, $2.032\,cm$ ($0.8''$) wide and $0.15875\,cm$ ($0.0625''$) thick. One pair of piezoceramic patches was surface-bonded opposite one other and they were electrically coupled to create one sensor/actuator as shown in Figure 6.5. The piezoceramics had the following geometric dimensions; the length was $6.35\,cm$, the width was $2.032\,cm$ and the thickness was $0.0254\,cm$. An accelerometer was also affixed to each beam near the midspan (this could be used to corroborate our findings with the accumulated strain data used in (6.7)).

We focused on vibration tests in which a maximum of the first three vibrational modes would be excited simultaneously. Since the Euler-Bernoulli theory with Kelvin-Voigt damping was employed, our beam model was inadequate for use in analyzing vibration data containing high frequencies. For the chosen beam dimension and material, the first three natural modes are under 100 Hz and the Euler-Bernoulli model is appropriate as was verified in Section 5.4.

The piezoceramic patch pair located $5.08\,cm$ from the clamped end was used as both a sensor and actuator for the damage detection experiments. A switching mechanism permitted the dual use of the patches. The switch could be engaged by applying 12 VDC during the excitation period and it could be disengaged afterward, thus allowing the piezoceramic to be used as a sensor. When used as an actuator, the ceramic was connected in series to an amplifier ($\times 10$) with a voltage range of ± 50 V. When used as a sensor, the piezoceramic

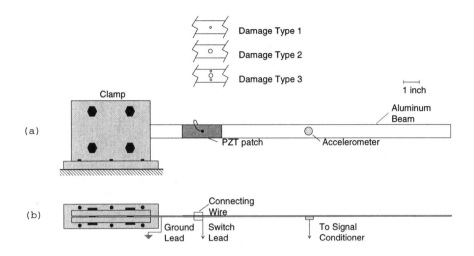

Figure 6.5: Diagram of a typical beam experiment, (a) Side View, (b) Top View.

was connected in series to a high impedance low-pass filter. The device had a 10 MΩ impedance and a corner frequency of 10,000 Hz, i.e., any frequencies higher than 10,000 Hz were cut off to avoid aliasing. These parameters were chosen such that the output voltage of the ceramic was proportional to strain over the frequency range of interest (between 1 Hz and 100 Hz). Figure 6.6 contains a diagram of the test set-up.

To ensure reliable measurement of small frequency changes, it was essential to maintain the same experimental conditions throughout tests. This includes the maintenance of boundary clamps for both the undamaged and damaged structures and external inputs to the beam. In vibration tests, an impulse hammer is commonly used to excite multiple modes. The disadvantage in using an impulse hammer is the associated poor repeatability of input. Moreover, it is difficult to simulate the resulting input force accurately. For the same reason, a random force input is also not suitable. To have repeatable input with frequency content over a desired bandwidth, a Schroeder-phased signal [52] was adopted. The excitation voltage to the piezoceramic patches was generated by a DSpace computer where the MATLAB Simulab Toolbox was utilized for implementing the Schroeder-phased signal. Each signal contained 1024 points and excited the beam for 4 seconds. The frequency content of each signal was chosen to excite certain modes of the beam. Signal 1 was broadband while signals 2 and 3 were narrowband (see Table 6.1). Schroeder-phased inputs exhibit a flat power spectrum over the specified frequency range, i.e., the power in the input signal is equally distributed over the range. This characteristic permitted very repeatable input time histories for all of the cases studied.

	Frequency Range (Hz)		
	0-100	0-20	20-50
Signal 1	x	-	-
Signal 2	-	x	-
Signal 3	-	-	x

Table 6.1: Frequency content of the Schroeder-phased inputs.

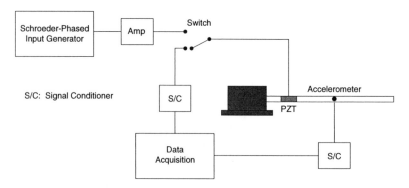

Figure 6.6: Test setup for experiments with damaged and undamaged beams.

The data acquisition system was a Fourier analyzer, Tektronix Analyzer 2600, with a PC (IBM AT) connected to it to display and record the data. The sampling rate of the analyzer was set at 512 Hz, and a total of 4096 data points (8 seconds) were recorded for each vibration response. The analyzer was set to start recording 3.6 seconds after initiation of the input signal. Hence 0.4 seconds of the forced vibration was recorded since the excitation lasted for 4 seconds. The rest of the data contained free decay vibration information.

Two separate series of experiments were performed with the location of the damage differing for each series. To conduct studies on the sensitivity of the method to the distances between the damage, the clamped end, and the sensor location, one defect was introduced near the clamped end and one near midspan. In Experiment D1, the damage was centered 2.54 cm from the clamped end of the beam while for Experiment D2 the damage was located 22.86 cm from the clamped end. An accelerometer was located at 25.4 cm in both experiments for monitoring the vibration and corroborating the data collected from the piezoceramic patches. Table 6.2 lists the damage type and location for each experiment. The damage types are depicted in Figure 6.5 and their dimensions are given in Table 6.3. The damage state column in Table 6.3 lists the ratio of the material removed to the material of the undamaged beam. All damage geometries were symmetric to the neutral axis of the beams.

Exp. #	Damage Cases			
	a	b	c	d
D1	No damage	DT1-1	DT2-1	DT3-1
D2	No damage	DT1-9	DT2-9	DT3-9

DT#-* refers to damage type (#) and location (*) in inches.

Table 6.2: Damage types and locations for the two experiments. Damage types are depicted in Figure 6.5.

Type	Description	State
DT1	one 3.175 mm hole	0.08%
DT2	one 6.350 mm hole	0.32%
DT3	one 6.350 mm hole + two 3.175 mm holes	0.47%

Table 6.3: Damage description and state (ratio of removed mass to the undamaged beam mass) for each damage type.

The procedures for each of the two series of experiments were identical. First, the undamaged beam was tested by exciting it with the Schroeder-phased inputs and measuring the response of the piezoceramics and the accelerometer. Responses were measured separately for each excitation signal. After the undamaged beam was tested, the beam was damaged by drilling a hole (type 1) through the structure at a specified location. After testing, the beam was again drilled to produce a larger hole (type 2). Again, after testing, this beam was further drilled to produce damage type 3. The holes were drilled without removing the beam from the clamp. This eliminated variation in the response due to changes in the boundary condition. Hence, each series of experiments was carried out on the same beam with the level of damage varying from no damage to the most severe damage. The excitation pattern was repeated for each damage case.

6.3.2 Identification results

The dimension N of the approximation space was set to 14 for the same reason (i.e., no changes in eigenvalues for the first three modes for larger N) stated in Section 5.3. A comparison with the selection of $N = 10$ for undamaged beams revealed that more basis functions were required to capture the small changes in the structure due to the damage than were needed to study the undamaged beam.

As might be expected from our earlier discussions, we readily observed in the experimental data that the frequency information is not sufficient to use as damage detection information. In Table 6.4, we list the frequency change in percentage for beams with different damage type and location versus undamaged beam. Note that in each case the change in frequency was *at most* 2.22%.

	Undamaged Beam	Damaged Beams					
		DT1		DT2		DT3	
Mode	f (Hz)	f (Hz)	$\Delta f/f$ (%)	f (Hz)	$\Delta f/f$ (%)	f (Hz)	$\Delta f/f$ (%)
1	6.000	6.000	0	5.867	2.22	5.867	2.22
2	33.867	33.867	0	33.600	0.79	33.200	1.97

(i) Experiment D1

	Undamaged Beam	Damaged Beams					
		DT1		DT2		DT3	
Mode	f (Hz)	f (Hz)	$\Delta f/f$ (%)	f (Hz)	$\Delta f/f$ (%)	f (Hz)	$\Delta f/f$ (%)
1	6.133	6.133	0	6.133	0	6.133	0
2	34.400	34.400	0	34.267	0.39	34.000	1.16

(ii) Experiment D2

Table 6.4: Frequency shift for $48.26 \times 2.032 \times 0.1651\, cm^3$ beams with holes at (i) $2.54\, cm$ and (ii) $22.86\, cm$ from clamp.

The piezoceramic parameters $(\mathcal{K}^B, E_{pe}, \rho_{pe})$, undamaged beam parameters (E, ρ) and damping parameters were estimated first from data on the undamaged beam vibrations. The results were then employed as fixed values in the damage detection procedures.

To remain consistent with the formulation for an Euler-Bernoulli beam, as we have already noted, the geometry of the damage was assumed to be symmetric about the neutral axis of the beam. For the same reason, we did not attempt to identify the shape of the damage since we use a 1-D equation to describe a 3-D structure, and many different 3-D shapes could be represented by a 1-D function (the level of nonuniqueness would be too high in such an endeavor). Instead, the possibility of using a fixed damage shape in the mathematical model was investigated. A reasonable shape for hole type damages is a circular function

$$S(y) = 2\sqrt{r^2 - (y - y_c)^2}, \qquad (6.13)$$

in which r is the radius, y_c is the center of the circle, and the factor of 2 in the equation is due to the symmetry of the beam about its neutral axis. In this case, the shape functions $S_A(y)$ and $S_I(y)$ become

$$S_A(y) = h\,S(y)$$
$$S_I(y) = \frac{1}{12}\,h^3\,S(y).$$

To obtain a good initial guess for the shape function parameters in searching for the optimal size and location of the hole, a series of simulated responses were computed for different sizes and locations. The parameter sets $\{(r, y_c)\}$ which yield smaller residual (comparing the numerical solutions with the experimental data) among all the integration runs were selected as initial guesses.

An IMSL routine of the Levenberg-Marquardt algorithm with a finite difference Jacobian algorithm (ZXSSQ in IMSL9) was used to solve the approximating finite dimensional least squares minimization problems. Further discussion concerning the Levenberg-Marquardt algorithm can be found in Section 5.3 and in [77, 96].

A summary of the estimated parameters for the undamaged beam is given in Table 6.5. The parameter identification result for beam I is shown in Figure 6.7 where the power spectrum of the data is compared to that of the model response corresponding to the best-fit parameters. The data fit results for Beam II are not presented here since they are very similar to those for Beam I.

		Given	Beam I	Beam II
E	beam	$7.300 + 10$	$7.341 + 10$	$6.971 + 10$
(N·m^2)	PZT	$6.300 + 10$	$7.151 + 10$	$6.829 + 10$
ρ	beam	$2.766 + 3$	$3.001 + 3$	$2.766 + 3$
(kg/m^3)	PZT	$7.600 + 3$	$9.171 + 3$	$8.987 + 3$
c_D	beam	—	$1.339 + 6$	$1.052 + 6$
$(\text{N·m}^2\text{·s})$	PZT	—	$1.219 + 6$	$1.219 + 6$
γ $(\text{N·m}^2\text{·s})$		—	$0.956 - 2$	$1.172 - 2$

Table 6.5: Given and estimated structural parameters for undamaged beams I and II.

The results of the damage estimation computations using the experimental data are graphically depicted in Figures 6.8 and 6.9. A summary of analytical damage detection results for damage types and locations b,c,d of Table 6.2 is given in Table 6.6. The radius for damage case (d) is an equivalent radius, i.e., the removed area of a circle with the radius is the same as the actual removed area due to three holes. Typical comparisons between the experimental data and model responses are displayed in Figures 6.10–6.11. In Figure 6.10, the top plot is time history and the bottom one is the power spectrum of the damaged beam DT1-1 response to the broadband excitation Signal 1 (see Table 6.1). The response to the narrowband excitation Signal 3 of the same beam is depicted in Figure 6.11. As illustrated by Figures 6.10 and 6.11, accurate models fits were obtained with the *same* parameters for quite different excitations. This highlights an advantage of the physics-based PDE method for damage detection.

| Beam | Damage Center & Radius | Damaged Beams | | | | | |
| | | b | | c | | d | |
		Actual	Est.	Actual	Est.	Actual	Est.
I	$\bar{y}_c$ (cm)	2.540	2.822	2.540	2.780	2.540	2.594
	$\bar{r}$ (cm)	0.159	0.230	0.318	0.442	0.355	0.675
II	$\bar{y}_c$ (cm)	22.866	23.817	22.866	24.175	22.866	24.255
	$\bar{r}$ (cm)	0.159	0.476	0.318	0.635	0.355	0.826

Table 6.6: Given and estimated damage for beams I and II. The damage types and locations b,c,d are summarized in Table 6.2.

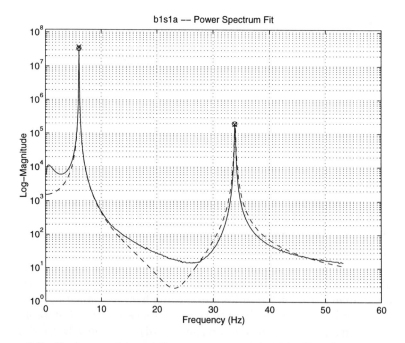

Figure 6.7: Undamaged beam I response to excitation Signal 1, x —— experimental data, o – – – model response with optimal parameters.

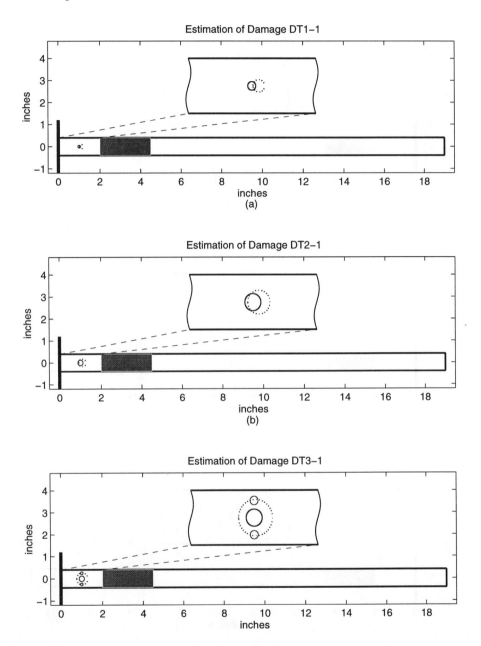

Figure 6.8: Estimated damages for Experiment D1, where solid lines are the actual damage and dotted lines are estimated damages.

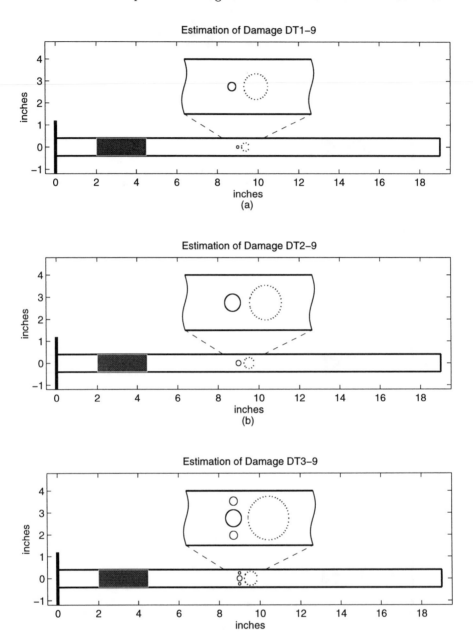

Figure 6.9: Estimated damages for Experiment D2, where solid lines are the actual damage and dotted lines are estimated damages.

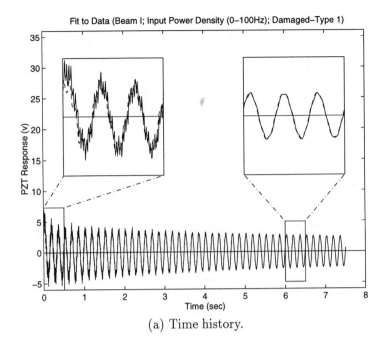

(a) Time history.

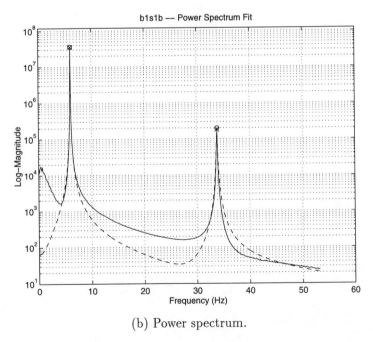

(b) Power spectrum.

Figure 6.10: Damaged beam (DT1-1) response to excitation Signal 1, x ———
experimental data (with $y_c = 2.54\,cm$, $r = 1.59\,mm$), o - - - model response
with optimal parameters ($\bar{y}_c = 2.82\,cm$, $\bar{r} = 2.30\,mm$).

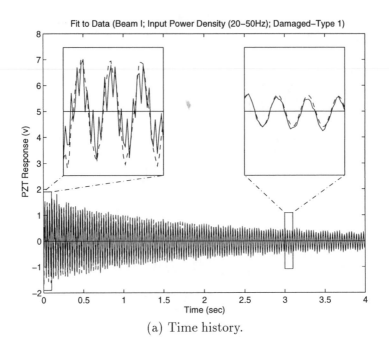

(a) Time history.

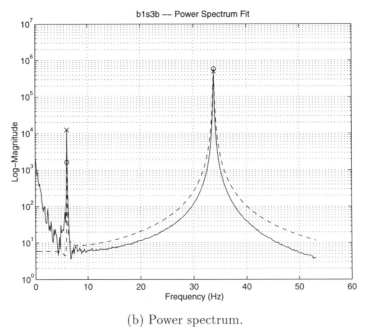

(b) Power spectrum.

Figure 6.11: Damaged beam (DT1-1) response to excitation Signal 3, x ——— experimental data (with $y_c = 2.54\,cm$, $r = 1.59\,mm$), o – – – model response with optimal parameters ($\bar{y}_c = 2.82\,cm$, $\bar{r} = 2.30\,mm$).

6.3.3 Discussion

We first point out that the parameter identification results of Table 6.5 for the undamaged beams demonstrate once again the consistency of our estimation methods for structures without damage. Two sets of estimated structural parameters obtained by using response data recorded from two identically made beams are extremely close in the values. This is significant since from Table 6.4, one can see that the damped natural frequencies of the two undamaged beams are different. This difference is caused in part by experimental limitations, e.g., it is very difficult to obtain consistent clamping in several experiments. To partially alleviate these limitations, the optimization procedure yields slightly different stiffness and mass density parameters for the two experiments. As one might expect, the ratio of stiffness to mass density for Beam I is lower than the one for Beam II since the damped natural frequencies for Beam I are smaller than those for Beam II.

Excellent agreement is obtained between the estimated damage location and the actual ones. Our parameter estimation method is sensitive to small changes due to the damage - the smallest hole is $3.175\,mm$ in diameter which is only 0.08% damage (ratio of removed mass to undamaged beam mass). However, the size of the damages are consistently over estimated. We suspect that the beams' characteristics were slightly changed during the drilling process in a manner not modeled in our equations (e.g., changes in mass density around the holes due to shearing and stress, etc.). Simulations of the numerical solutions with actual damage in the model yield less frequency response change than is present in the experimental data. The estimation of damage location for damage type (d) is as good as the other two even though the assumed damaged shape (one circle) in the mathematical model is very different from actual shape (Damage Type 3 in Figure 6.5).

Comparing the damaged beams DT#-1 to DT#-9, we find that the experimental frequency response changes are less for a damage location further away from the clamped end. Even so, our results demonstrate that the method is sensitive to the different locations of the damage. Even though the estimated locations for Experiment D2 are not as good as those obtained from Experiment D1, they are within 6% error from the actual locations.

For these experiments, attempts at using a fixed damage shape were successful. The estimation of damage location for Damage Type 3 is as good as for Damage Type 1 and 2 although we matched the experimental data for DT3 with a numerical solution to a mathematical model with a singular circular hole. In many situations, it is more important to determine whether there are damages and where they are than what shape they possess. In these cases, we can proceed to estimate the locations and the approximate sizes characterized by radii. This significantly reduces computation time since we only estimate one parameter, the radius, as opposed to a function characterized by many unknown coefficients.

In this chapter we have discussed a theoretical, numerical and experimental investigation of the use of smart structures, parameterized partial differential equations and Galerkin approximation techniques to detect and locate damage. As we have already noted, the idea of using vibration testing as a basis for damage detection in structures is not a new one. However, most methods to date are based on modal information. The approach presented here is independent of modal information from the structure, and changes in damping, mass and stiffness properties of the structure are estimated using time histories of the input and vibration response of the structure, generated and measured by the piezoceramic patches internal to the structure.

Although results have been obtained only for aluminum beams, the ideas and framework can be readily applied to plate, shell and beam-like structures. However, in the cases of crack damage and delaminations, the PDE model developed here can not be applied directly since the physical parameters ρ, EI and $c_D I$ can, of course, no longer be used to simply describe the damage. Nevertheless, once a proper model describing the dynamics of a particular structure with cracks is developed, we suggest that the same framework and convergence arguments could be applied with some modifications.

The framework developed is also valid for composite material structures since our theoretical and computational methods can include weaker (than Kelvin-Voigt) and more complex damping operators.

Our efforts with variable geometry and variable elasticity, mass and damping parameters promise to alleviate some difficulties encountered with modal methods. This, among other of our findings, strongly supports our own belief that geometry-based partial differential equation techniques can play a very important role in damage detection modeling in the emerging technology of adaptive or smart material structures. Our conclusions are based on the methodology developed here which contains theory, computational and experimental tests, all of which are consistent.

Chapter 7

Infinite Dimensional Control Problem and Galerkin Approximation

A feedback control methodology for the infinite dimensional systems discussed in Chapters 2 and 3 is presented here. These control results are most naturally formulated for first-order systems, and the abstract Cauchy formulation presented in Chapter 4 for the PDE models of Chapters 2 and 3 provides such a framework. In this Cauchy form, control and approximation results for the finite time and infinite horizon problems can be discussed.

The emphasis throughout this discussion will be on unbounded (discontinuous) control input operators since this is the case when elements such as piezoceramic or electrostrictive patches are used as control actuators. Since a number of the systems of interest to us are subjected to harmonic disturbances, various results for control systems with a periodic exogenous force $\tilde{g}$ are also included in the discussion. For convenience, the abstract first-order system derived in Chapter 4 for structural systems with unbounded input terms and periodic exogenous disturbances is recalled in Section 7.1. Throughout this discussion, emphasis is placed on constructing operators and solution semigroups which are compatible with the unbounded control input terms.

Control results for these problems are discussed in Sections 7.2–7.5. A complete theory exists for the unbounded control input problem with no exogenous input ($\tilde{g} \equiv 0$) in which full state measurements are available. The fundamentals of this theory are outlined in Section 7.2. A complete approximation framework for the unbounded linear quadratic regulator (LQR) problem is summarized in Section 7.3. The theory in both sections utilizes a formulation of the first-order system in terms of sesquilinear forms and associated linear operators. Taken in combination, Sections 7.1–7.3 provide a rigorous formulation of the infinite dimensional LQR problem with unbounded input term and $\tilde{g} \equiv 0$ and summarize the convergence theory necessary for approximating the gains.

The control problem for a system driven by a persistent disturbance is discussed briefly in Section 7.4. Depending upon the application, this distur-

bance can be treated either as a nonhomogeneous forcing term in the abstract
Cauchy problem modeling the system or as a trajectory to be tracked by one
of the states. The control and convergence theory for this problem is less complete than that for problems with undisturbed systems, and open questions
are indicated.

Results in the first four sections depend upon having knowledge of the full
state in order to calculate control gains. In almost all distributed parameter
problems, however, only partial state information is available, and the state
must be estimated or reconstructed from observations and a dynamic output
feedback law (compensator) developed for controlling the system dynamics.
While a theory for compensator design based on sesquilinear forms has been
developed for the case of bounded input/output operators, this theory has not
yet been extended to the unbounded operator regime. In Section 7.5, we outline instead a purely operator-based approach for obtaining finite dimensional
compensators for the infinite dimensional problem when the input/output operators are unbounded. This analysis is currently limited to the problem with
no exogenous disturbance ($\tilde{g} \equiv 0$) and open questions regarding the extension of the theory to systems with exogenous disturbances as well as issues
concerning order reduction in the compensators are discussed.

The emphasis throughout this chapter is on a careful discussion of the infinite dimensional control problems as well as related convergence issues when
the LQR and output feedback problems are approximated. The implementation of this control methodology is considered in Chapter 8. That discussion
begins with a finite dimensional control system, obtained either through application of the approximation theory of this chapter or other discretization
techniques, and finite dimensional compensator theory is summarized.

Unlike other chapters in this monograph, the discussions here are not to-
tally self-contained. To establish with rigor the results discussed here would
require a rather lengthy digression. Instead, we provide references to the research literature for those results which can be proven.

7.1   Abstract Formulation

In Chapter 4 it was demonstrated that the structural models discussed in
Chapters 2 and 3 can be written in the weak form

$$\langle \ddot{w}(t), \eta \rangle_{V^*,V} + \sigma_2(\dot{w}(t), \eta) + \sigma_1(w(t), \eta) = \langle Bu(t) + \tilde{g}(t), \eta \rangle_{V^*,V} \qquad (7.1)$$

for all $\eta \in V$ where V is a Hilbert space of test functions with corresponding
inner product $\langle \cdot, \cdot \rangle_V$. Here w denotes displacement in a Hilbert space H (state
space with inner product $\langle \cdot, \cdot \rangle_H$), B is a control operator with input u, and $\tilde{g}$
is an exogenous external force or moment applied to the structure. We have
the usual Gelfand triple construction $V \hookrightarrow H \cong H^* \hookrightarrow V^*$ with the duality
product on the pivot space H denoted by $\langle \cdot, \cdot \rangle_{V^*,V}$. We note that w may be
vector-valued and can, for example, incorporate longitudinal, circumferential

and transverse displacements. The input u is considered in a Hilbert space U. In a typical application involving the control of structural vibrations using piezoceramic actuators, u denotes the voltage to an actuator and $B \in \mathcal{L}(U, V^*)$ is unbounded due to the discontinuous geometry of the patches which leads to external applied line moments in the structure. The exogenous force $\tilde{g}$ is commonly assumed to be periodic so as to model oscillatory disturbances to the system.

As discussed in Chapter 4, σ_1 incorporates stiffness components of the structural model and σ_2 contains the damping terms (such as Kelvin-Voigt, viscous or others) for the model. Recalling our presentation of Chapter 4, we restrict our discussions here to the case where $V_2 = V$ (i.e., both the stiffness and damping forms are defined on the same space V). The symmetry, continuity and coercivity conditions satisfied by the sesquilinear forms are discussed in Chapter 4, and it was noted there that due to the continuity (boundedness), one can define operators $A_1, A_2 \in \mathcal{L}(V, V^*)$ by

$$\langle A_i \phi, \eta \rangle_{V^*, V} = \sigma_i(\phi, \eta) \quad , \quad i = 1, 2 \ .$$

The weak form (7.1), taken with initial conditions, can then be written equivalently as the second-order system

$$\ddot{w}(t) + A_2 \dot{w}(t) + A_1 w(t) = Bu(t) + \tilde{g}(t) \quad \text{in } V^*$$
$$w(0) = w_0 \ , \ w_t(0) = w_1 \ . \tag{7.2}$$

For consideration of the control problem, it is advantageous to write the system in first-order form. To this end, we define the product spaces $\mathcal{H} = V \times H$ and $\mathcal{V} = V \times V$ with the norms

$$|(\phi_1, \phi_2)|^2_{\mathcal{H}} = |\phi_1|^2_V + |\phi_2|^2_H$$
$$|(\phi_1, \phi_2)|^2_{\mathcal{V}} = |\phi_1|^2_V + |\phi_2|^2_V \ .$$

We point out that $\mathcal{V} \hookrightarrow \mathcal{H} \simeq \mathcal{H}^* \hookrightarrow \mathcal{V}^*$ again forms a Gelfand triple with $\mathcal{V}^* = V \times V^*$. The state in $\mathcal{H}$ is denoted by $z(t) = (w(t), \dot{w}(t))$ (we caution the reader that the use of the variable z to denote the state for the first-order problem, here and in Chapter 4, should not be confused with the symbol's use as a coordinate in Chapters 2 and 3). The stiffness and damping components are combined in the sesquilinear form $\sigma : \mathcal{V} \times \mathcal{V} \to \mathbb{C}$ given by

$$\sigma(\Phi, \Psi) = -\langle \phi_2, \psi_1 \rangle_V + \sigma_1(\phi_1, \psi_2) + \sigma_2(\phi_2, \psi_2)$$

where $\Phi = (\phi_1, \phi_2)$ and $\Psi = (\psi_1, \psi_2)$. Finally, the product space forcing terms are formulated as

$$g(t) = \begin{bmatrix} 0 \\ \tilde{g}(t) \end{bmatrix} \quad , \quad \mathcal{B}u(t) = \begin{bmatrix} 0 \\ Bu(t) \end{bmatrix} \ .$$

The weak form of the system equations can then be written as the first-order equation

$$\langle \dot{z}(t), \Psi \rangle_{\mathcal{V}^*, \mathcal{V}} + \sigma(z(t), \Psi) = \langle \mathcal{B}u(t) + g(t), \Psi \rangle_{\mathcal{V}^*, \mathcal{V}} \tag{7.3}$$

for $\Psi \in \mathcal{V}$. As discussed in Chapter 4, this is formally equivalent to the strong form of the equation

$$\dot{z}(t) = \mathcal{A}z(t) + \mathcal{B}u(t) + g(t) \quad \text{in } \mathcal{V}^*$$

$$z(0) = z_0 = \begin{bmatrix} w_0 \\ w_1 \end{bmatrix} \tag{7.4}$$

where $\mathcal{A}$ is given by

$$\text{dom}\mathcal{A} = \{(\phi_1, \phi_2) \in \mathcal{H} \,|\, \phi_2 \in V, A_1\phi_1 + A_2\phi_2 \in H\}$$

$$\mathcal{A} = \begin{bmatrix} 0 & I \\ -A_1 & -A_2 \end{bmatrix}. \tag{7.5}$$

We recall that $\mathcal{A}$ is the negative of the restriction to $\text{dom}\mathcal{A}$ of the operator $\tilde{\mathcal{A}} \in \mathcal{L}(\mathcal{V}, \mathcal{V}^*)$ defined by $\sigma(\Phi, \Psi) = \langle \tilde{\mathcal{A}}\Phi, \Psi \rangle_{\mathcal{V}^*, \mathcal{V}}$ so that $\sigma(\Phi, \Psi) = \langle -\mathcal{A}\Phi, \Psi \rangle_{\mathcal{H}}$ for $\Phi \in \text{dom}\mathcal{A}$, $\Psi \in \mathcal{V}$.

In Chapter 4, it was demonstrated that when σ_2 is V-elliptic (which is the case when a structural model includes strong Kelvin-Voigt damping), the product space sesquilinear form σ is $\mathcal{V}$-elliptic and $\mathcal{A}$ generates an analytic semigroup $\mathcal{T}(t)$ on $\mathcal{V}, \mathcal{H}$ and $\mathcal{V}^*$. (In this case, the $\text{dom}\mathcal{A}$ defined in (7.5) is actually $\text{dom}_{\mathcal{H}}(\mathcal{A})$, the domain of $\mathcal{A}$ as a generator in $\mathcal{H}$.) The use of one symbol to denote the semigroups (or general operators) defined on each of the Gelfand triple spaces is common in the literature and should not cause ambiguity; we merely warn the reader to be prepared for this convention.

In weakly damped systems (such as a structural acoustic system in which a vibrating structure is coupled with an adjacent, undamped acoustic field), however, σ_2 may be only H-semielliptic and σ is *not* $\mathcal{V}$-elliptic. In this case, $\mathcal{A}$ generates a semigroup $\mathcal{T}(t)$ on $\mathcal{H}$ which is strongly continuous but is *not* analytic. To define a solution compatible with the unbounded input terms in $\{0\} \times V^*$, this semigroup must be extended through extrapolation space techniques to a space which includes the input terms.

It is convenient at this point to interpret (7.4) in the mild form with a solution given by the variation-of-constants formula

$$z(t) = \mathcal{T}(t)z_0 + \int_0^t \mathcal{T}(t-s) \left[\mathcal{B}u(s) + g(s)\right] ds \tag{7.6}$$

where $\mathcal{B}u + g \in L^2(0, T; \mathcal{V}^*)$. The necessity of extrapolating the semigroup to either $\mathcal{V}^*$ or a space containing $\mathcal{V}^*$ to accommodate the input $\mathcal{B}u$ is readily apparent and motivates the previously mentioned extensions. In Chapter 4, the mild solution (7.6) was shown to be equivalent to the strong and weak solutions when σ_2 is $\mathcal{V}$-elliptic and $\mathcal{B}u + g$ is sufficiently smooth. Moreover, the well-posedness of the problem (7.4) or equivalently (7.2) was established. Indeed, the well-posedness of (7.6) as a mild solution of (7.4), and equivalence to (7.2), was established under even weaker assumptions on σ_2, but that will not be of concern to us in this chapter.

7.2 Infinite Dimensional LQR Control Problem with No Exogenous Input

In this section we consider the infinite dimensional control problem in which there is no exogenous disturbance ($g \equiv 0$) on the system. Moreover, it is assumed here that full state ($z = (w, \dot{w})$) information is available for calculation of the feedback control. Whereas only partial state observations are available in current distributed parameter applications, the consideration of the full state feedback case is a necessary first step in the development of a state estimator and compensator feedback system. Moreover, the full state feedback theory may eventually be applied directly as sensors distributed over the entire structure (e.g., optical and laser) are perfected.

We also consider observations of the state

$$z_{ob}(t) = \mathcal{C}z(t) \tag{7.7}$$

where $\mathcal{C}$ denotes an observation operator mapping into the observation space Y. When physically implementing the controller, the observation operator is typically unbounded on the state space $\mathcal{H}$ due to the discrete nature of the measurement devices and hence $\mathcal{C} \in \mathcal{L}(\mathcal{V}, Y)$. The technical details associated with the unbounded operator $\mathcal{C}$ tend to obscure an initial exposition, however, and we will concentrate here primarily on the continuous measurement case $\mathcal{C} \in \mathcal{L}(H, Y)$. The reader is then referred to various results in [17] or [110] for analysis extending the bounded observation operator results to the unbounded case (see also [128]).

Finally, we restrict our theoretical discussions to the case when σ_2 is V-elliptic and hence σ is V-elliptic and $\mathcal{T}(t)$ is an analytic semigroup. Thus we are assuming for our discussion in this chapter that hypotheses (H1)-(H3), (H4$'$) and (H5$'$) of Chapter 4 hold along with the ensuing results of Theorem 4.8. As discussed previously, this includes structural models that incorporate strong damping (e.g., Kelvin-Voigt damping). Although the unbounded control input methodology has been numerically demonstrated as viable and effective for many weakly damped systems, the theory has not yet been completely extended to that regime. To guide the reader interested in results requiring extension to weakly damped systems, comments regarding such systems will be interspersed throughout the discussion even though much of the analysis discussed here strictly applies only to the strongly damped problem.

To guarantee well-defined trajectories in the three spaces $\mathcal{V}, \mathcal{H}, \mathcal{V}^*$, the following assumptions are typically made [17, 110, 157].

(C0) The semigroup $\mathcal{T}(t)$ is strongly continuous on $\mathcal{H}$ and $\mathcal{V}^*$.

(C1) For every $u(\cdot) \in L^2(0, T; U)$, $\int_0^T \mathcal{T}(T - s)\mathcal{B}u(s)ds \in \mathcal{H}$ and there exists $b > 0$ such that

$$\left| \int_0^T \mathcal{T}(T - s)\mathcal{B}u(s)ds \right|_{\mathcal{H}} \leq b\,|u|_{L^2(0,T;U)}.$$

(C2) For every $\phi \in \mathcal{H}$ and $\mathcal{C} \in \mathcal{L}(\mathcal{H}, Y)$, there exists $c > 0$ such that

$$|\mathcal{C}\mathcal{T}(t)\phi|_{L^2(0,T;Y)} \le c|\phi|_{\mathcal{V}^*}$$

for all $\phi \in \mathcal{H}$.

(C3) Consider the Hilbert space $\mathcal{W} = \text{dom}_{\mathcal{V}^*}(\mathcal{A})$, the domain of $\mathcal{A}$ as an infinitesimal generator in $\mathcal{V}^*$, with the graph norm of $\mathcal{A}$ on $\mathcal{V}^*$. Then $\mathcal{W}$ is continuously and densely embedded in $\mathcal{H}$; that is, $\mathcal{W} \hookrightarrow \mathcal{H}$.

The following observations can be made regarding these criteria.

Remark 7.1

(0) The existence of a strongly continuous semigroup $\mathcal{T}$ on $\mathcal{H}$ follows from the symmetry, continuity and coercivity properties of the sesquilinear forms σ_1 and σ_2. For the strongly damped case under consideration, the extension to $\mathcal{V}^*$, as required in (C0), follows directly from Theorem 4.8.

(1) When σ_2 is V-elliptic, (C1) follows directly from the equivalence and continuous dependence (e.g., see Corollary 4.3 and Theorem 4.14) results of Chapter 4. For any $u \in L^2(0, T; U)$ and initial value $z_0 \in \mathcal{H}$, this hypothesis implies that the mild solution $z(\cdot)$ defined by (7.6) is continuous on $[0, T]$ with values in $\mathcal{H}$. In this case, the output can be expressed as

$$z_{ob}(t) = \mathcal{C}\mathcal{T}(t)z_0 + \mathcal{C}\int_0^t \mathcal{T}(t - s)\mathcal{B}u(s)ds \qquad (7.8)$$

with well-defined values in Y. If, on the other hand, $z_0 \in \mathcal{V}^*$, then z may have values in $\mathcal{V}^*$ and (C2) must be invoked to obtain a well-defined output.

(2) The bound of (C2) also follows from standard estimates obtained when $\mathcal{A}$ is defined by (7.5) and σ_2 generating A_2 is V-elliptic (see (3.10) and (6.6) of [17]). This property implies that the operator mapping $z_0 \in \mathcal{H}$ to $\mathcal{C}\mathcal{T}(\cdot)z_0$ can be extended through continuity to define an L^2 (in time) function $\mathcal{C}\mathcal{T}(\cdot)z_0$ for $z_0 \in \mathcal{V}^*$. In this manner, the output given by (7.8) can be expressed as a well-defined function in $L^2(0, T; Y)$ for $z_0 \in \mathcal{V}^*$.

(3) As noted in [157], the assumption (C3) is not very restrictive and is satisfied in many systems of interest through correct choices of $\mathcal{H}$ and $\mathcal{V}^*$. For the case under consideration in which σ_2 is V-elliptic, the domain of $\mathcal{A}$ defined on $\mathcal{V}^*$ satisfies (see (2.9) of [17])

$$\text{dom}_{\mathcal{V}^*}(\mathcal{A}) = \{\phi \in \mathcal{V}|\mathcal{A}\phi \in \mathcal{V}^*\} = \mathcal{V} .$$

7.2.1   Finite Horizon Control Problem

For the control problem on a finite time interval $[0, T]$, we consider the quadratic performance index

$$J_T(u, z_0) = \int_0^T \left\{ |\mathcal{C}z(t)|_Y^2 + |\mathcal{R}^{1/2}u(t)|_U^2 \right\} dt + \langle z(T), \mathcal{G}z(T) \rangle_{\mathcal{V}^*, \mathcal{V}} \qquad (7.9)$$

where $\mathcal{G} \in \mathcal{L}(\mathcal{V}^*, \mathcal{V})$ is a nonnegative operator which is self-adjoint in the sense that its restriction to $\mathcal{H}$ is self-adjoint. The positive, self-adjoint operator $\mathcal{R} \in \mathcal{L}(U)$ can be used to weight various components of the control (in the applications of Chapter 8, it is used to weight the controlling voltage to specific piezoceramic patches). The minimization is performed over $u \in L^2(0, T; U)$ subject to z satisfying (7.4) with $g \equiv 0$.

As noted in [17, 110], the determination of an optimal control is facilitated by formulating the output solution $z_{ob}(t)$ of (7.8) as

$$z_{ob}(t) = \mathcal{M}z_0 + \mathcal{J}u(t)$$

in $L^2(0, T; Y)$. The operators $\mathcal{M} : \mathcal{H} \to L^2(0, T; Y)$ and $\mathcal{J} : L^2(0, T; U) \to L^2(0, T; Y)$ are defined by

$$(\mathcal{M}z_0)(t) = \mathcal{C}T(t)z_0$$

$$(\mathcal{J}u)(t) = \mathcal{C} \int_0^t T(t - s)\mathcal{B}u(s)ds$$

for $t \in (0, T)$. From (C1)-(C4), it follows that $\mathcal{M}$ can be uniquely extended to $\mathcal{V}^*$ so as to be compatible with forces and data in the applications discussed in subsequent chapters. In the spirit of the semigroup operator notation, we will let $\mathcal{M}$ also denote this extension so that $\mathcal{M} \in \mathcal{L}(\mathcal{V}^*, L^2(0, T; Y))$. For fixed terminal time $T > 0$, we also define the operators $\mathcal{J}_T : L^2(0, T; U) \to \mathcal{H}$ and $\mathcal{M}_T : \mathcal{V}^* \to \mathcal{H}$ by

$$\mathcal{J}_T u = \mathcal{G}^{1/2} \int_0^T T(T - s)\mathcal{B}u(s)ds$$

$$\mathcal{M}_T z_0 = \mathcal{G}^{1/2} T(T)z_0$$

with $\mathcal{G}^{1/2} \in \mathcal{L}(\mathcal{V}^*, \mathcal{H})$ (further details regarding the characterization $\mathcal{G} = (\mathcal{G}^{1/2})^* \mathcal{G}^{1/2}$ can be found in Corollary 2.1 of [110]). In terms of these operators, the performance index (7.9) can be written as

$$J_T(u, z_0) = |\mathcal{M}z_0 + \mathcal{J}u|_{L^2(0,T;Y)}^2 + |\mathcal{R}^{1/2}u|_{L^2(0,T;U)}^2 + |\mathcal{M}_T z_0 + \mathcal{J}_T u|_{\mathcal{H}}^2 . \qquad (7.10)$$

The optimal control $\overline{u}_T$ is then specified in the following theorem which is Theorem 3.1 in [17] or Theorem 2.2 in [110] (the results also follow from Theorem 2.7 of [157]).

Theorem 7.2 *Let σ be bounded and $\mathcal{V}$-elliptic and $\mathcal{C} \in \mathcal{L}(\mathcal{H}, Y)$ be a bounded observation operator. For finite T and $z_0 \in \mathcal{H}$, the optimal control $\overline{u}_T$ which minimizes (7.10) is given by*

$$\overline{u}_T = -\left(I + \mathcal{J}^*\mathcal{J} + \mathcal{J}_T^*\mathcal{J}_T\right)^{-1} \left(\mathcal{J}^*\mathcal{M} + \mathcal{J}_T^*\mathcal{M}_T\right) z_0 .$$

The performance of the optimal control can be specified in terms of the self-adjoint Riccati operator $\Pi_T \in \mathcal{L}(\mathcal{V}^*, \mathcal{V})$ defined by

$$\Pi_T = (\mathcal{M}^*, \mathcal{M}_T^*) \left(\begin{bmatrix} I & 0 \\ 0 & I \end{bmatrix} + \begin{bmatrix} \mathcal{J} \\ \mathcal{J}_T \end{bmatrix} [\mathcal{J}^*, \mathcal{J}_T^*] \right)^{-1} \begin{pmatrix} \mathcal{M} \\ \mathcal{M}_T \end{pmatrix}$$

which satisfies

$$\langle \Pi_T z_0, z_0 \rangle_{\mathcal{V}, \mathcal{V}^*} = J_T(\overline{u}_T, z_0) = \min_u J_T(u, z_0)$$

(we point out that Π_T is self-adjoint in the sense that its restriction to $\mathcal{H}$ is self-adjoint). Moreover, if we let $\Pi_T(t) \equiv \Pi_{T-t}, t \leq T$, then the optimal control is given by the feedback law

$$\overline{u}_T(t) = -\mathcal{R}^{-1} \mathcal{B}^* \Pi_T(t) \overline{z}(t)$$

where $\Pi_T(t)$ satisfies the differential Riccati equation

$$\left(\frac{d}{dt} \Pi_T(t) + \mathcal{A}^* \Pi_T(t) + \Pi_T(t) \mathcal{A} - \Pi_T(t) \mathcal{B} \mathcal{R}^{-1} \mathcal{B}^* \Pi_T(t) + \mathcal{C}^* \mathcal{C} \right) z = 0 \ ,$$

for all $z \in \mathcal{V}$, and $\overline{z}(t)$ denotes the corresponding optimal trajectory.

Proofs and discussion regarding this theorem can be found in [17, 110, 157].

7.2.2   Infinite Horizon Control Problem

While control over a specified finite time interval is important in some applications, it is more common that the time interval is indefinite in length and controls applicable for an unbounded time interval are sought. For the infinite horizon problem, a control u is sought which minimizes the quadratic cost functional

$$J(u, z_0) = \int_0^\infty \left\{ |\mathcal{C}z(t)|_Y^2 + |\mathcal{R}^{1/2} u(t)|_U^2 \right\} dt \tag{7.11}$$

subject to (7.4) with $g \equiv 0$. The following definitions and theorem detail the form of the optimal control for this case.

Definition 7.3 The pair $(\mathcal{A}, \mathcal{B})$ is said to be stabilizable if there exists an operator $\mathcal{K} \in \mathcal{L}(\mathcal{V}^*, U)$ such that $\mathcal{A} - \mathcal{B}\mathcal{K}$ generates an exponentially stable semigroup on $\mathcal{V}^*$ (i.e., $|e^{t(\mathcal{A}-\mathcal{B}\mathcal{K})}|_{\mathcal{L}(\mathcal{V}^*)} \leq Me^{-\omega t}$ for $M \geq 1, \omega > 0$).

Definition 7.4 The pair $(\mathcal{A}, \mathcal{C})$ is said to be detectable if there exists an operator $\mathcal{F} \in \mathcal{L}(Y, \mathcal{V}^*)$ such that $\mathcal{A} - \mathcal{F}\mathcal{C}$ generates an exponentially stable semigroup on $\mathcal{V}^*$.

Theorem 7.5 *If $(\mathcal{A}, \mathcal{B})$ is stabilizable and $(\mathcal{A}, \mathcal{C})$ is detectable, then the algebraic Riccati equation*

$$\left(\mathcal{A}^*\Pi + \Pi\mathcal{A} - \Pi\mathcal{B}\mathcal{R}^{-1}\mathcal{B}^*\Pi + \mathcal{C}^*\mathcal{C}\right) z = 0 \qquad \text{for all } z \in \mathcal{V}$$

has a unique nonnegative solution $\Pi \in \mathcal{L}(\mathcal{V}^*, \mathcal{V})$, $\mathcal{A} - \mathcal{B}\mathcal{R}^{-1}\mathcal{B}^*\Pi$ generates an exponentially stable closed loop semigroup $\mathcal{S}(t)$ on $\mathcal{H}, \mathcal{V}$ and $\mathcal{V}^*$, and the optimal control that minimizes (7.11) is given by

$$\bar{u}(t) = -\mathcal{R}^{-1}\mathcal{B}^*\Pi\bar{z}(t)$$

where $\bar{z}(t) = \mathcal{S}(t)z_0$ for $z_0 \in \mathcal{V}^*$.

Remark 7.6

(1) Theorem 7.5 is essentially Theorem 3.4 in [17] or Theorem 2.3 in [110], and further discussion and proofs can be found therein. A full discussion of analysis leading to this theorem can be found in Theorems 3.3, 3.4 and Remark 3.5 of [157].

(2) The conditions of stabilizability and detectability are often referred to as exponential stabilizability and detectability in the literature.

(3) We point out that the stabilizability and detectability of the system can be defined in terms of operators $\mathcal{K} \in \mathcal{L}(\mathcal{H}, U)$ and $\mathcal{F} \in \mathcal{L}(Y, \mathcal{H})$ leading to the generation of exponentially stable semigroups on $\mathcal{H}$. Lemma 3.3 of [17] can then be invoked to obtain the desired exponential stability of the semigroups on $\mathcal{V}$ and $\mathcal{V}^*$.

As noted in the proof of Theorem 3.3 of [157], the solution to the algebraic Riccati equation is derived as the limit of solutions to finite-time integral Riccati solutions as $T \to \infty$. In that theorem it is noted that in order to obtain convergence of operators and controls in this limit, the admissibility hypothesis

(C4) For every $z_0 \in \mathcal{V}^*$ there exists a control $u_{z_0}(\cdot) \in L^2(0, \infty; U)$ such that $J(u_{z_0}) < \infty$

along with (C0)-(C3) is required. While yielding the existence of a Riccati operator, this hypothesis alone does not indicate specifically when one can expect to obtain a *unique* Riccati solution or an exponentially stable closed loop semigroup. In Theorem 3.4 of [157], it was then proved that when the hypothesis

(C5) If $z_0 \in \mathcal{V}^*$ and $u(\cdot) \in L^2(0, \infty; U)$ are such that $J(u) < \infty$, then $z(\cdot) \in L^2(0, \infty; \mathcal{V}^*)$ where $z(t), t \geq 0$, is given by (7.6)

is included with (C0)-(C4), the uniqueness of the Riccati solution $\Pi \in \mathcal{L}(\mathcal{V}^*, \mathcal{V})$ and exponential stability of the closed loop semigroup can be concluded.

The weakness of Theorem 3.4 of [157] is that the required hypotheses (C4) and (C5) are difficult to check in actual applications. A practical alternative is to employ the stabilizability and detectability assumptions which are stronger than (C4) and (C5) but are more easily verified and, as discussed in Remark 3.5 of [157], can be used to obtain (C4) and (C5). Specifically, the hypothesis of stabilizability yields (C4) and hence can be used to guarantee the existence of a Riccati solution $\Pi \in \mathcal{L}(\mathcal{V}^*, \mathcal{V})$. Hypothesis (C5) can be obtained from the assumption of detectability. This then yields the uniqueness of the Riccati solution and exponential stability of the closed loop semigroup.

7.3   Approximate LQR Control Problem with No Exogenous Input

The control results discussed thus far have been for the infinite dimensional problem and controls have been defined in terms of operators and functions satisfying appropriate smoothness constraints. In order for these controls to be implemented, the problem must be discretized and a sequence of finite dimensional LQR problems considered. This discretization is considered here in the framework of Galerkin approximations and approximate solutions are sought in finite dimensional subspaces $\mathcal{V}^N \subset \mathcal{V} \subset \mathcal{H}$. The bases for these subspaces can consist of modes, splines or finite elements which satisfy convergence criteria to be discussed in this section. We point out that the inclusion of $\mathcal{V}^N$ in $\mathcal{V}$ may be too restrictive for some approximation methods such as finite differences, certain spectral and collocation approximations. In such cases, a relaxation of hypotheses in the manner discussed in [25] for the bounded control input analysis can be employed. It is also noted that $\mathcal{V}^N$ is not required to be in $\mathrm{dom}_{\mathcal{H}}(\mathcal{A})$. This is important when choosing a basis for $\mathcal{V}^N$ and permits the use of linear splines in second-order problems and cubic splines in fourth-order systems.

We first assume that the approximation method satisfies the following convergence condition:

(C1N) For any $z \in \mathcal{V}$, there exists a sequence $\tilde{z}^N \in \mathcal{V}^N$ such that $|z - \tilde{z}^N|_V \to 0$ as $N \to \infty$.

This assumption is standard and is satisfied by most reasonable approximation methods.

The operator $\mathcal{A}^N : \mathcal{V}^N \to \mathcal{V}^N$ which approximates $\mathcal{A}$ is defined by restricting σ to $\mathcal{V}^N \times \mathcal{V}^N$; hence

$$\left\langle -\mathcal{A}^N \Phi, \Psi \right\rangle_{\mathcal{H}} = \sigma(\Phi, \Psi) \qquad \text{for all}\ \ \Phi, \Psi \in \mathcal{V}^N .$$

For each N, the C_0 semigroup on $\mathcal{V}^N$ that is generated by $\mathcal{A}^N$ is denoted by $T^N(t)$. The control operator $\mathcal{B} \in \mathcal{L}(U, \mathcal{V}^*)$ is approximated by $\mathcal{B}^N \in \mathcal{L}(U, \mathcal{V}^N)$ which is defined through duality by

$$\left\langle \mathcal{B}^N u, \Psi \right\rangle_{\mathcal{H}} = \left\langle u, \mathcal{B}^* \Psi \right\rangle_U \qquad \text{for all } u \in U \ , \ \Psi \in \mathcal{V}^N \ . \tag{7.12}$$

The observation operator $\mathcal{C}^N$ is simply obtained by restricting $\mathcal{C}$ to $\mathcal{V}^N$. Finally, we let P^N denote the usual orthogonal projection of $\mathcal{H}$ onto $\mathcal{V}^N$ which by definition satisfies

(i) $P^N \Phi \in \mathcal{V}^N$ for $\Phi \in \mathcal{H}$

(ii) $\left\langle P^N \Phi - \Phi, \Psi \right\rangle_{\mathcal{H}} = 0$ for all $\Psi \in \mathcal{V}^N$.

This projection can be extended to $P^N \in \mathcal{L}(\mathcal{V}^*, \mathcal{V}^N)$ by replacing the $\mathcal{H}$-inner product $\langle \Phi, \Psi \rangle_{\mathcal{H}}$ in the definition (ii) by $\langle \Phi, \Psi \rangle_{\mathcal{V}^*, \mathcal{V}}$ and considering $\Phi \in \mathcal{V}^*$. The approximate problem corresponding to (7.3) with $g \equiv 0$ can then be formulated as

$$\frac{d}{dt} \left\langle z^N(t), \Psi \right\rangle_{\mathcal{H}} + \sigma(z^N(t), \Psi) = \left\langle \mathcal{B}^N u(t), \Psi \right\rangle_{\mathcal{H}} \qquad \text{for all } \Psi \in \mathcal{V}^N$$
$$z^N(0) = P^N z_0 \tag{7.13}$$

with the solution

$$z^N(t) = T^N(t) P^N z_0 + \int_0^t T^N(t-s) \mathcal{B}^N u(s) ds \ . \tag{7.14}$$

If (C1N) is satisfied, z denotes the solution to (7.3), and $e^N(t) \equiv z^N(t) - z(t)$ denotes the error, then one can expect the convergence

$$\left| e^N(t) \right|_{\mathcal{H}} \to 0$$

$$\int_0^t \left| e^N(s) \right|_{\mathcal{V}}^2 ds \to 0$$

as $N \to \infty$ (see, for example, Chapter III in [137] for a proof of this result).

7.3.1   Approximate Control Gains in the Finite Horizon Problem

In Section 7.2 we considered the minimization of the functional (7.9) to obtain the optimal control over a finite time interval $(0, T)$. The corresponding N^{th} approximate problem in $\mathcal{V}^N$ concerns the minimization of the cost functional

$$J_T^N(u, z_0) = \int_0^T \left\{ \left| \mathcal{C}^N z^N(t) \right|_Y^2 + \left| \mathcal{R}^{1/2} u(t) \right|_U^2 \right\} dt + \left\langle \mathcal{G}^N z^N(T), z^N(T) \right\rangle_{\mathcal{H}} \tag{7.15}$$

subject to

$$\frac{d}{dt} z^N(t) = \mathcal{A}^N z^N(t) + \mathcal{B}^N u(t) \ , \quad 0 < t < T$$
$$z^N(0) = P^N z_0 \ . \tag{7.16}$$

The symmetric operator $\mathcal{G}^N$ is defined by $\mathcal{G}^N = P^N \mathcal{G} P^N$.

Since the trajectories of (7.16) evolve in $\mathcal{V}^N$, finite dimensional control theory can be used to determine the optimal control which minimizes (7.15). The main question of interest then concerns the convergence of the finite dimensional Riccati operators and approximate controls to their infinite dimensional counterparts. Convergence properties of the controls and Riccati operators for the finite-time problem are summarized in the following theorem.

Theorem 7.7 *Suppose $\mathcal{A}$ is defined by (7.5) and $\mathcal{B} \in \mathcal{L}(U, \mathcal{V}^*), \mathcal{C} \in \mathcal{L}(\mathcal{V}, Y),$ $\mathcal{G} \in \mathcal{L}(\mathcal{V}^*, \mathcal{V})$ are the previously discussed control, observation and terminal weight operators. Consider an approximation scheme satisfying (C1N) with the corresponding approximate operators $\mathcal{A}^N, \mathcal{B}^N, \mathcal{C}^N, \mathcal{G}^N$. Moreover, let $\Pi_T^N(t), t \leq T$ denote the solution to the Riccati equation*

$$\frac{d}{dt}\Pi_T^N(t) + \mathcal{A}^{N^*}\Pi_T^N(t) + \Pi_T^N(t)\mathcal{A}^N - \Pi_T^N(t)\mathcal{B}^N\mathcal{R}^{-1}\mathcal{B}^{N^*}\Pi_T^N(t) + \mathcal{C}^{N^*}\mathcal{C}^N = 0$$

$$\Pi_T^N(T) = P^N\mathcal{G}P^N = \mathcal{G}^N$$

in $\mathcal{V}^N$, and let $\bar{u}_T^N$ denote the optimal control for the Nth approximate problem (7.15)-(7.16). The following convergence is then obtained:

(i) $|\bar{u}_T^N - \bar{u}_T|_{L^2(0,T;U)} \to 0$ *for all* $z_0 \in \mathcal{H}$

(ii) $|\Pi_T^N(t)P^N z - \Pi_T(t)z|_{\mathcal{H}} \to 0$ *uniformly in* $t \in [0,T]$ *for all* $z \in \mathcal{H}$.

The optimal control $\bar{u}_T^N$ to the Nth approximate problem is given by

$$\bar{u}_T^N(t) = -\mathcal{R}^{-1}\mathcal{B}^{N^*}\Pi_T^N(t)z^N(t) .$$

Furthermore, if $\mathcal{C}$ is bounded (i.e., $\mathcal{C} \in \mathcal{L}(\mathcal{H}, Y)$) and (C2) is satisfied, then (ii) can be strengthened to yield convergence in the uniform operator topology

(ii′) $\Pi_T^N(t)$ *converges to* $\Pi_T(t)$ *in* $\mathcal{L}(\mathcal{V}^*, \mathcal{V})$ *uniformly for t in $[0,T]$.*

The reader is referred to Theorems 4.5 and 4.6 (as well as Remark 3.2(1), Lemma 6.1 and the discussion following the lemma) in [17] and Theorem 2.4 in [110] for further discussion and proofs of these results.

7.3.2   Approximate Control for the Infinite Horizon Problem

When implementing the method over an indefinite time period, one typically must consider the convergence problem for controls determined with $T = \infty$. The optimization problem in this case consists of finding $u \in L^2(0, \infty; U)$ which minimizes

$$J^N(u, z_0) = \int_0^\infty \left\{ \left|\mathcal{C}^N z^N(t)\right|_Y^2 + \left|\mathcal{R}^{1/2}u(t)\right|_U^2 \right\} dt \qquad (7.17)$$

subject to z^N satisfying the evolution equation

$$\frac{d}{dt}z^N(t) = \mathcal{A}^N z^N(t) + \mathcal{B}^N u(t) \quad , \quad t > 0$$
$$z^N(0) = P^N z_0 . \tag{7.18}$$

An example of a physical situation in which this control formulation might be of interest could consist of a structure with surface-mounted piezoceramic patches that starts from an initial deformation and then oscillates to an equilibrium state. Due to the piecewise constant nature of the actuators, the approximate control operators $\mathcal{B}^N$ must converge to an unbounded operator $\mathcal{B}$. For the discussion here, we will assume that full state (e.g., in the case of a structure, distributed displacement and velocity) measurements are available so that $\mathcal{C}^N$ is the restriction of $\mathcal{C} \in \mathcal{L}(\mathcal{H}, Y)$ to $\mathcal{V}^N$. The assumption of a bounded observation operator simplifies the presentation. Moreover, it provides a framework from which to consider the unbounded observation operator problem that arises in most applications when only a limited number of discrete state measurements are available. It also provides a basis for constructing a state estimator or compensator that can be used to reconstruct the state from the discrete measurements.

An examination of the functional J^N of (7.17) reveals that two limits are involved in the convergence process, namely $N \to \infty$ and $T \to \infty$. As detailed in the proof of Theorem 2.2 of [25] for the bounded control operator and Theorem 2.6 of [110] for an unbounded operator, convergence arguments typically start with the consideration of the finite-time minimization problem

$$\text{minimize} \int_0^T \left\{ \left| \mathcal{C}^N z^N(t) \right|_Y^2 + \left| \mathcal{R}^{1/2} u(t) \right|_U^2 \right\} dt$$

subject to (7.18).

The strategy is to then bound the observed trajectory and control as T increases in a manner which guarantees the existence of nonnegative self-adjoint solutions $\Pi^N \in \mathcal{L}(\mathcal{V}^*, \mathcal{V})$ to the Nth approximate algebraic Riccati equation in $\mathcal{V}^N$

$$\mathcal{A}^{N^*}\Pi^N + \Pi^N \mathcal{A}^N - \Pi^N \mathcal{B}^N \mathcal{R}^{-1} \mathcal{B}^{N^*} \Pi^N + \mathcal{C}^{N^*} \mathcal{C}^N = 0 \tag{7.19}$$

for N sufficiently large. This is "Step 1" in the proof outlined in [17] and detailed in [110].

In "Step 2" of the previously mentioned proofs, the uniqueness of the approximate Riccati solutions is established and the exponential stability of the approximate closed loop semigroup $\mathcal{S}^N(t)$ generated by $\mathcal{A}^N - \mathcal{B}^N \mathcal{R}^{-1} \mathcal{B}^{N^*} \Pi^N$ is proved. The convergence of the Riccati operators and hence controls then follows from the standard estimates described in (A.6), (A.7) of [110] and the Appendix of [25]. Thus in a general sense, if one assumes the existence of unique Π and Π^N solving the respective infinite dimensional and approximate Riccati equations and is able to show that $\mathcal{S}(t)$ and $\mathcal{S}^N(t)$ are exponentially stable, then the desired convergence can be obtained directly (see, for example,

Theorem 2.2 of [25] for the bounded control case). Due to the nature of these assumptions (e.g., difficulty in verifying directly in problems), however, such a result is less than desirable, and conditions that can be verified in applications and that yield these assumptions are sought.

We recall from the discussion of the corresponding infinite dimensional problem in Section 7.2 that the condition of stabilizability of $(\mathcal{A}, \mathcal{B})$ was used to guarantee the existence of a Riccati solution $\Pi \in \mathcal{L}(\mathcal{V}^*, \mathcal{V})$ to the algebraic Riccati equation of Theorem 7.5. The uniqueness of the solution and exponential stability of the closed loop semigroup were obtained using the hypothesis of detectability of $(\mathcal{A}, \mathcal{C})$. As discussed in [17] and [110], the analogous concepts of uniform stabilizability of $(\mathcal{A}^N, \mathcal{B}^N)$ and uniform detectability of $(\mathcal{A}^N, \mathcal{C}^N)$ can be used to prove the existence of a unique Riccati solution Π^N and exponential stability of the closed loop semigroup $\mathcal{S}^N(t)$. We state these results more precisely.

Definition 7.8 The pair $(\mathcal{A}^N, \mathcal{B}^N)$ is said to be uniformly stabilizable if there exist constants $M_1 \geq 1, \omega_1 > 0$ independent of N and a sequence of operators $\mathcal{K}^N \in \mathcal{L}(\mathcal{V}^N, U)$ such that $\sup_N |\mathcal{K}^N| < \infty$ and

$$\left| e^{t(\mathcal{A}^N - \mathcal{B}^N \mathcal{K}^N)} P^N z \right|_{\mathcal{H}} \leq M_1 e^{-\omega_1 t} |z|_{\mathcal{H}}$$

for $z \in \mathcal{H}$.

Definition 7.9 The pair $(\mathcal{A}^N, \mathcal{C}^N)$ is said to be uniformly detectable if there exist constants $M_2 \geq 1, \omega_2 > 0$ independent of N and a sequence of operators $\mathcal{F}^N \in \mathcal{L}(Y, \mathcal{V}^N)$ such that $\sup_N |\mathcal{F}^N| < \infty$ and

$$\left| e^{t(\mathcal{A}^N - \mathcal{F}^N \mathcal{C}^N)} P^N z \right|_{\mathcal{H}} \leq M_2 e^{-\omega_2 t} |z|_{\mathcal{H}}$$

for $z \in \mathcal{H}$.

To treat $z \in \mathcal{V}^*$, the projections can be extended to $P^N \in \mathcal{L}(\mathcal{V}^*, \mathcal{V}^N)$ with norms altered accordingly.

Convergence results for the infinite horizon problem are then summarized in the following theorem whose proof can be obtained from the discussions of Theorem 4.8 and Section 6.1 in [17] or Theorem 2.6 in [110] along with the observation that for second-order systems, the operator $\mathcal{B}$ has the form $\mathcal{B} = [0, B]^T$. To obtain the desired result, the following assumption is also required.

(C6) The injection $i : V \hookrightarrow H$ is compact.

Theorem 7.10 *Suppose (C6) holds. Let the sesquilinear form σ associated with the first-order system (7.3) be continuous and $\mathcal{V}$-elliptic. Assume that the operators $\mathcal{A}, \mathcal{B}, \mathcal{C}$ of (7.4), (7.7) satisfy: $(\mathcal{A}, \mathcal{B})$ is stabilizable and $(\mathcal{A}, \mathcal{C})$ is detectable where $\mathcal{B} \in \mathcal{L}(U, \mathcal{V}^*)$ is unbounded and $\mathcal{C} \in \mathcal{L}(\mathcal{H}, Y)$ is bounded. Consider an approximation method which satisfies (C1N). Finally, suppose that for fixed N_0 and $N > N_0$, the pair $(\mathcal{A}^N, \mathcal{B}^N)$ is uniformly stabilizable and $(\mathcal{A}^N, \mathcal{C}^N)$ is uniformly detectable.*

Then for N sufficiently large, there exists a unique nonnegative self-adjoint solution $\Pi^N \in \mathcal{L}(\mathcal{V}^*, \mathcal{V})$ to the Nth approximate algebraic Riccati equation (7.19) in $\mathcal{V}^N$. There also exist constants $M_3 \geq 1$ and $\omega_3 > 0$ independent of N such that $\mathcal{S}^N(t) = e^{(\mathcal{A}^N - \mathcal{B}^N \mathcal{R}^{-1} \mathcal{B}^{N^*} \Pi^N)t}$ satisfies

$$\left| \mathcal{S}^N(t) \right|_{\mathcal{V}^N} \leq M_3 e^{-\omega_3 t} \quad , \quad t > 0$$

or equivalently

$$\left| e^{t(\mathcal{A}^N - \mathcal{B}^N \mathcal{R}^{-1} \mathcal{B}^{N^*} \Pi^N)} P^N z_0 \right|_{\mathcal{H}} \leq M_3 e^{-\omega_3 t} |z_0|_{\mathcal{H}} \quad , \quad t > 0 \, , \, z_0 \in \mathcal{H} \, .$$

Additionally, the convergences

$$\Pi^N P^N z \xrightarrow{s} \Pi z \ \text{ in } \mathcal{V} \ \text{ for every } z \in \mathcal{V}^*$$

$$\left| \mathcal{B}^{N^*} \Pi^N P^N - \mathcal{B}^* \Pi \right|_{\mathcal{L}(\mathcal{H}, U)} \to 0 \, ,$$

as $N \to \infty$, of the Riccati and control operators are obtained. Moreover, the feedback system operator $\mathcal{A} - \mathcal{B} \mathcal{R}^{-1} \mathcal{B}^{N^*} \Pi^N$ generates an exponentially stable analytic semigroup on $\mathcal{H}$ and for every $z_0 \in \mathcal{H}$,

$$J\left(-\mathcal{B}^{N^*} \Pi^N z(\cdot), z_0 \right) - J(\bar{u}, z_0) \leq \varepsilon(N) |z_0|_{\mathcal{H}}^2$$

where $\varepsilon(N) \to 0$ as $N \to \infty$.

Remark 7.11

(1) Theorem 7.10 is the second-order system analogue of Theorem 4.8 of [17], which is for first-order systems. To obtain Theorem 7.10 directly from Theorem 4.8 in [17], one would require that the embedding $\mathcal{V} \hookrightarrow \mathcal{H}$ be compact. This is not true for our second-order systems where $\mathcal{V} = V \times V$ and $\mathcal{H} = V \times H$. However, one can use arguments similar to those employed in Theorem 4.8 of [17] along with (C6) and the special structure of our second-order systems written in first-order form to prove Theorem 7.10.

(2) The final condition of Theorem 7.10 implies that the finite dimensional control yields an exponentially stable semigroup when applied to the infinite dimensional system. This is of great practical importance since it represents the case when computed controls are applied to the actual physical systems of interest.

(3) For the analysis here, the system is assumed to be strongly damped or essentially parabolic in nature; hence σ_2 is V-elliptic and σ is $\mathcal{V}$-elliptic. As discussed in Lemmas 4.2 and 4.3 of [17], this leads to resultant estimates for $\mathcal{A}^N$ and bounds and estimates for the open loop semigroups $\mathcal{T}(t)$ and $\mathcal{T}^*(t)$ in both the $\mathcal{H}$ and $\mathcal{V}$ norms and hence convergence of the approximating and adjoint approximating semigroups $\mathcal{T}^N$ and $\mathcal{T}^{N^*}$ (see (i)-(iii) of Lemma 4.3 in [17]). Moreover, the convergence results summarized in Lemma 4.4 of [17] for $\mathcal{M}^N$ and $\mathcal{M}^{N^*}$ are typical of the consistency criteria that can be obtained.

(4) Conditions such as the required adjoint convergence $\mathcal{T}^{N^*}(t)P^N z \to \mathcal{T}^*(t)z$, as $N \to \infty$, may appear on first encounter as purely mathematical with little physical relevance to the actual problem. However, as demonstrated by examples in [6, 7, 21, 31, 115, 116] in which the calculation of feedback gains for a delay equation are considered, failure of approximation methods to satisfy adjoint convergence can lead to gains which do not strongly converge.

(5) Care must also be taken when choosing an approximation method to guarantee that the uniform stabilizability margins are preserved under approximation. As illustrated in the context of delay equations [21] and weakly damped PDE systems [19], approximation methods which are quite adequate for open loop simulations and/or parameter estimation (e.g., finite differences and finite elements) may not preserve these margins. This can slow or even prevent convergence of the feedback gains and illustrates the care that must be taken when choosing an approximation methodology that is suitable for control calculations.

While Theorem 7.10 yields the desired convergence results, the hypotheses of uniform stabilizability of $(\mathcal{A}^N, \mathcal{B}^N)$ and uniform detectability of $(\mathcal{A}^N, \mathcal{C}^N)$ are often difficult to verify directly, and we now discuss more readily checked assumptions which yield the uniform stabilizability and detectability conditions.

For general first-order systems, uniform stabilizability is obtained using the following lemma which is Lemma 4.7 in [17] and is the unbounded input analogue of the (POES) condition in [7, 25].

Lemma 7.12 *Suppose $(\mathcal{A}, \mathcal{B})$ is stabilizable and the injection $\mathcal{V} \hookrightarrow \mathcal{H}$ is compact. Then there exists a positive integer N_0 such that for all $N > N_0$, the pair $(\mathcal{A}^N, \mathcal{B}^N)$ is uniformly stabilizable.*

For truly first-order systems, the compact injection condition is quite often satisfied through natural choices for the state and test function space (e.g., $\mathcal{H} = L^2(\Omega), \mathcal{V} = H^1(\Omega)$) due to standard Sobolev embedding theorems. In such cases, Lemma 7.12 can be combined with the first-order analogue of

Theorem 7.10 to yield the desired convergence results with hypotheses that can be verified in physical applications (the reader is referred to [7] for examples demonstrating the verification of these hypotheses when $\mathcal{B}$ is bounded).

On the other hand, in second-order systems that are written in first-order form through a product space formulation, the embedding $i : V \hookrightarrow \mathcal{H}$ is *not* compact and hence Lemma 7.12 is not useful. Lemma 7.13 below contains conditions used (along with Theorem 7.10) to obtain uniform exponential bounds on the approximating semigroups $\mathcal{S}^N(t)$ on $\mathcal{H}^N \subset \mathcal{H} = V \times H$ for the second-order systems of interest to us.

Lemma 7.13 *Assume that (C6) holds. Moreover, suppose that the damping sesquilinear form can be decomposed as $\sigma_2 = \delta\sigma_1 + \tilde{\sigma}_2$, for some $\delta > 0$, where the continuous sesquilinear form $\tilde{\sigma}_2$ satisfies for some $\mu \in \mathbb{R}$*

$$Re\ \tilde{\sigma}_2(\phi, \phi) \geq -\frac{\delta}{2}|\phi|_V^2 - \mu|\phi|_H^2 \quad \text{for all } \phi \in V \ .$$

Finally, suppose that the operator $A_1^{-1}\tilde{A}_2$, where $\tilde{A}_2 \in \mathcal{L}(V, V^*)$ is defined by $\left\langle \tilde{A}_2\phi, \eta \right\rangle_{V^*,V} = \tilde{\sigma}_2(\phi, \eta)$, is compact on V.

Let $\mathcal{T}$ denote the open loop semigroup generated by the product space operator $\mathcal{A}$ and let $\mathcal{T}^N$ be generated by $\mathcal{A}^N$. If for some $\omega \in \mathbb{R}$ and $M \geq 1$

$$|\mathcal{T}(t)|_{\mathcal{L}(\mathcal{H})} \leq Me^{\omega t} \quad , \quad t \geq 0 \ ,$$

then for any $\varepsilon > 0$ there exists an integer N_ε such that for $N \geq N_\varepsilon$

$$|\mathcal{T}^N(t)P^N|_{\mathcal{L}(\mathcal{H})} \leq \widetilde{M}e^{(\omega+\varepsilon)t} \quad , \quad t \geq 0$$

for some constant $\widetilde{M} > 0$ independent of N.

This is Lemma 6.2 in [17] and the proof can be found therein. When combined with Theorem 7.10, one then obtains convergence of the Riccati and control operators for the special first-order systems corresponding to second-order problems.

To apply the lemma, one must first verify that for the system under consideration, the operator $A_1^{-1}\tilde{A}_2$ is compact in V. In considering this proposition, it can be advantageous to use the property that the product of a compact linear operator and a bounded linear operator yields a compact linear operator (see, for example, Lemma 8.3-2 of [122]). For many problems of interest, A_1 is a differential operator whose inverse is compact, whereas $\tilde{A}_2$ is quite often bounded. This is illustrated in the following example in the context of the Euler-Bernoulli beam model from earlier chapters discussed once again in a control setting.

Example 7.14 We consider here the cantilever Euler-Bernoulli beam having length ℓ with fixed-end at $y = 0$ given at the beginning of Chapter 4. We assume Kelvin-Voigt internal damping and viscous or air external damping. Control is effected through a pair of identical piezoceramic patches as discussed in Chapter 4.

The density, Young's modulus, moment of inertia, viscous damping and Kelvin-Voigt damping coefficients are denoted by ρ, E, I, γ and c_D, respectively, and are assumed constant. This assumption would be reasonable for the case of a homogeneous beam or a beam with piezoceramic patches embedded in a manner so that material properties do not vary across the region of the patches. We consider only the transverse displacement which is denoted by w. An external moment is applied through an applied voltage $u(t)$ to a pair of embedded piezoceramic patches operated out-of-phase. As discussed in Chapter 4, the equation describing the transverse beam motion is

$$\rho \frac{\partial^2 w}{\partial t^2} + \gamma \frac{\partial w}{\partial t} + \frac{\partial^2}{\partial y^2}(bM_y) = \frac{\partial^2}{\partial y^2}\left(\mathcal{K}^B u(t) \chi_{pe}(y)\right)$$

$$w(t,0) = \frac{\partial}{\partial y}w(t,0) = 0 \, , \quad bM_y(t,\ell) = \frac{\partial}{\partial y}(bM_y)(t,\ell) = 0$$

where the derivatives $\frac{\partial}{\partial y}, \frac{\partial^2}{\partial y^2}$ are taken in the distributional sense and the internal moment bM_y is given by (4.2) with $\widetilde{EI}, \widetilde{c_D I}$ replaced by $EI, c_D I$, respectively. We recall that $\mathcal{K}^B$ is a patch constant while $\chi_{pe}(y)$ denotes the characteristic function over the region containing the patch pair.

The state space $H = L^2(0,\ell)$, the space of test functions $V = H_L^2(0,\ell)$, and their respective inner products were described in Section 4.5. For these choices, the compact embedding of V in H follows immediately from standard Sobolev theory. We also noted (see (4.8), (4.44), (4.45)) that the equation of motion can then be written in the weak form

$$\langle \ddot{w}(t), \eta \rangle_{V^*,V} + \sigma_2(\dot{w}(t), \eta) + \sigma_1(w(t), \eta) = \langle Bu(t), \eta \rangle_{V^*,V} \quad \text{for all } \eta \in V$$

where

$$\sigma_1(w, \eta) = \int_0^\ell EI w'' \eta'' dy$$

$$\sigma_2(w, \eta) = \int_0^\ell c_D I w'' \eta'' dy + \gamma \int_0^\ell w \eta \, dy$$

are continuous, V-elliptic and symmetric. Thus the results of Chapter 4 concerning variational solutions and semigroup generation (in particular, Theorems 4.8 and 4.14) hold for this example. The first-order form of the equation along with the associated generator $\mathcal{A}$ are precisely defined as in (4.27), (4.28) or (7.4), (7.5).

For the constant coefficient case under consideration here, the damping form σ_2 can be decomposed as $\sigma_2 = \delta \sigma_1 + \tilde{\sigma}_2$ where δ and $\tilde{\sigma}_2$ are given by

$\delta = \frac{c_D}{E}$ and $\tilde{\sigma}_2(w, \eta) = \gamma \int_0^\ell w\eta \, dy$. Clearly, $\tilde{\sigma}_2$ satisfies

$$Re \, \tilde{\sigma}_2(\phi, \phi) = \frac{\gamma}{\rho}|\phi|_H^2 \geq -\frac{\delta}{2}|\phi|_V^2$$

for all $\phi \in V$. The operator $\tilde{A}_2$ generated by $\tilde{\sigma}_2$ is given by $\tilde{A}_2 = \gamma I$ and is therefore bounded. Since the injections $i : V \hookrightarrow H$, $i^* : H \hookrightarrow V^*$ are compact, we have that $A_1^{-1} \in \mathcal{L}(V^*, V)$ can be written as an operator on $V \to V$ by $A_1^{-1} = A_1^{-1} i^* i$. Thus A_1^{-1} is compact on V and hence $A_1^{-1} \tilde{A}_2$ is compact on V.

Finally, the exponential stability of the open loop semigroup $T(t)$ generated by $\mathcal{A}$ is guaranteed by the following theorem which is discussed in [16] and summarized in [7]:

Theorem 7.15 *Suppose σ_1 is V-elliptic, continuous and symmetric and σ_2 is H-elliptic, continuous and symmetric. Then $\mathcal{A}$ defined via (7.5) is the infinitesimal generator of a C_0-semigroup $T(t)$ in $\mathcal{H} = V \times H$ that is exponentially stable (i.e., $|T(t)z_0|_{\mathcal{H}} \leq M e^{-\omega t}|z_0|_{\mathcal{H}}$).*

From Theorem 7.15, it also follows that $(\mathcal{A}, \mathcal{B})$ is stabilizable and $(\mathcal{A}, \mathcal{C})$ is detectable for the example under discussion.

The hypotheses for Lemma 7.13 are satisfied for Example 7.14 and one therefore obtains uniform bounds on the approximating semigroups. When combined with the conclusions of Theorem 7.10, this then leads to the convergence of the Riccati and control operators in this example. We point out that these arguments are not unique to the Euler-Bernoulli beam model and other structural models (e.g., the plate model discussed in Chapters 5 and 8) can be readily analyzed in a similar manner.

As we have seen, the version of Lemma 7.13 given above is readily used in the case of constant coefficient models. It is possible to prove an analogue of Lemma 7.13 which can be used to treat models with spatially varying coefficients. The details are rather technical and the most direct analogue requires some smoothness on the coefficients (H^1 in the case of beams such as in Example 7.14). Thus this analogue is not immediately applicable to the case of piecewise constant coefficients such as the situation where piezoceramic patches are bonded to the surface of a beam. While numerical results have demonstrated the veracity of the conclusions of Lemma 7.13 and convergence of the feedback gains for various structural models with discontinuous variable physical parameters and unbounded control input operators, general theoretical results encompassing these models are still under development.

The convergence of Theorem 7.10 is obtained under the assumption of a bounded observation operator $\mathcal{C} \in \mathcal{L}(\mathcal{H}, Y)$. As noted previously, however, the observation operator in most current applications is unbounded (discontinuous) due to the spatially discrete nature of commonly-used measurement devices (e.g., accelerometers, laser vibrometers, proximity sensors, piezoelectric sensors, strain gauges, microphones). The reader is referred to Sections 5 and 6.2 of [17] for an extension of the theory for Theorem 7.10 to the un-bounded observation operator case, $C \in \mathcal{L}(\mathcal{V}, Y)$. The arguments for this case are somewhat more technical in that they involve intermediate spaces, but the-oretical results quite similar to those outlined above for bounded observation operators can be obtained. The theory concerning the form of the observa-tion operator takes on practical significance when designing an estimator or observer for state reconstruction from partial state measurements, and further discussion concerning this topic is given in Section 7.5.

The analysis outlined in Sections 7.2 and 7.3 was developed in the context of a sesquilinear functional formulation of the control problem. This provided a natural framework in which to incorporate the unbounded input and output operators. Other approaches involving primarily operator-based theory have also been developed for analysis of the LQR problems with unbounded input and output operators, and the reader is referred to [129, 131] and references therein for development of theory in the context of this latter methodology.

7.4   LQR Control Problems for Systems with Exogenous Inputs

The discussion in Sections 7.2 and 7.3 concerned systems in which the only applied force consisted of the control input. Models of this type arise when considering physical systems with no exogenous force, and the controls in these systems must attenuate only transient state responses.

On the other hand, a large number of physical systems are driven by a persistent exogenous force and in such cases, the feedback control law must be modified to accommodate both the transient and the nontrivial steady state behavior of the forced state. For the discussion here, the exogenous force is assumed to be periodic in time. This is an appropriate assumption in many mechanical, electromechanical, structural acoustic and fluid/structure systems in which the persistent disturbance is due to rotating or oscillating components. For other systems in which the exogenous force has an underlying periodic component that is diluted by white noise, H^∞/MinMax analysis can in some cases be coupled effectively with the theory described here to provide controllers.

The exogenous input can enter a control system $\dot{z}(t) = \mathcal{A}z(t) + \mathcal{B}u(t)$ in a variety of forms depending upon the application under consideration. To illustrate, consider an acoustic system in which ϕ denotes an acoustic potential and $p = \frac{\partial \phi}{\partial t}$ is acoustic pressure (see, for example the system discussed in Chapter 9). Let $\hat{p}$ model a periodic exogenous pressure field which, for example, could be generated by aircraft engines or the electromagnetic field inside a transformer.

One means of attenuating the offending pressure field $\hat{p}$ is through the use of acoustic sources (e.g., speakers) which generate a canceling secondary pressure field p. For $z = (p, \frac{\partial p}{\partial t})$ and $\hat{z} = (\hat{p}, \frac{\partial \hat{p}}{\partial t})$, an appropriate control might be obtained by minimizing the functional

$$J_\tau(u, z_0) = \frac{1}{2} \int_0^\tau \left\{ |\mathcal{C}\,[z(t) + \hat{z}(t)]\,|_Y^2 + |\mathcal{R}^{1/2} u(t)|_U^2 \right\} dt$$

subject to z satisfying modeling equations for the interior field (see [23, 24]). In formulating this functional, τ is chosen so as to be commensurate with all frequencies present in the exogenous signal. This is an example of a tracking problem since the external disturbance can be negated to yield a prescribed trajectory along which the controlled trajectory should track. Hence the disturbance appears in the cost functional and the observed combined state $|\mathcal{C}[z(t) + \hat{z}(t)]|$ is minimized subject to constraints on the control u.

In other applications, the exogenous disturbance may not be directly related to a state being minimized, in which case, it does not provide a trajectory to be tracked. Motivating this concept with a structural acoustic model discussed in Chapter 9, we consider a system in which an external acoustic field $\hat{p}$ impinges upon an enclosed structure. Due to structural acoustic coupling, this then leads to structural vibrations and unwanted interior sound pressure levels. With w, ϕ and $p = \frac{\partial \phi}{\partial t}$ denoting structural displacement, acoustic potential and acoustic pressure for the interior field, the control objective may be the reduction of $p, w, \frac{\partial w}{\partial t}$ or a combination of all three depending upon the application (e.g., the reduction of $\frac{\partial w}{\partial t}$ might be emphasized to reduce structural fatigue). The exogenous pressure $\hat{p}$ enters as a force to the structure (recall that pressure = force/area) and hence does not provide a trajectory to be tracked directly. With $z = (\phi, w, p, \frac{\partial w}{\partial t})$ denoting the state for the associated first-order system, an appropriate penalty functional might be

$$J_\tau(u, z_0) = \frac{1}{2} \int_0^\tau \left\{ |\mathcal{C}z(t)|_Y^2 + |\mathcal{R}^{1/2} u(t)|_U^2 \right\} dt$$

where z satisfies equations modeling the dynamics of the coupled structural acoustic system driven by the exogenous disturbance $\hat{p}$. It should be noted that in this case, the disturbance is not included in the penalty functional since it is not a trajectory to be tracked by some component of the state (even if the interior pressure p was the only state being minimized, $\hat{p}$ would not provide a good trajectory to track because differences between the interior and exterior fields means that minimization of $|\mathcal{C}[p(t) + \hat{p}(t)]|$ would not lead to optimal reduction in interior sound pressure levels).

While the mechanisms leading to the tracking formulation and system with persistent excitation differ, the final control laws are nearly identical. In the remainder of this section, a brief discussion regarding the tracking problem and control of a system with persistent excitation is given. The latter problem arises more commonly in smart material applications and the primary discussion is directed toward that case. The theory for this problem is less complete than that for a system with no exogenous input, discussed in Section 7.2, and we concentrate here on the bounded control input and observation case. While numerical investigations have demonstrated that the techniques can be extended to the unbounded operator regime which encompasses many smart material systems (e.g., piezoceramic or electrostrictive actuators and sensors) driven by exogenous forces, a complete theory for such systems with unbounded operators is still being developed.

7.4.1   Bounded Control Input $\mathcal{B}$ and Periodic Exogenous Force g

We consider here the problem of controlling a system of the form

$$\dot{z}(t) = \mathcal{A}z(t) + \mathcal{B}u(t) + g(t)$$
$$z(0) = z(\tau) \tag{7.20}$$

where $\mathcal{B} \in \mathcal{L}(U, \mathcal{H})$ is bounded and $g \in L^2(0, \tau; \mathcal{H})$ is τ-periodic. As in the previous discussion, the state space $\mathcal{H}$ is a Hilbert space with inner product $\langle \cdot, \cdot \rangle_{\mathcal{H}}$. For this discussion, we also assume that the observation operator $\mathcal{C} \in \mathcal{L}(\mathcal{H}, Y)$ is bounded. Due to periodicity, the penalty functional to be minimized is

$$J_\tau(u, z_0) = \frac{1}{2} \int_0^\tau \left\{ |\mathcal{C}z(t)|_Y^2 + |\mathcal{R}^{1/2}u(t)|_U^2 \right\} dt$$

subject to z satisfying (7.20).

Using standard calculus of variations arguments, it follows that the optimal control $\bar{u}$ and optimal state $\bar{z}$ must satisfy the optimality conditions

$$\dot{\bar{z}}(t) = \mathcal{A}\bar{z}(t) + \mathcal{B}\bar{u}(t) + g(t) \quad , \qquad \bar{z}(0) = \bar{z}(\tau)$$
$$\dot{\lambda}(t) = -\mathcal{A}^*\lambda(t) - \mathcal{C}^*\mathcal{C}\bar{z}(t) \quad , \qquad \lambda(0) = \lambda(\tau)$$
$$\bar{u}(t) = -\mathcal{R}^{-1}\mathcal{B}^*\lambda(t)$$

(see [75]). Formal dynamic programming arguments can then be invoked to argue that the adjoint solution λ can be expressed in the form

$$\lambda = \Pi\bar{z} - r \tag{7.21}$$

where Π and r solve the operator equations

$$\mathcal{A}^*\Pi + \Pi\mathcal{A} - \Pi\mathcal{B}\mathcal{R}^{-1}\mathcal{B}^*\Pi + \mathcal{C}^*\mathcal{C} = 0 \tag{7.22}$$

and

$$\dot{r}(t) = -\left[\mathcal{A}^* - \Pi\mathcal{B}\mathcal{R}^{-1}\mathcal{B}^*\right] r(t) + \Pi g(t)$$
$$r(0) = r(\tau) , \tag{7.23}$$

respectively. We point out that in the discussion of [75], the operators $\mathcal{A}, \mathcal{B}$ and $\mathcal{C}$ are time-dependent and τ-periodic. This then yields time-dependent Riccati solutions satisfying a differential Riccati equation with periodic boundary conditions. For the applications considered here, $\mathcal{A}, \mathcal{B}$ and $\mathcal{C}$ are time invariant which implies that Π is also time invariant.

In order to justify the formal arguments leading to the solution form (7.21), conditions guaranteeing the existence of a unique solution to the Riccati equation (7.22) must be determined. Moreover, the dynamic programming arguments must be justified for the infinite dimensional system under consideration. As was the case for the unforced system considered in Section 7.2, existence of a Riccati solution is obtained through a stabilizability assumption whereas detectability is used to guarantee the uniqueness of the solution. The stabilizability and detectability hypotheses used in this case are summarized in (C7) and (C8):

(C7) There exists an operator $\mathcal{K} \in \mathcal{L}(\mathcal{H}, U)$ such that $\mathcal{A} - \mathcal{B}\mathcal{K}$ generates an exponentially stable semigroup on $\mathcal{H}$.

(C8) There exists an operator $\mathcal{F} \in \mathcal{L}(Y, \mathcal{H})$ such that $\mathcal{A} - \mathcal{F}\mathcal{C}$ generates an exponentially stable semigroup on $\mathcal{H}$.

The conditions guaranteeing the existence of a unique control for the infinite dimensional system (7.20) with persistent excitation are summarized in the following theorem.

Theorem 7.16 *Consider the system (7.20) which is assumed to have a unique mild solution. A bounded control operator $\mathcal{B} \in \mathcal{L}(U, \mathcal{H})$ is considered and $g \in L^2(0, \tau; \mathcal{H})$ is taken to be τ-periodic. Moreover, we suppose that the stabilizability and detectability conditions (C7) and (C8) hold.*

The Riccati equation (7.22) then has a unique nonnegative solution Π. Furthermore, if r denotes the τ-periodic tracking solution of (7.23) and $\bar{z}$ is the closed loop solution of

$$\dot{\bar{z}}(t) = \left[\mathcal{A} - \mathcal{B}\mathcal{R}^{-1}\mathcal{B}^*\Pi\right]\bar{z}(t) - \mathcal{B}\mathcal{R}^{-1}\mathcal{B}^*r(t) + g(t)$$

$$\bar{z}(0) = \bar{z}(\tau) \,,$$

then the optimal control is given by

$$\bar{u}(t) = -\mathcal{R}^{-1}\mathcal{B}^*[\Pi\bar{z}(t) - r(t)] \,. \tag{7.24}$$

To better understand the form of the control $\bar{u}$ defined in (7.24), it is useful to compare it with the corresponding expression in Theorem 7.5 obtained with $g \equiv 0$. In the latter case, the state vibrations are transient in nature and the control is directly proportional to the state. When driven by a persistent excitation, however, the state response exhibits both transient and steady state components. The control of the transient dynamics is accomplished primarily by the state component of (7.24) whereas much of the steady state information is provided through the tracking variable r (the state does provide some steady state information since $\bar{z}$ is driven by g).

Further discussion regarding these results and a proof Theorem 7.16 can be found in [75]. While this theory does provide conditions leading to the existence of a control for the infinite dimensional problem, it does not provide convergence results for the corresponding finite dimensional approximate control problems nor does it address the cases of unbounded control input $\mathcal{B}$ or unbounded observation $\mathcal{C}$. Although extensive numerical simulations have demonstrated the effectiveness and convergence of controls corresponding to (7.24) with unbounded $\mathcal{B}$ and $\mathcal{C}$ and periodic g, a comprehensive theory including convergence results has not yet been completely developed.

7.4.2   Persistent State Excitation and a Tracking Problem

Motivated by the previously described acoustic problem, we illustrate the tracking problem by considering the cancellation of a periodic, exogenous acoustic pressure $\hat{p}(t)$ through the generation of a secondary field $p(t)$. A suitable control $u \in L^2(0, \tau; U)$ is determined by minimizing the tracking functional

$$J_\tau(u, z_0) = \int_0^\tau \left\{ |\mathcal{C}[z(t) + \hat{z}(t)]|_Y^2 + |\mathcal{R}^{1/2} u(t)|_U^2 \right\} dt \qquad (7.25)$$

subject to

$$\dot{z}(t) = \mathcal{A}z(t) + \mathcal{B}u(t) \quad , \quad 0 \le t \le \tau$$
$$z(0) = z(\tau) \qquad\qquad (7.26)$$

where $z = (p, \frac{\partial p}{\partial t})$ and $\hat{z} = (\hat{p}, \frac{\partial \hat{p}}{\partial t})$. The parameter τ is again chosen to be commensurate with all frequencies in the periodic disturbance $\hat{p}$.

As discussed in [23, 24], under stabilizability and detectability assumptions, the optimal control which minimizes (7.25) is

$$\overline{u}(t) = -\mathcal{R}^{-1}\mathcal{B}^*[\Pi\overline{z}(t) - r(t)] \qquad (7.27)$$

where $\overline{z}(t)$ is the optimal, periodic, closed loop trajectory and Π and r solve the operator equations

$$\mathcal{A}^*\Pi + \Pi\mathcal{A} - \Pi\mathcal{B}\mathcal{R}^{-1}\mathcal{B}^*\Pi + \mathcal{C}^*\mathcal{C} = 0$$

and

$$\dot{r}(t) = -\left[\mathcal{A}^* - \Pi\mathcal{B}\mathcal{R}^{-1}\mathcal{B}^*\right] r(t) + \mathcal{C}^*\mathcal{C}\hat{z}(t)$$
$$r(0) = r(\tau) \, ,$$

respectively. A comparison of the optimal control (7.27) with that specified in Theorem 7.16 for the system with persistent exogenous input reveals a similar form with the only difference occurring in the forcing terms in the tracking equations.

The control defined in (7.27) was derived for the system (7.26) with periodic boundary conditions. For many physical systems starting from rest, however, it is more realistic to consider the minimization of (7.25) subject to the initial value problem

$$\dot{z}(t) = \mathcal{A}z(t) + \mathcal{B}u(t) \quad , \quad t \ge 0$$
$$z(0) = 0 \, . \qquad\qquad (7.28)$$

As discussed in [23, 24], an effective control strategy for (7.28) is obtained through the application of the control (7.27) which yields the closed loop system

$$\dot{z}(t) = \left[\mathcal{A} - \mathcal{B}\mathcal{R}^{-1}\mathcal{B}^*\Pi\right] z(t) - \mathcal{B}\mathcal{R}^{-1}\mathcal{B}^* r(t) \quad , \quad t \geq 0$$

$$z(0) = 0 .$$

This strategy is asymptotically stable in the sense that the state settles into the optimal state $\bar{z}(t)$ in a stable manner. Moreover, the control (7.27) is optimal for (7.28) in that it minimizes the penalty functional

$$\tilde{J}(u, z_0) = \limsup_{T \to \infty} \frac{1}{T} \int_0^T \left\{ |\mathcal{C}[z(t) + \hat{z}(t)]|_Y^2 + |\mathcal{R}^{1/2} u(t)|_U^2 \right\} dt$$

subject to (7.28).

Convergence analysis for this tracking problem can be found in [24]. It is demonstrated there that if the approximation method is chosen so that the approximate control systems are uniformly stabilizable and detectable, the convergence of the Riccati solutions, optimal trajectories, tracking solutions and controls is obtained.

7.4.3 Smart Material Applications with Unbounded Operators

In the majority of applications involving smart material structures, the exogenous disturbance is naturally formulated as an input force to the modeling equations rather than as a trajectory to be tracked by one of the state variables. This is due to the coupling and interaction between components that occurs in many such systems. Hence the control formulation discussed in Section 7.4.1 will typically be employed with the optimal control specified by Theorem 7.16. If the application does warrant the use of a tracking formulation, a nearly identical control law is used with the tracking equation modified slightly to incorporate the difference in formulations.

The current theory for problems modeled by infinite dimensional systems with exogenous forces is limited to bounded control input and observation operators. Convergence analysis for the tracking problem has been developed but it is still limited to the bounded operator case. On the other hand, the use of discrete actuators and sensors (e.g., piezoceramic or electrostrictive patches) in smart material applications typically leads to unbounded control input and observation operators. As discussed in Chapter 8, extensive numerical investigations, coupled in some cases with experimental validation, have demonstrated the validity of extending the control methods to the unbounded operator situation. However, a comprehensive theory, including convergence analysis, for the unbounded operator case for systems with persistent excitation is still being developed.

7.5   Control with Partial State Measurements

The theory in the previous sections of this chapter was developed under the assumption that full state information is available for calculation of the feedback control $u(t)$. While such a theory is a necessary step in the development of an infinite dimensional control methodology, measurement of the full state typically is not feasible in current applications. Instead, partial state measurements are obtained using physical devices such as accelerometers, laser vibrometers, proximity sensors, piezoelectric sensors, strain gauges, or microphones, and from these observations, the state must be reconstructed. Based on the reconstructed state variable, a dynamic output feedback law is developed for the purpose of controlling the system dynamics. Although the original systems are infinite dimensional, it is necessary for applications to consider a sequence of finite dimensional observers and compensators of sufficiently low order so as to permit real time implementation. Moreover, these finite dimensional observers and compensators must be constructed in a manner which preserves dynamical and asymptotic stability properties of the original infinite dimensional system so as to provide effective controllers when fed back into that system.

Dynamic compensators for finite dimensional systems have been studied extensively with much of the initial work being attributed to Luenberger [140]. These concepts have also been extended to infinite dimensional systems with bounded and unbounded input/output operators. A comprehensive discussion of the linear quadratic Gaussian (LQG) problem in the context of flexible structures can be found in [95]. There it is demonstrated that the problem of designing a compensator can be separated into an LQR control problem and the dual state estimation problem. Issues associated with the approximation of the optimal infinite dimensional compensator by a sequence of finite dimensional compensators are discussed and conclusions are illustrated through a numerical example involving a flexible beam. The input and output operators throughout the analysis in [95] are assumed to be bounded.

As will be noted in ensuing discussions, the direct approximation of an infinite dimensional system by traditional approximation methods leads to compensators whose order is dictated by the discretization index. For models having two and three spatial variables, such compensators are often too large (e.g., on the order of 1000-10,000) to implement in real time. This necessitates the use of model/controller reduction techniques or direct construction of low-order compensators. Direct construction of such low (and fixed) order compensators for infinite dimensional systems with bounded input and output operators was considered in [165] and discussed and refined in [54]. The reduction of the compensator order was also considered in [109]. In this latter work, a sesquilinear form representation of the Cauchy system, similar to that described in Section 7.2, was used as a basis for developing Galerkin approximations of the compensator. For the analysis in [109], the system was assumed to satisfy a "spectrum determined growth condition" and the input/output operators were taken to be bounded.

As in the theory regarding the LQR control problem, analysis regarding the unbounded input/output case is less complete than its bounded operator counterpart. The existence and design of finite dimensional compensators for parabolic and retarded functional differential systems with unbounded input operators is considered in [73, 74]. Because this analysis is based upon eigen-representations of the systems, it can be difficult to directly implement in complex systems. This is due to the necessity of numerically approximating eigenvalues and eigenfunctions. A Galerkin approach was taken in [58] where the problem of constructing fixed (low) order compensators for stabilizing Burgers' equation was considered. In this work, the optimal projection method of [54] involving a pair of modified Riccati equations as well as a pair of modified Lyapunov equations was extended to the unbounded operator regime. While a general theoretical analysis was not included, numerical examples clearly demonstrate the feasibility of applying fixed (low) order compensators to nonlinear infinite dimensional systems having unbounded input/output operators. Finally, a general theoretical framework for the compensator problem with unbounded input/output operators can be found in [127, 128]. The analysis in [127] concerns hyperbolic-type systems whereas systems generating analytic semigroups are considered in [128].

The analysis in [127, 128] depends on a purely operator-based framework as compared with the sesquilinear functional framework employed in the analysis discussed in Sections 7.2 and 7.3. The compensator analysis in [109] more closely resembles that used in the control discussion of this chapter but, as pointed out previously, that analysis has not yet been extended to the unbounded operator regime. We outline here the compensator analysis for unbounded operators that is discussed in [127, 128]; this will provide a basis for comparison with the sesquilinear framework in earlier sections and yields a framework in which to consider the finite dimensional problem as discussed in Chapter 8. This discussion is not intended to be exhaustive and is meant to give the reader an overview of the framework and appreciation for the issues. Details can be found in the references.

7.5.1   The Operator $\mathcal{A}$ Generates a C_0 Semigroup; $g \equiv 0$

We consider here the system

$$\dot{z}(t) = \mathcal{A}z(t) + \mathcal{B}u(t) \quad \text{on } (\text{dom}\mathcal{A}^*)^*$$
$$z(0) = z_0 \in \mathcal{H} \tag{7.29}$$
$$z_{ob}(t) = \mathcal{C}z(t)$$

in which $\mathcal{A}$ is assumed to generate a C_0 semigroup which is not necessarily analytic. Such a system would arise when modeling a weakly damped structural system or a structural system with a coupled hyperbolic component (e.g., a structural acoustic system). It is also assumed that $\mathcal{B}$ is unbounded and $\mathcal{C}$

is bounded with $\mathcal{B} : U \rightarrow (\text{dom}\mathcal{A}^*)^*$ and $\mathcal{C} \in \mathcal{L}(\mathcal{H}, Y)$. We point out that the treatment of $\mathcal{B}u$ in $(\text{dom}\mathcal{A}^*)^*$ was considered in the analysis of the general system in Chapter 4 where $\mathcal{A}$ did not generate an analytic semigroup. This should be compared with the consideration of $\mathcal{B} : U \rightarrow \mathcal{V}^* \subset (\text{dom}\mathcal{A}^*)^*$ in Section 7.2 where it was assumed that $\mathcal{A}$ generated an analytic semigroup.

To guarantee that the system (7.29) is well-posed on $\mathcal{H}$, it is also assumed that $\mathcal{B}$ satisfies the regularity constraint

(C9) $\displaystyle \int_0^T \left| \mathcal{B}^* e^{\mathcal{A}^* t} z \right|_U^2 dt \leq M_T |z|_{\mathcal{H}}^2$, $z \in \text{dom}\mathcal{A}^*$

where

$$\langle \mathcal{B}u, \Psi \rangle_{\mathcal{H}} = \langle u, \mathcal{B}^* \Psi \rangle_U \quad , \quad u \in U \ , \ \Psi \in \text{dom}\mathcal{B}^* \subset \text{dom}\mathcal{A}^* \ .$$

We note that hypothesis (C9) is essentially (H1*) in [157] and is equivalent to the statement dual to (C1) in Section 7.2.

As discussed in [127], if there exist operators $\mathcal{K} \in \mathcal{L}(\mathcal{H}, U)$ and $\mathcal{F} \in \mathcal{L}(Y, \mathcal{H})$ such that $\mathcal{A} - \mathcal{B}\mathcal{K}$ and $\mathcal{A} - \mathcal{F}\mathcal{C}$ generate exponentially stable semigroups, then the *infinite dimensional* estimator or observer

$$\begin{aligned} \dot{z}_c(t) &= \mathcal{A}z_c(t) + \mathcal{B}u(t) + \mathcal{F}[z_{ob}(t) - \mathcal{C}z_c(t)] \\ z_c(0) &= z_{c0} \end{aligned} \tag{7.30}$$

with the dynamic feedback law

$$u(t) = -\mathcal{K}z_c(t) \tag{7.31}$$

exponentially stabilizes the original system (7.29) (the combination of the estimator (7.30) and the feedback law (7.31) is sometimes referred to as the compensator [95, 127]). This statement is expressed more precisely in the following theorem.

Theorem 7.17 *Let $\tilde{\mathcal{H}} \equiv \mathcal{H} \times \mathcal{H}$ and define $\mathcal{A}_c : \text{dom}\mathcal{A}_c \subset \tilde{\mathcal{H}} \rightarrow \tilde{\mathcal{H}}$ by*

$$\mathcal{A}_c \equiv \begin{bmatrix} \mathcal{A} & -\mathcal{B}\mathcal{K} \\ \mathcal{F}\mathcal{C} & \mathcal{A} - \mathcal{B}\mathcal{K} - \mathcal{F}\mathcal{C} \end{bmatrix}$$

$$\text{dom}\mathcal{A}_c \equiv \big\{ (\Phi, \Psi) \in \tilde{\mathcal{H}} \,|\, \mathcal{A}\Phi - \mathcal{B}\mathcal{K}\Psi \in \mathcal{H} \,, \tag{7.32}$$

$$\mathcal{F}\mathcal{C}\Phi + (\mathcal{A} - \mathcal{B}\mathcal{K} - \mathcal{F}\mathcal{C})\Psi \in \mathcal{H} \big\} \ .$$

If there exist operators $\mathcal{K} \in \mathcal{L}(\mathcal{H}, U), \mathcal{F} \in \mathcal{L}(Y, \mathcal{H})$ such that $\mathcal{A} - \mathcal{B}\mathcal{K}$ and $\mathcal{A} - \mathcal{F}\mathcal{C}$ generate exponentially stable semigroups, then the semigroup generated by $\mathcal{A}_c$ is exponentially stable. Moreover, the reconstruction error satisfies the bound

$$|z(t) - z_c(t)|_{\mathcal{H}} \leq M e^{-\omega t} |z_0 - z_{c0}|_{\mathcal{H}}$$

where M and ω are positive constants.

To construct feedback gains $\mathcal{K}$ in terms of the solution to an LQR problem, we consider minimization of the functional

$$J(u, z_0) = \int_0^\infty \{\langle \mathcal{Q}z(t), z(t)\rangle_{\mathcal{H}} + \langle \mathcal{R}u(t), u(t)\rangle_U\}\, dt \qquad (7.33)$$

subject to the evolution equation (7.29). The nonnegative, self-adjoint operator $\mathcal{Q} \equiv \mathcal{D}^*\mathcal{D} \in \mathcal{L}(\mathcal{H})$ can be specified to weight state components, window certain frequencies, et cetera. The control weight $\mathcal{R} = (\mathcal{R}^{1/2})^2 \in \mathcal{L}(U)$ is again assumed to be positive and self-adjoint. Under the assumption that $(\mathcal{A}, \mathcal{B})$ is stabilizable and $(\mathcal{A}, \mathcal{D})$ is detectable, the optimal solution to (7.33) is given by $u(t) = -\mathcal{K}z(t)$ where $\mathcal{K} = \mathcal{R}^{-1}\mathcal{B}^*\Pi$ and Π solves the algebraic Riccati equation

$$\left(\mathcal{A}^*\Pi + \Pi\mathcal{A} + \mathcal{Q} - \Pi\mathcal{B}\mathcal{R}^{-1}\mathcal{B}^*\Pi\right)z = 0\,, \quad z \in \mathrm{dom}\mathcal{A}\,.$$

Similar analysis for the dual problem yields the expression $\mathcal{F} = \hat{\Pi}\mathcal{C}^*\hat{\mathcal{R}}^{-1}$ for the Kalman filter or observer gain. In this case, $\hat{\Pi}$ solves the dual Riccati equation

$$\left(\mathcal{A}\hat{\Pi} + \hat{\Pi}\mathcal{A}^* + \hat{\mathcal{Q}} - \hat{\Pi}\mathcal{C}^*\hat{\mathcal{R}}^{-1}\mathcal{C}\hat{\Pi}\right)\Psi = 0\,, \quad \Psi \in \mathrm{dom}\mathcal{A}^*$$

where $\hat{\mathcal{Q}} \equiv \hat{\mathcal{D}}\hat{\mathcal{D}}^* \in \mathcal{L}(\mathcal{H})$ and $\hat{\mathcal{R}} \equiv \hat{\mathcal{E}}\hat{\mathcal{E}}^* \in \mathcal{L}(Y)$, like $\mathcal{Q}$ and $\mathcal{R}$, are typically specified according to design criteria for the application under consideration (specific examples for the corresponding finite dimensional problem are detailed in Chapter 8). The existence of the the unique Riccati solution $\hat{\Pi}$ is guaranteed by the assumptions that $(\mathcal{A}, \mathcal{C})$ is detectable and $(\mathcal{A}, \hat{\mathcal{D}})$ is stabilizable. The four stabilizability and detectability assumptions are summarized in Condition (C10)

(C10)   (a) $(\mathcal{A}, \mathcal{B})$ and $(\mathcal{A}, \hat{\mathcal{D}})$ are stabilizable on $\mathcal{H}$

   (b) $(\mathcal{A}, \mathcal{C})$ and $(\mathcal{A}, \mathcal{D})$ are detectable on $\mathcal{H}$.

The feedback operator $\mathcal{K}$ is in general unbounded, and it is assumed at certain points in the analysis of [127] that the design operator $\mathcal{D}$ satisfies the regularity property

(C11) $\displaystyle\int_0^T \left|\mathcal{D}^*\mathcal{D}e^{\mathcal{A}t}\mathcal{B}u\right|_{\mathcal{H}}\, dt \leq M_T |u|_U$.

While this yields a bounded feedback gain $\mathcal{K}$, it can be too restrictive in certain applications, and alternative assumptions are discussed at the end of this section.

Of more practical interest in applications than the infinite dimensional compensator is the corresponding problem of constructing a sequence of finite dimensional compensators which approximate (7.30) and yield a stable control system when fed back into the infinite dimensional system. To this end, we consider a sequence of finite dimensional subspaces $\mathcal{V}^N$ with orthogonal projections P^N mapping $\mathcal{H}$ onto $\mathcal{V}^N$. We also consider finite dimensional operators

$$\mathcal{A}^N : \mathcal{V}^N \to \mathcal{V}^N \quad , \quad \mathcal{K}^N : \mathcal{V}^N \to U \quad , \quad \mathcal{B}^N : U \to \mathcal{V}^N \quad , \quad \mathcal{F}^N : Y \to \mathcal{V}^N$$

$$\mathcal{C}^N : \mathcal{V}^N \to Y \quad , \quad \mathcal{Q}^N : \mathcal{V}^N \to \mathcal{V}^N \quad , \quad \hat{\mathcal{Q}}^N : \mathcal{V}^N \to \mathcal{V}^N$$

and the discrete Riccati equations

$$\mathcal{A}^{N^*}\Pi^N + \Pi^N \mathcal{A}^N + \mathcal{Q}^N - \Pi^N \mathcal{B}^N \mathcal{R}^{-1} \mathcal{B}^{N^*}\Pi^N = 0$$

$$\mathcal{A}^N \hat{\Pi}^N + \hat{\Pi}^N \mathcal{A}^{N^*} + \hat{\mathcal{Q}}^N - \hat{\Pi}^N \mathcal{C}^{N^*} \hat{\mathcal{R}}^{-1} \mathcal{C}^N \hat{\Pi}^N = 0 \tag{7.34}$$

where $\Pi^N \in \mathcal{L}(\mathcal{V}^N), \hat{\Pi}^N \in \mathcal{L}(\mathcal{V}^N)$. Recall that if the original infinite dimensional system is defined in terms of a sesquilinear form σ, $\mathcal{A}^N$ can be obtained by restricting σ to $\mathcal{V}^N \times \mathcal{V}^N$. The approximate control operator is defined through the duality relation (7.12) while $\mathcal{C}^N$ again denotes the restriction of $\mathcal{C}$ to $\mathcal{V}^N$. Finally, $\mathcal{Q}^N$ and $\hat{\mathcal{Q}}^N$ are defined in the usual manner as $\mathcal{Q}^N = P^N \mathcal{Q} P^N$ and $\hat{\mathcal{Q}}^N = P^N \hat{\mathcal{Q}} P^N$.

To guarantee that the finite dimensional compensator retains the asymptotic stability attributes of the original infinite dimensional system, the following approximation hypotheses are assumed.

(C2N)   (a) $\left| e^{\mathcal{A}^N t} \right|_{\mathcal{L}(\mathcal{H})} \leq M e^{\omega_0 t} \quad , \quad M, \omega > 0$

(b) $\int_0^T \left| \mathcal{B}^{N^*} e^{\mathcal{A}^{N^*} t} z^N \right|_U^2 dt \leq M_T \left| z^N \right|_{\mathcal{H}}^2 .$

(C3N)   Let $R_\lambda(\mathcal{A}) \equiv (\lambda I - \mathcal{A})^{-1}$ and $R_\lambda(\mathcal{A}^N) \equiv (\lambda I - \mathcal{A}^N)^{-1}$ for some $\lambda > 0$. The following convergence holds for $N \to \infty$

(a) $\left| \left[R_\lambda(\mathcal{A}^N) P^N - R_\lambda(\mathcal{A}) \right] z \right|_{\mathcal{H}} \to 0 \quad , \quad z \in H$

(b) $\left| \left[R_\lambda(\mathcal{A}) - R_\lambda(\mathcal{A}^N) \right] P^N \mathcal{B} u \right|_{\mathcal{H}} \to 0 \quad , \quad u \in U$

(c) $\left| R_\lambda(\mathcal{A}) \left[\mathcal{B}^N - \mathcal{B} \right] u \right|_{\mathcal{H}} \to 0 \quad , \quad u \in U$

(d) $\left| \left[R_\lambda(\mathcal{A}^N) - R_\lambda(\mathcal{A}) \right] \mathcal{B}^N u \right|_{\mathcal{H}} \to 0 \quad , \quad u \in U.$

(C4N)   (a) $(\mathcal{A}^N, \mathcal{B}^N)$ and $(\mathcal{A}^N, \hat{\mathcal{D}}^N)$ are uniformly stabilizable

(b) $(\mathcal{A}^N, \mathcal{C}^N)$ and $(\mathcal{A}^N, \mathcal{D}^N)$ are uniformly detectable.

(C5N) $\int_0^T \left| \mathcal{D}^{N^*} \mathcal{D}^N e^{\mathcal{A}^N t} \mathcal{B}^N u \right|_{\mathcal{H}} dt \leq M_T |u|_U .$

We point out that alternative, and in some cases equivalent, approximation hypotheses can be used to obtain the desired results. For example, strong semigroup convergence yields the resolvent condition (C3N)(a) in most approximation methods. The given hypotheses yield the desired convergence results, however, and illustrate issues which must be considered when designing finite dimensional compensators for infinite dimensional systems.

The following theorem, compiled from results in [127], summarizes the stability and convergence properties of a finite dimensional compensator for the infinite dimensional system (7.29).

Theorem 7.18 *Let* $\mathcal{A}_c : \mathrm{dom}\,\mathcal{A}_c \subset \mathcal{H} \times \mathcal{H} \to \mathcal{H} \times \mathcal{H}$ *be given by (7.32) and define* $\mathcal{A}_c^N : \mathcal{H} \times \mathcal{V}^N \to \mathcal{H} \times \mathcal{V}^N$ *by*

$$\mathcal{A}_c^N \equiv \begin{bmatrix} \mathcal{A} & -\mathcal{B}\mathcal{K}^N \\ \mathcal{F}^N \mathcal{C} & \mathcal{A}^N - \mathcal{B}^N \mathcal{K}^N - \mathcal{F}^N \mathcal{C}^N \end{bmatrix} .$$

We then consider the control systems

$$\frac{d}{dt} \begin{bmatrix} z(t) \\ z_c(t) \end{bmatrix} = \mathcal{A}_c \begin{bmatrix} z(t) \\ z_c(t) \end{bmatrix}$$

and

$$\frac{d}{dt} \begin{bmatrix} z(t) \\ z_c^N(t) \end{bmatrix} = \mathcal{A}_c^N \begin{bmatrix} z(t) \\ z_c^N(t) \end{bmatrix} .$$

Suppose that the regularity hypotheses (C9), (C11), stabilizability and detectability hypothesis (C10) and approximation hypotheses (C2N)–(C5N) are satisfied. Moreover, assume that the observation operator $\mathcal{C}$ *is compact.*

Then there exist constants M *and* $\omega < 0$ *independent of* N *such that*

$$\left| e^{\mathcal{A}_c^N t} \right|_{\mathcal{L}(\mathcal{H})} \le M e^{\omega t}$$

with $\mathcal{K}^N, \mathcal{F}^N$ *given by* $\mathcal{K}^N = \mathcal{R}^{-1} \mathcal{B}^{N*} \Pi^N$ *and* $\mathcal{F}^N = \hat{\Pi}^N \mathcal{C}^{N*} \hat{\mathcal{R}}^{-1}$ *where* $\Pi^N, \hat{\Pi}^N$ *satisfy (7.34). One also obtains the semigroup and controller convergence*

$$\sup_{t \ge 0} \left| \left[e^{\mathcal{A}_c^N t} P^N - e^{\mathcal{A}_c t} \right] z \right|_{\mathcal{H}} \to 0 \quad , \quad z \in \mathcal{H}$$

$$\left| u^N - u \right|_{L^2(0,\infty;U)} \to 0$$

where u *and* u^N *are given by*

$$u(t) = -\mathcal{R}^{-1} \mathcal{B}^* \Pi z_c(t)$$

$$u^N(t) = -\mathcal{R}^{-1} \mathcal{B}^{N*} \Pi^N z_c^N(t) .$$

The reader is referred to [127] for proofs of the assertions in this theorem.

For systems with no exogenous force g, Theorem 7.18 provides a formal framework that can be used for constructing finite dimensional compensators for some smart structure applications. While verification of the theorem hypotheses is illustrated in [127] for various models, it is also acknowledged that some of the hypotheses may be overly restrictive for certain applications. For example, the regularity assumption (C11) requires that $Q = \mathcal{D}^*\mathcal{D}$ be sufficiently smoothing so as to yield a continuous operator $\mathcal{D}^*\mathcal{D}e^{\mathcal{A}t}\mathcal{B}$ whereas $\mathcal{B}$ itself is often discontinuous. As noted in Chapter 8, a natural choice for $Q = \mathcal{D}^*\mathcal{D}$ in many applications is the mass matrix (identity with respect to the $\mathcal{H}$-inner product) in which case, this condition may not be satisfied. In Theorem 2.4 of [127], it is noted that the regularity assumptions (C11) and (C5N) can be replaced by the assumptions that $P^N\mathcal{A} = \mathcal{A}P^N$ and $\mathcal{B}^N \equiv P^N\mathcal{B}$. These latter conditions are satisfied in modal approximations due to the truncation nature of the expansions. For more general Galerkin methods, however, the commutativity of $\mathcal{A}$ cannot be guaranteed, and another mode of analysis must be employed.

As discussed in [128], several of the conditions can be relaxed if the model is parabolic, or more generally, $\mathcal{A}$ generates an analytic semigroup. This category includes a large number of smart material applications since purely structural models which include strong (Kelvin-Voigt) damping generally have operators which generate analytic semigroups. The analysis of [128] also includes a case of unbounded observation operators which makes it applicable in a variety of applications involving spatially discrete measurement devices. As noted both in [128] and [130], the conditions of uniform stabilizability and detectability can be replaced by a uniform analyticity condition when $\mathcal{A}$ generates an analytic semigroup. This can provide a condition which is more readily verified in some applications.

Finally, the analysis outlined here yields finite dimensional compensators whose order is dictated by the approximation index. For applications involving models with two and three spatial dimensions, the high order of such compensators may prevent real-time implementation (e.g., compensators designed in this manner for 3-D structural acoustic models may be on the order of 1000-10,000). For such cases, ideas such as those discussed in [58, 109] must be melded with unbounded operator analysis to yield applicable fixed (and low) order compensators.

7.5.2   Compensators for Systems with Exogenous Inputs

In Section 7.4, infinite dimensional control systems with exogenous inputs g were discussed. It was noted there that in many such systems, the hypothesis that g is periodic in time is appropriate due to the oscillatory nature of components contributing to the persistent disturbance. The modeling of such systems in smart material applications leads to abstract systems of the form

$$\dot{z}(t) = \mathcal{A}z(t) + \mathcal{B}u(t) + g(t)$$
$$z(0) = z(\tau) \qquad\qquad (7.35)$$
$$z_{ob}(t) = \mathcal{C}z(t)$$

where $\mathcal{B}$ and $\mathcal{C}$ are in general unbounded. Issues regarding the development of LQR feedback laws for the bounded operator case were discussed in Section 7.4 while the design of observers and compensators for the case $g \equiv 0$ were outlined in Section 7.5.1. A general theory regarding the design of finite dimensional compensators for the infinite dimensional system (7.35) is still in the development stage for both the bounded and unbounded input/output cases.

In the next chapter, a typical Galerkin approximation method is used to discretize a system of the form (7.35) which models a structure with piezoceramic actuators and discrete sensors (accelerometers). Due to the nature of the actuators and sensors, both $\mathcal{B}$ and $\mathcal{C}$ are unbounded in the model. Existing finite dimensional theory is then used to construct a compensator. Numerical results are provided which demonstrate the effectiveness of the compensator when fed back into the discretized model while experimental results verify that the compensator yields controls which effectively attenuate structural vibrations when fed back into the experimental system. This demonstrates the feasibility and validity of using Galerkin techniques to construct effective finite dimensional compensators for infinite dimensional control systems of the form (7.35) with unbounded input and output operators even though a complete theoretical foundation has yet to be developed.

Chapter 8

Implementation of Finite-Dimensional Compensators

The discussion in Chapter 7 focused on the problem of designing control laws for infinite dimensional systems which arise when a system of partial differential equations (PDE) is used to model a physical application. The applications of interest to us consist of smart material systems which, due to the spatially discrete nature of the actuators (e.g., piezoceramic or electrostrictive patches) and sensors (e.g., accelerometers or piezoelectric films), lead to unbounded input and output operators. That chapter concluded with a discussion concerning the design of finite dimensional dynamic output feedback laws for such infinite dimensional systems with emphasis placed on the design of compensators that can be implemented in physical applications.

In this chapter, we focus on implementation issues regarding the design of compensators for smart material applications. While the compensator discussion in Chapter 7 focused on the LQG problem for systems with no exogenous input, we include here the design of H^∞/MinMax compensators for these systems as well as LQG and MinMax compensators for systems driven by a persistent disturbance. Summaries for the various control systems, state estimators or observers, and compensators are given in Section 8.1. Throughout this section, the temporally continuous nature of the systems is retained with system dynamics approximated by matrix differential equations. Discrete approximations of the control problem and compensators are considered in Section 8.2 with the emphasis placed on discretization techniques which facilitate implementation. In the following section, experimental results concerning the control of circular plate vibrations are presented to demonstrate the effectiveness of the methods in physical application and illustrate issues which arise when implementing PDE-based compensators. A PDE model for the plate system is developed and a Galerkin approximation method consistent with the criteria discussed in Chapter 7 is outlined. Specific choices for design parameters in the compensator are discussed and illustrated through a set of experimental examples. These experimental results demonstrate the feasibility and effectiveness of the PDE-based compensator for attenuating transient and persistent periodic steady state plate vibrations. They also indicate directions for future research. The issues and various open questions regarding the implementation of PDE-based compensators are further summarized in a concluding section.

8.1   Continuous-Time Compensator Problem

Throughout Chapter 7, superscript N's were used to denote the level of discretization as is standard in Galerkin approximation theory for infinite dimensional systems. In finite dimensional control theory, however, the level of discretization is typically fixed and these superscript N's are typically omitted to simplify notation. We shall follow this convention throughout this chapter so as to remain consistent with standard control notation.

All operators, states and controls in this chapter are assumed to be finite dimensional; hence z, which denoted the infinite dimensional state in Chapter 7, represents here a finite dimensional state. Roman fonts are used throughout to represent the matrix approximations to the general operators of Chapter 7 which were set in calligraphic fonts. Hence the operator A in this chapter denotes the matrix representation for the finite dimensional operator $\mathcal{A}^N$ which approximates $\mathcal{A}$. Similar representations are used for the other operators. Finally, Euclidean inner products and norms are used exclusively throughout this chapter with the dimension to be understood from the context.

8.1.1   LQG Control Law – No Exogenous Disturbance

In Section 7.5.1, the problem of constructing a compensator for the infinite dimensional control system (7.29) was considered. It was noted there that in many smart material applications, the control operator $\mathcal{B}$ and observation operator $\mathcal{C}$ are unbounded due to the spatially discrete nature of the actuators and sensors. Feedback gains and controls u were determined by minimizing the quadratic functional (7.33) subject to (7.29). The operators $Q \equiv \mathcal{D}^*\mathcal{D}$ and $\mathcal{R} \equiv \mathcal{E}^*\mathcal{E}$ in that functional were taken to be design parameters which, in the applications described here, weight various components of the state and control, respectively.

Discretization of the system yields the following optimization problem: minimize

$$
\begin{aligned}
J(u, z_0) &= \int_0^\infty |z_{out}(t)|^2 \, dt \\
&= \int_0^\infty \left\{ \langle Qz(t), z(t) \rangle + \langle Ru(t), u(t) \rangle \right\} dt
\end{aligned}
\tag{8.1}
$$

subject to the N-dimensional system

$$\dot{z}(t) = Az(t) + Bu(t) \quad , \quad z(0) = z_0$$

$$z_{ob}(t) = Cz(t) \tag{8.2}$$

$$z_{out}(t) = Dz(t) + Eu(t) .$$

Here z_{ob} denotes observations in $\mathbb{R}^p$ and C is a $p \times N$ observation matrix whose structure is determined by the manner and number of observations being used (the specific C matrix used in the plate experiments is described in Section 8.3.3). Moreover, $z_{out} \in \mathbb{R}^r$ denotes the performance output obtained under the assumption that E and D are time invariant matrices satisfying $D^T E = 0$. The $N \times N$ matrix $Q \equiv D^T D$ can be chosen to satisfy various design criteria including frequency windowing, the weighting of various state components, or minimization of certain energy measures. For the applications described in Section 8.3.3, Q was chosen using energy considerations and the $s \times s$ matrix $R \equiv E^T E$ was used to weight voltages to various patches or patch pairs.

In the ideal case that $p = N$ and C is the identity, the optimal control u minimizing (8.2) is obtained from standard finite dimensional LQR theory. It is most usual in applications, however that one has only partial state information, $p < N$, and the state must be estimated or reconstructed from observations before a control u can be determined. In designing a state estimator, the asymptotic and dynamic properties of the original system must be retained in a manner which guarantees the convergence of the state estimates to the finite and infinite dimensional states as $t \to \infty$.

As discussed in Section 7.5.1, such a state estimator is given by

$$\dot{z}_c(t) = Az_c(t) + Bu(t) + F\left[z_{ob}(t) - Cz_c(t)\right]$$

$$z_c(0) = z_{c0}$$

(the reader is referred to [140] for a complete discussion of state estimators for finite dimensional systems). When coupled with the dynamic feedback law

$$u(t) = -Kz_c(t) ,$$

this yields a compensator for determining controls to be fed back into the original infinite dimensional system. Under the stabilizability and detectability hypotheses (C10) and (C4N) of Section 7.5.1, the optimal feedback (regulator) gain K and observer (Kalman filter) gain F are given by

$$K = R^{-1}B^T \Pi$$

$$F = \hat{\Pi}C^T \hat{R}^{-1} \tag{8.3}$$

where Π and $\hat{\Pi} \in \mathbb{R}^N$ are unique nonnegative-definite solutions to the following feedback and observer algebraic Riccati equations

$$\Pi A + A^T \Pi - \Pi B R^{-1} B^T \Pi + Q = 0$$

$$\hat{\Pi}A^T + A\hat{\Pi} - \hat{\Pi}C^T \hat{R}^{-1} C\hat{\Pi} + \hat{Q} = 0 , \tag{8.4}$$

respectively. As was the case with the matrices Q and R, the matrices $\hat{Q} \equiv \hat{D}^T\hat{D}$ and $\hat{R} \equiv \hat{E}^T\hat{E}$ are design criteria for the specific control application under consideration (recall that $\hat{D}$ and $\hat{E}$ are duality variables - see Chapter 7, and (C4N) in particular). Specific choices used in the plate experiments are summarized in Section 8.3.3.

In terms of the component matrices and control voltage, the compensator can be expressed as

$$\dot{z}_c(t) = \left[A - BR^{-1}B^T\Pi\right] z_c(t) + \hat{\Pi}C^T\hat{R}^{-1}C\left[z_{ob}(t) - z_c(t)\right] \ .$$

The observed system, performance index, state estimator and compensator for the system are summarized in Algorithm 8.1.

The state estimator and output feedback control law described here are full-order in the sense that they have the same order (N) as the finite dimensional control system (8.2). As demonstrated in later sections, such a compensator can be used to control plate vibrations when the system matrices $A, B, C, Q, R, \hat{Q}, \hat{R}$, the Riccati solutions $\Pi, \hat{\Pi}$ and the gains K, F are computed offline and loaded as datafiles for online computations. For the plate application, the order of the state estimator is less than 100, and the online approximation of the state estimator and calculation of the output feedback control can be performed in real time.

For models of complex two and three dimensional systems, however, the order can be sufficiently large (over 1000 variables) so as to prohibit direct implementation, even when component matrices and Riccati solutions are computed offline. In such cases, model/controller reduction techniques or direct construction of reduced (low) order estimators and compensators of the type discussed in [54, 58] must be employed in order to attain real-time implementation.

8.1.2   LQG Control Law – Periodic Exogenous Disturbance

The finite dimensional problem described in the last section approximates the dynamics of a system starting from an initial, nonzero state and evolving with no exogenous force throughout the time interval of interest. The output feedback is directly proportional to the state (or the estimated state) and must control only transient dynamics. In many applications, however, a persistent exogenous force drives the system and both transient and steady state dynamics must be attenuated by the controller. Throughout this discussion, the exogenous force is taken to be periodic in time which is a natural assumption for systems in which disturbances are created by oscillating components or periodic sources. A discussion of the corresponding LQR problem for infinite dimensional systems with bounded input and output operators can be found in

---

(1) Observed System

$$\dot{z}(t) = Az(t) + Bu(t) \qquad\qquad z \in \mathbb{R}^N, u \in \mathbb{R}^s$$
$$z_{ob}(t) = Cz(t) \qquad\qquad\qquad z_{ob} \in \mathbb{R}^p$$
$$z_{out}(t) = Dz(t) + Eu(t) \qquad\quad z_{out} \in \mathbb{R}^r \ , \ D^T E = 0$$

---

(2) Performance Index

$$J(u, z_0) = \int_0^\infty |z_{out}(t)|^2 \, dt$$
$$= \int_0^\infty \{\langle Qz(t), z(t)\rangle + \langle Ru(t), u(t)\rangle\} \, dt \ \ , \ \ Q = D^T D \ , \ R = E^T E$$

---

(3) Compensator

$$\dot{z}_c(t) = A_c z_c(t) + F z_{ob}(t)$$
$$A_c = A - FC - BK$$

$$K = R^{-1} B^T \Pi$$
$$F = \hat{\Pi} C^T \hat{R}^{-1}$$

$$\Pi A + A^T \Pi - \Pi B R^{-1} B^T \Pi + Q = 0$$
$$\hat{\Pi} A^T + A\hat{\Pi} - \hat{\Pi} C^T \hat{R}^{-1} C \hat{\Pi} + \hat{Q} = 0 \ , \qquad \hat{Q} = \hat{D}\hat{D}^T \ , \ \hat{R} = \hat{E}\hat{E}^T$$

Feedback Control
$$u(t) = -K z_c(t)$$

Algorithm 8.1: LQG control law for systems with no exogenous forces.

Section 7.4. As noted there, knowledge of the periodic driving force can be incorporated in the control through an additional nonhomogeneous term in the optimal trajectory in combination with a tracking component. The design of a compensator for the corresponding finite dimensional problem is considered here.

Discretization of the infinite dimensional system (7.20), with bounded input $\mathcal{B}$ or (7.35), where $\mathcal{B}$ is assumed unbounded, yields the matrix equation

$$\dot{z}(t) = Az(t) + Bu(t) + g(t) \ \ , \quad z(0) = z(\tau)$$
$$z_{ob}(t) = Cz(t) \tag{8.5}$$

where $g(t) \in \mathbb{R}^N$ is assumed periodic with period τ. In the case of multiple frequencies, it is assumed that τ can be chosen so as to be commensurate with all frequencies present in the driving signal. The control in this case is determined by minimizing the quadratic functional

$$J_\tau(u, z_0) = \frac{1}{2} \int_0^\tau \{\langle Qz(t), z(t) \rangle + \langle Ru(t), u(t) \rangle\} \, dt$$

subject to (8.5).

With K and F defined in (8.3), the reconstructed state in this case satisfies the system

$$\dot{z}_c(t) = Az_c(t) + Bu(t) + F\left[z_{ob}(t) - Cz_c(t)\right] + g(t)$$
$$z_c(0) = z_c(\tau) \tag{8.6}$$

with the optimal voltage given by

$$u(t) = -Kz_c(t) + R^{-1}B^T r(t) . \tag{8.7}$$

Here r is a tracking variable defined by the system

$$\dot{r}(t) = -[A - BK]^T r(t) + \Pi g(t)$$
$$r(0) = r(\tau) , \tag{8.8}$$

where Π solves the first of the algebraic Riccati equations (8.4). The continuous-time system and compensator when g is periodic are summarized in Algorithm 8.2.

As was the case in the corresponding infinite dimensional LQR problem discussed in Section 7.4.1, the control in systems driven by an exogenous force contains two components. Transient information is provided by the estimated state while the influence due to the periodic force is manifested through both the estimated state and the tracking variable.

We note that the compensator described here is again full-order. For models of complex two and three dimensional physical systems, such a compensator may not be feasible for implementation due to the large order of the state estimator and tracking equations. Again, for such applications, model/controller reduction techniques or direct construction of low-order compensators must be considered.

---

(1) Observed System

$$\dot{z}(t) = Az(t) + Bu(t) + g(t) \qquad\qquad z \in \mathbb{R}^N, u \in \mathbb{R}^s$$

$$z(0) = z(\tau)$$

$$z_{ob}(t) = Cz(t) \qquad\qquad z_{ob} \in \mathbb{R}^p$$

$$z_{out}(t) = Dz(t) + Eu(t) \qquad\qquad z_{out} \in \mathbb{R}^r \ , \ D^T E = 0$$

---

(2) Performance Index

$$J_\tau(u, z_0) = \frac{1}{2} \int_0^\tau |z_{out}(t)|^2 \, dt$$

$$= \frac{1}{2} \int_0^\tau \{\langle Qz(t), z(t)\rangle + \langle Ru(t), u(t)\rangle\} \, dt \ , \qquad Q = D^T D \ , \ R = E^T E$$

---

(3) Compensator

$$\dot{z}_c(t) = A_c z_c(t) + F z_{ob}(t) + B_r r(t) + g(t)$$

$$A_c = A - FC - BK$$

$$B_r = BR^{-1}B^T$$

$$K = R^{-1}B^T \Pi$$

$$F = \hat{\Pi}C^T \hat{R}^{-1}$$

$$\Pi A + A^T \Pi - \Pi B R^{-1} B^T \Pi + Q = 0$$

$$\hat{\Pi}A^T + A\hat{\Pi} - \hat{\Pi}C^T \hat{R}^{-1} C\hat{\Pi} + \hat{Q} = 0 \ , \qquad \hat{Q} = \hat{D}\hat{D}^T \ , \ \hat{R} = \hat{E}\hat{E}^T$$

Tracking Equation

$$\dot{r}(t) = A_{tr}r(t) + \Pi g(t)$$

$$A_{tr} = -[A - BK]^T$$

Feedback Control

$$u(t) = -Kz_c(t) + R^{-1}B^T r(t)$$

Algorithm 8.2: LQR Control law for systems with periodic exogenous forces.

8.1.3 H^∞/MinMax Compensators

In the previous sections, LQG theory was invoked to yield compensators for a system with no persistent disturbance as well as the corresponding system driven by a periodic exogenous force. While these compensators are constructed to stabilize the underlying infinite dimensional systems, they may lack robustness with respect to certain state and measurement uncertainties and disturbance. To obtain additional robustness with regard to certain types of modeling errors and system or observation noise, an H^∞/MinMax compensator of the type discussed in [51, 117] may be considered.

To incorporate such uncertainties into the finite dimensional model, we take $v(t) \in \mathbb{R}^q$ and let $\hat{D}v$ and $\hat{E}v$ denote input and output disturbances (the reader is cautioned not to confuse this use of the symbol v with the circumferential displacements discussed in previous chapters). The system with no exogenous force is then

$$\dot{z}(t) = Az(t) + Bu(t) + \hat{D}v(t) \quad , \quad z(t) = z_0$$

$$z_{ob}(t) = Cz(t) + \hat{E}v(t) \tag{8.9}$$

$$z_{out}(t) = Dz(t) + Eu(t)$$

(compare with the infinite dimensional system (7.29) or the finite dimensional system (8.1)). To simplify the discussion, it is assumed that the input and output disturbances are independent and hence $\hat{D}\hat{E}^T = 0$. This is a matter of convenience and, as discussed in [117], only slight modifications are necessary to extend the theory to the dependent case.

The disturbance is also incorporated into the quadratic penalty functional

$$
\begin{aligned}
J_\gamma(u, v, z_0) &= \int_0^\infty \left\{ |z_{out}(t)|^2 - \gamma^2 |v(t)|^2 \right\} dt \\
&= \int_0^\infty \left\{ \langle Qz(t), z(t) \rangle + \langle Ru(t), u(t) \rangle - \gamma^2 \langle v(t), v(t) \rangle \right\} dt .
\end{aligned}
$$

The associated Min/Max optimization problem consists of finding a control $\bar{u} \in U \equiv L^2(0, \infty; \mathbb{R}^s)$ and disturbance $\bar{v} \in W \equiv L^2(0, \infty; \mathbb{R}^q)$ such that

$$\bar{J}_\gamma = \inf_{u \in U} \sup_{v \in W} J_\gamma(u, v, z_0) = J_\gamma(\bar{u}, \bar{v}, z_0) .$$

This is motivated by game theoretic considerations where one wishes to minimize the output performance in the presence of maximum (i.e., worst case) disturbances. As noted in [51], the results from this optimization problem yield a bound γ for the H^∞ norm of the transfer function from the disturbance $L(v)$ to the performance output $L(z_{out})$ where L denotes the Laplace transform.

For a given attenuation level γ, one obtains the existence of unique (minimal) positive definite solutions to the algebraic feedback and observer Riccati equations

$$\Pi A + A^T \Pi - \Pi \left[BR^{-1}B^T - \gamma^{-2}\hat{Q} \right] \Pi + Q = 0$$
$$\hat{\Pi}A^T + A\hat{\Pi} - \hat{\Pi} \left[C^T\hat{R}^{-1}C - \gamma^{-2}Q \right] \hat{\Pi} + \hat{Q} = 0$$

(8.10)

provided (A, B), $(A, \hat{D})$ are stabilizable and (A, C), (A, D) are detectable. Moreover, if the spectral radius ρ of $\hat{\Pi}\,\Pi$ satisfies the condition

$$\rho(\hat{\Pi}\,\Pi) < \gamma^2, \quad \text{or} \quad \Pi - \gamma^2\hat{\Pi}^{-1} < 0,$$

then there exists a unique optimal control given by

$$\bar{u}(t) = -R^{-1}B^T\Pi z_c(t) \ .$$

With A_c and F given by

$$A_c = A - BK - FC + \gamma^{-2}\hat{Q}\Pi$$
$$F = \left[I - \gamma^{-2}\hat{\Pi}\,\Pi \right]^{-1} \hat{\Pi}C^T\hat{R}^{-1} \ ,$$

the compensator, z_c satisfies the matrix differential equation

$$\dot{z}_c(t) = A_c z_c(t) + F z_{ob}(t)$$
$$z_c(0) = z_{c_0} \ .$$

(8.11)

The H^∞/MinMax compensator is summarized in Algorithm 8.3.

A comparison of the compensator equation (8.11) with that obtained using LQG theory shows that while the form of the systems are the same, the component matrices differ with the H^∞/MinMax estimator incorporating attenuation terms that are absent in the LQG estimator. Similarly, the form of the output feedback law is the same for both theories but the Riccati matrices differ due to the attenuation components in (8.10).

When implementing the method, the attenuation bound γ, Riccati solutions and component matrices can be calculated offline and then loaded as datafiles or filters for online computations. Hence the actual online controller can run at the same rate as that obtained using the LQG methodology.

The extension of the H^∞/MinMax compensator described here to systems driven by a periodic exogenous force follows in a manner analogous to that described for the LQG compensator in Section 8.1.2, and the resulting control law is summarized in Algorithm 8.4. We also refer the reader to [10] where details concerning the design of an H^∞/MinMax compensator for a structural acoustic system that is subjected to a periodic exogenous force is considered.

---

(1) Observed System

$$\dot{z}(t) = Az(t) + Bu(t) + \hat{D}v(t) \qquad z \in \mathbb{R}^N, u \in \mathbb{R}^s, v \in \mathbb{R}^q$$

$$z_{ob}(t) = Cz(t) + \hat{E}v(t) \qquad z_{ob} \in \mathbb{R}^p, \ \hat{D}\hat{E}^T = 0$$

$$z_{out}(t) = Dz(t) + Eu(t) \qquad z_{out} \in \mathbb{R}^r, \ D^T E = 0$$

$$Q = D^T D \geq 0, \ R = E^T E > 0$$

$$\hat{Q} = \hat{D}\hat{D}^T \geq 0, \ \hat{R} = \hat{E}\hat{E}^T > 0$$

---

(2) Performance Index

$$J_\gamma(u, v, z_0) = \int_0^\infty \left\{ |z_{out}(t)|^2 - \gamma^2 |v(t)|^2 \right\} dt$$

$$= \int_0^\infty \left\{ \langle Qz(t), z(t) \rangle + \langle Ru(t), u(t) \rangle - \gamma^2 \langle v(t), v(t) \rangle \right\} dt$$

---

(3) Compensator

$$\dot{z}_c(t) = A_c z_c(t) + F z_{ob}(t)$$

$$A_c = A - FC - BK + \gamma^{-2}\hat{Q}\Pi$$

$$K = R^{-1}B^T \Pi$$

$$F = \left[I - \gamma^{-2}\hat{\Pi}\,\Pi \right]^{-1} \hat{\Pi} C^T \hat{R}^{-1}$$

$$\Pi A + A^T \Pi - \Pi \left[BR^{-1}B^T - \gamma^{-2}\hat{Q} \right] \Pi + Q = 0$$

$$\hat{\Pi} A^T + A\hat{\Pi} - \hat{\Pi} \left[C^T \hat{R}^{-1} C - \gamma^{-2}Q \right] \hat{\Pi} + \hat{Q} = 0$$

Feedback Control

$$u(t) = -K z_c(t)$$

Note: $\rho(\hat{\Pi}\,\Pi) < \gamma^2$ or $\Pi - \gamma^2 \hat{\Pi}^{-1} < 0$

Algorithm 8.3: H^∞/MinMax control law for systems with no exogenous forces.

(1) Observed System

$$\dot{z}(t) = Az(t) + Bu(t) + \hat{D}v(t) + g(t) \qquad z \in \mathbb{R}^N, u \in \mathbb{R}^s, v \in \mathbb{R}^q$$
$$z(0) = z(\tau)$$

$$z_{ob}(t) = Cz(t) + \hat{E}v(t) \qquad\qquad z_{ob} \in \mathbb{R}^p \ , \ \hat{D}\hat{E}^T = 0$$
$$z_{out}(t) = Dz(t) + Eu(t) \qquad\qquad z_{out} \in \mathbb{R}^r \ , \ D^T E = 0$$
$$\qquad\qquad\qquad\qquad\qquad Q = D^T D \geq 0 \ , \ R = E^T E > 0$$
$$\qquad\qquad\qquad\qquad\qquad \hat{Q} = \hat{D}\hat{D}^T \geq 0 \ , \ \hat{R} = \hat{E}\hat{E}^T > 0$$

(2) Performance Index

$$J_\gamma(u, v, z_0) = \frac{1}{2} \int_0^\tau \left\{ |z_{out}(t)|^2 - \gamma^2 |v(t)|^2 \right\} dt$$
$$= \frac{1}{2} \int_0^\tau \left\{ \langle Qz(t), z(t) \rangle + \langle Ru(t), u(t) \rangle - \gamma^2 \langle v(t), v(t) \rangle \right\} dt$$

(3) Compensator

$$\dot{z}_c(t) = A_c z_c(t) + F z_{ob}(t) + B_r r(t) + g(t)$$
$$A_c = A - FC - BK + \gamma^{-2}\hat{Q}\Pi$$
$$B_r = BR^{-1}B^T - \gamma^{-2}\hat{Q}$$

$$K = R^{-1}B^T\Pi$$
$$F = \left[I - \gamma^{-2}\hat{\Pi}\Pi \right]^{-1} \hat{\Pi}C^T\hat{R}^{-1}$$

$$\Pi A + A^T\Pi - \Pi\left[BR^{-1}B^T - \gamma^{-2}\hat{Q} \right]\Pi + Q = 0$$
$$\hat{\Pi}A^T + A\hat{\Pi} - \hat{\Pi}\left[C^T\hat{R}^{-1}C - \gamma^{-2}Q \right]\hat{\Pi} + \hat{Q} = 0$$

Tracking Equation

$$\dot{r}(t) = A_{tr}r(t) + \Pi g(t)$$
$$A_{tr} = -\left[A - BK - \gamma^{-2}\hat{Q}\Pi \right]^T$$

Feedback Control

$$u(t) = -Kz_c(t) + R^{-1}B^T r(t)$$

Note: $\rho(\hat{\Pi}\,\Pi) < \gamma^2$ or $\Pi - \gamma^2\hat{\Pi}^{-1} < 0$

Algorithm 8.4: H^∞/MinMax control law for systems with periodic forces.

8.2   Discrete-Time Compensator Problem

The state estimators and feedback control laws summarized in the last section were derived under the assumption of continuous-time modeling and sampling of observed state and force data. Moreover, it was assumed that exact integration of the state, state estimator and tracking equations could be performed to obtain reconstructed states and controls throughout the time interval of interest. In numerical and experimental applications, however, neither the continuous sampling of data nor the exact solution of the differential equations are typically feasible. Instead, data is digitally processed and hence the observations z_{ob} and force measurements g are available only at discrete times t_j. The sample rate, and hence times t_j, is governed by the data acquisition system, the software being used, and the number of calculations required between samples.

The solutions to the modeling state equations, state estimator and tracking equations must also be numerically approximated in most applications. The state equations are typically used to simulate open loop system dynamics, and the integration can usually be performed using any sufficiently accurate ODE routine which is efficient for the system under consideration. In many cases, a variable order, variable stepsize method is preferred since the semidiscrete systems arising in structural applications are quite often stiff.

If the goal when approximating the compensator and tracking equations is solely to perform simulations, this can be easily accomplished using the same ODE solver used to integrate the state equation (indeed, the state and estimator equations can be combined into a single system and integrated simultaneously). This is not practical when experimentally implementing the method, however, and one must typically approximate the compensator and tracking equations subject to the following criteria. The method must be sufficiently efficient so as to guarantee that the tracking component $r(t_j)$, compensator $z_c(t_j)$ and control $u(t_j)$ are calculated before the arrival of data at time $t_{j+1} = t_j + \Delta t$. It must also be sufficiently accurate so as to resolve system dynamics. Like the state equations, the systems are quite often stiff and this implies that either a-stability or α-stability is important. Finally, the difficulties in storing past data make it prohibitive to use many popular multistep methods.

For the experiments performed with the circular plate reported on below, the sample rate was sufficiently fast (and hence Δt was sufficiently small) that a modified backward Euler or trapezoidal method produced adequate results. We illustrate here such a modified backward Euler method.

8.2.1   Discrete Estimator and Compensator for Systems with No Exogenous Input

For the system with no primary exogenous force, only the equation for the compensator must be numerically approximated when calculating the output feedback control. For a time step $\Delta t = t_{j+1} - t_j$ dictated by the sample rate, an a-stable modified backward Euler approximate to the solution at time t_{j+1} is given by

$$\begin{aligned} z_{c_{j+1}} &= (I - \Delta t A_c)^{-1} z_{c_j} + (I - \Delta t A_c)^{-1} F z_{ob}(t_j) \\ &= R(A_c) z_{c_j} + R(A_c) F z_{ob}(t_j) \, . \end{aligned}$$

The method is modified in the sense that current observation values $z_{ob}(t_j)$ are used as input since futures values at t_{j+1} are unknown. We point out that the matrix $R(A_c) \equiv (I - \Delta t A_c)^{-1}$ and vector $R(A_c)F \equiv (I - \Delta t A_c)^{-1} F$ can be computed offline and then loaded as datafiles for the online computations. Hence the implicit nature of the method, which is necessary to ensure stability, does not slow the implementation. The discrete time implementation of the method is summarized in Algorithm 8.5 which follows. Details regarding the component matrices can be found in Section 8.3.1.

Offline   (i)   **Construct matrices $A, B, C, Q, R, \hat{Q}, \hat{R}$**

           (ii)   **Solve Riccati equations (8.4) for Π and $\hat{\Pi}$**

           (iii)   **Construct $K = R^{-1} B^T \Pi$**

$$F = \hat{\Pi} C^T \hat{R}^{-1}$$
$$A_c = A - BK - FC$$

           (iv)   **Choose appropriate Δt (determined by sample rate)**

           (v)   **Construct $R(A_c) = (I - \Delta t A_c)^{-1}$**

$$R(A_c)F = (I - \Delta t A_c)^{-1} F$$

Online   (i)   **Collect data $a(t_j)$**

           (ii)   **Process data to obtain $z_{ob}(t_j)$**

           (iii)   **Time step the discrete compensator system**

$$z_{c_{j+1}} = R(A_c) z_{c_j} + R(A_c) F z_{ob}(t_j)$$

           (iv)   **Calculate the voltage $u(t_{j+1}) = -K z_{c_{j+1}}$**

Algorithm 8.5: Discrete compensator for systems with no exogenous forces.

8.2.2   Discrete Estimator and Compensator for Systems Driven by a Periodic Exogenous Force

For systems driven by a periodic exogenous force, approximations for both compensator and tracking components must be obtained before an output feedback control can be calculated. While a variety of techniques and strategies can be used to obtain approximate values of $r(t_j)$, which are then used when computing $z_c(t_j)$, these calculations must ultimately be performed in real time when implementing the method. In the experiments detailed below involving the circular plate, the exogenous force was measured for several periods and the solutions to the tracking equation were approximated and stored over a time period commensurate with the driving frequency. These stored tracking values were then used as a filter when approximating the estimated state during the remainder of the experiment.

Illustrating with a modified backward Euler discretization, the approximate to the tracking solution was determined from the difference equation

$$
\begin{aligned}
r_{j+1} &= (I - \Delta t A_{tr})^{-1} r_j + (I - \Delta t A_{tr})^{-1} \Pi g(t_j) \\
&= R(A_{tr}) r_j + R(A_{tr}) \Pi g(t_j)
\end{aligned}
$$

subject to the final condition $r(\tau) = 0$ (when implementing the method, one can simply search for a "suitable" zero crossing to start the approximation).

This approximation was continued throughout several periods of the driving force with r_j being stored in a circular buffer. This buffer was then treated as a filter when estimating the state using the difference equations

$$
\begin{aligned}
z_{c_{j+1}} &= (I - \Delta t A_c)^{-1} z_{c_j} + (I - \Delta t A_c)^{-1} F z_{ob}(t_j) \\
&+ (I - \Delta t A_c)^{-1} B_r r_j + (I - \Delta t A_c)^{-1} g(t_j) \\
&= R(A_c) z_{c_j} + R(A_c) F z_{ob}(t_j) + R(A_c) B_r r_j + R(A_c) g(t_j) \; .
\end{aligned}
$$

The offline construction of matrices and online approximation of the LQG tracking solutions and compensators are summarized in Algorithm 8.6 (similar expressions for the H^∞ approximates are obtained simply by augmenting the component matrices). For the full-order compensator described here, the algorithm was sufficiently fast to implement with the circular plate system described in the next section. For more complex systems involving two and three dimensional components, however, the full order of the system may prevent real-time implementation. In such cases, low-order compensators again may need to be employed to attain the rates necessary for successful implementation.

8.2.3   Higher-Order Approximations

In the previous discussion, a modified backward Euler method was used to discretize the compensator and tracking equations. As indicated, by numerical simulations reported in [40] and experimental results in the next section, for small Δt this provides sufficient accuracy to calculate an effective feedback voltage. If more accuracy is needed, a trapezoid rule or hybrid method of the nature discussed in [124, page 225] can be used. These provide increased accuracy without adding complexity during implementation since the components $R(A_{tr})$, $R(A_c)$, $R(A_c) B_r$ and $R(A_c) F$ can still be computed offline.

Offline (i) **Construct matrices** $A, B, C, Q, R, \hat{Q}, \hat{R}$

(ii) **Solve Riccati equations (8.4) for** Π **and** $\hat{\Pi}$

(iii) **Construct** $K = R^{-1}B^T\Pi$

$$F = \hat{\Pi}C^T\hat{R}^{-1}$$
$$A_c = A - BK - FC$$
$$A_{tr} = -[A - BK]^T$$
$$B_r = BR^{-1}B^T$$

(iv) **Choose appropriate** Δt **(determined by sample rate)**

(v) **Construct** $R(A_c) = (I - \Delta t A_c)^{-1}$

$$R(A_c)B_r = (I - \Delta t A_c)^{-1} B_r$$
$$R(A_{tr}) = (I - \Delta t A_{tr})^{-1}$$
$$R(A_c)F = (I - \Delta t A_c)^{-1} F$$

Online (i) **Collect data** $a(t_j)$

(ii) **Process data to obtain** $z_{ob}(t_j)$

(iii) **Approximate and store tracking values**

$$r_{j+1} = R(A_{tr})r_j + R(A_{tr})\Pi g(t_j)$$
$$r(\tau) = 0$$

(iii) **Time step the discrete compensator system**

$$z_{c_{j+1}} = R(A_c)z_{c_j} + R(A_c)Fz_{ob}(t_j)$$
$$+ R(A_c)B_r r_j + R(A_c)g(t_j)$$
$$z_c(k\tau) = 0$$

(iv) **Calculate the voltage** $u(t_{j+1}) = -Kz_{c_{j+1}} + R^{-1}B^T r_{j+1}$

Algorithm 8.6: Discrete compensator for systems with periodic exogenous forces.

8.3 Control of Circular Plate Vibrations

The application of the previously described control methodology to an experimental system is considered here. The experimental structure consisted of a circular plate having a radius of $0.228\,m$ ($9''$) and a thickness of $0.00127\,m$ ($0.05''$) as depicted in Figure 8.1 (the same plate described in the identification experiments described in Section 5.5). Bonded to the plate center were a pair of surface-mounted piezoceramic patches having radius $0.019\,m$ ($0.75''$) and thickness $0.00018\,m$ ($0.007''$). In both the transient and steady state experiments, only one patch was used for control. In the steady state experiments, the opposite patch was used to drive the plate while it was allowed to remain

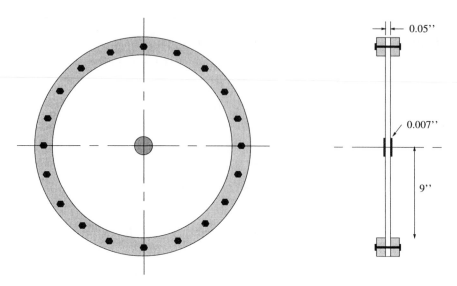

Figure 8.1: Clamped thin circular plate with surface-mounted piezoceramic patches.

uncharged in the transient case. In both sets of experiments, accelerometers were employed as sensors; hence both the input and output operators in the modeling system are unbounded. Finally, the plate was mounted in a wooden frame by circular aluminum collars bolted on each side of the plate (again see, Figure 8.1). This provided experimental boundary conditions which were sufficiently close to clamped-edge conditions so as to justify the boundary assumption of zero displacement and slope in the model.

The weak form of the modeling equations and the corresponding matrix system are summarized in the next section. This illustrates the first step in the development of a PDE-based control methodology, namely, the derivation of modeling equations. It also illustrates the manner through which a Galerkin method can be used to discretize that model. The estimation of physical parameters, in the context of the PDE-based control methodology, is summarized in Section 8.3.2 (the reader is referred to Chapter 5 for a complete discussion of this topic in the context of smart material systems). Compensator design and construction of matrices specific to the circular plate application are then detailed. The discussion in Section 8.3.4 centers on the techniques used to approximately integrate the acceleration data to obtain velocity state values. This is accomplished via a second-order approximate system which asymptotically matches the exactly integrated data and has the advantage of being robust with respect to inexact initial conditions and DC offsets in the data. Finally, experimental results demonstrating the attenuation of purely transient vibrations as well as the attenuation of transient and steady state vibrations generated by a periodic exogenous force are summarized in remaining sections. These results demonstrate the validity and effectiveness of using PDE-based compensators incorporating the physics of the system to control the system dynamics.

8.3.1 Circular Plate Model and Approximate System

To model the dynamics of the experimental structure, we consider a model for a clamped circular plate to which s pairs of piezoceramic patches are bonded (see Figure 3.4 of Chapter 3). The radius and thickness of the plate are denoted by a and h, respectively, while ρ, E, ν, c_D and γ represent the density, Young's modulus, Poisson ratio, Kelvin-Voigt and air damping parameters. The transverse displacement and external force to the plate are denoted by w and $\hat{g}$, respectively, while the region occupied by the unstrained neutral surface is indicated by Γ_0.

The space of test functions is taken to be the subspace of the Sobolev space $H^2(\Gamma_0)$ which satisfies the fixed-edge boundary conditions $w = \frac{\partial w}{\partial r} = 0$ at $r = a$. As detailed in Section 2.2.5 and Section 3.2.2, an appropriate weak form of the modeling equations for the system is

$$
\int_{\Gamma_0} \rho h \frac{\partial^2 w}{\partial t^2} \overline{\eta} d\gamma + \int_{\Gamma_0} \gamma \frac{\partial w}{\partial t} \overline{\eta} d\gamma - \int_{\Gamma_0} M_r \overline{\frac{\partial^2 \eta}{\partial r^2}} d\gamma
$$

$$
- \int_{\Gamma_0} \frac{1}{r^2} M_\theta \left[r \overline{\frac{\partial \eta}{\partial r}} + \overline{\frac{\partial^2 \eta}{\partial \theta^2}} \right] d\gamma - 2 \int_{\Gamma_0} \frac{1}{r^2} M_{r\theta} \left[r \overline{\frac{\partial^2 \eta}{\partial r \partial \theta}} - \overline{\frac{\partial \eta}{\partial \theta}} \right] d\gamma \quad (8.12)
$$

$$
= \int_{\Gamma_0} \sum_{i=1}^{s} \mathcal{K}_i^B u_i(t) \chi_i(r, \theta) \overline{\nabla^2 \eta} d\gamma + \int_{\Gamma_0} \hat{g} \overline{\eta} d\gamma
$$

for all $\eta \in V$. The overbar here denotes complex conjugation and the differential is $d\gamma = r d\theta dr$. Expressions for the internal moments M_r, M_θ and $M_{r\theta}$ in terms of E, ν, c_D are given in (3.14) while (3.16) can be used to analytically calculate the patch constants $\mathcal{K}_i^B$. We reiterate that while the expressions (3.14) and (3.16) incorporate the material contributions of the patches, the parameters $\rho, D = \frac{Eh^3}{12(1-\nu^2)}, \nu, c_D, \gamma$ and $\mathcal{K}_i^B$ must be estimated using data from the actual structure under consideration in order to guarantee an accurate model fit (this is necessitated by uncertainties, inhomogeneities and manufacturer variations in materials). The patch terms on the right-hand side model the moments generated by the application of applied out-of-phase voltages $u_i(t)$ to the patch pairs. The moments are localized to the regions covered by the patches by the characteristic functions $\chi_i(r, \theta)$ which leads to discontinuous control input operators.

To discretize the system (8.12), a basis satisfying the clamped boundary conditions and providing consistent approximations at the plate center is chosen, and a Galerkin approximate is formed. As discussed in [37, 170], an appropriate choice for the basis is $B_k^N(r, \theta) = r^{|\bar{m}|} B_n^m(r) e^{im\theta}$ where k denotes an ordering in which n varies for each fixed Fourier index m. The displacement

w is then approximated by

$$w^{\mathcal{N}}(t,r,\theta) = \sum_{k=1}^{\mathcal{N}} w_k^{\mathcal{N}}(t) B_k^{\mathcal{N}}(r,\theta) = \sum_{m=-\tilde{M}}^{\tilde{M}} \sum_{n=1}^{N^m} w_{mn}^{\mathcal{N}}(t) r^{|\hat{m}|} B_n^m(r) e^{im\theta} . \qquad (8.13)$$

Here $B_n^m(r)$ is the nth modified cubic spline satisfying $B_n^m(a) = \frac{dB_n^m(a)}{dr} = 0$ with the condition $\frac{dB_n^m(0)}{dr} = 0$ being enforced when $m = 0$ (this latter condition guarantees differentiability at the origin and implies that

$$N^m = \begin{cases} \tilde{N} & , \quad m = 0 \\ \tilde{N} + 1 & , \quad m \neq 0 \end{cases}$$

where $\tilde{N}$ denotes the number of modified cubic splines). The total number of plate basis functions is $\mathcal{N} = (2\tilde{M}+1)(\tilde{N}+1)-1$. As discussed in the [37, 170], the inclusion of the weighting term $r^{|\hat{m}|}$ with

$$\hat{m} = \begin{cases} 0 & , \quad m = 0 \\ 1 & , \quad m \neq 0 \end{cases}$$

is motivated by the asymptotic behavior of the Bessel functions, which make up the analytic plate solution, as $r \to 0$. It also serves to ensure the uniqueness of the solution at the origin. The Fourier coefficient in the weight is truncated to control the conditioning of the mass and stiffness matrices (see the examples in [37]).

To obtain a matrix system, the $\mathcal{N}$ dimensional approximating subspace is taken to be $H^{\mathcal{N}} = span\{B_k^{\mathcal{N}}\}$ and the product space for the usual corresponding first-order vector system is $\mathcal{H}^{\mathcal{N}} \times \mathcal{H}^{\mathcal{N}}$. The restriction of the first-order form for the infinite dimensional system (8.12) to the space $\mathcal{H}^{\mathcal{N}} \times \mathcal{H}^{\mathcal{N}}$ then yields the matrix equation

$$\begin{bmatrix} K_D & 0 \\ 0 & M \end{bmatrix} \begin{bmatrix} \dot{\vartheta}(t) \\ \ddot{\vartheta}(t) \end{bmatrix} = \begin{bmatrix} 0 & K_D \\ -K_D & -K_{c_D} \end{bmatrix} \begin{bmatrix} \vartheta(t) \\ \dot{\vartheta}(t) \end{bmatrix} + \begin{bmatrix} 0 \\ \tilde{B} \end{bmatrix} u(t) + \begin{bmatrix} 0 \\ \tilde{g}(t) \end{bmatrix}$$

$$\begin{bmatrix} K_D & 0 \\ 0 & M \end{bmatrix} \begin{bmatrix} \vartheta(0) \\ \dot{\vartheta}(0) \end{bmatrix} = \begin{bmatrix} g_1 \\ g_2 \end{bmatrix}$$

where $\vartheta(t) = [w_1^{\mathcal{N}}(t), w_2^{\mathcal{N}}(t), \cdots, w_{\mathcal{N}}^{\mathcal{N}}(t)]^T$ denotes the column $\mathcal{N}$ vector containing the generalized Fourier coefficients for the approximate displacement (see (8.13)). The component matrices and vectors are given by

$$K_D = K_{D1} + K_{D2} + K_{D3} + K_{D4} + K_{D5}$$

$$K_{c_D} = K_{c_D1} + K_{c_D2} + K_{c_D3} + K_{c_D4} + K_{c_D5} + \int_{\Gamma_0} \gamma B_k^{\mathcal{N}} \overline{B_\ell^{\mathcal{N}}} d\gamma$$

$$[M]_{\ell,k} = \int_{\Gamma_0} \rho h B_k^{\mathcal{N}} \overline{B_\ell^{\mathcal{N}}} d\gamma , \qquad\qquad\qquad (8.14)$$

$$[\tilde{g}(t)]_\ell = \int_{\Gamma_0} \hat{g} \overline{B_\ell^{\mathcal{N}}} d\gamma \quad , \quad [\tilde{B}]_{\ell,j} = \int_{j^{th} patch} \mathcal{K}_j \overline{\nabla^2 B_\ell^{\mathcal{N}}} d\gamma$$

$$[g_1]_\ell = \left\langle w_0, B_\ell^{\mathcal{N}} \right\rangle_V \quad , \quad [g_2]_\ell = \left\langle w_0, B_\ell^{\mathcal{N}} \right\rangle_H$$

where

$$[K_{D1}]_{\ell,k} = \int_{\Gamma_0} D \left[\frac{\partial^2 B_k^{\mathcal{N}}}{\partial r^2} + \frac{\nu}{r} \frac{\partial B_k^{\mathcal{N}}}{\partial r} + \frac{\nu}{r^2} \frac{\partial^2 B_k^{\mathcal{N}}}{\partial \theta^2} \right] \frac{\partial^2 \overline{B_\ell^{\mathcal{N}}}}{\partial r^2} \, d\gamma \, ,$$

$$[K_{D2}]_{\ell,k} = \int_{\Gamma_0} D \left[\frac{1}{r^2} \frac{\partial B_k^{\mathcal{N}}}{\partial r} + \frac{1}{r^3} \frac{\partial^2 B_k^{\mathcal{N}}}{\partial \theta^2} + \frac{\nu}{r} \frac{\partial^2 B_k^{\mathcal{N}}}{\partial r^2} \right] \frac{\partial \overline{B_\ell^{\mathcal{N}}}}{\partial r} \, d\gamma \, ,$$

$$[K_{D3}]_{\ell,k} = \int_{\Gamma_0} D \left[\frac{1}{r^3} \frac{\partial B_k^{\mathcal{N}}}{\partial r} + \frac{1}{r^4} \frac{\partial^2 B_k^{\mathcal{N}}}{\partial \theta^2} + \frac{\nu}{r^2} \frac{\partial^2 B_k^{\mathcal{N}}}{\partial r^2} \right] \frac{\partial^2 \overline{B_\ell^{\mathcal{N}}}}{\partial \theta^2} \, d\gamma \, ,$$

$$[K_{D4}]_{\ell,k} = 2 \int_{\Gamma_0} D(1-\nu) \left[\frac{1}{r^2} \frac{\partial^2 B_k^{\mathcal{N}}}{\partial r \partial \theta} - \frac{1}{r^3} \frac{\partial B_k^{\mathcal{N}}}{\partial \theta} \right] \frac{\partial^2 \overline{B_\ell^{\mathcal{N}}}}{\partial r \partial \theta} \, d\gamma \, ,$$

$$[K_{D5}]_{\ell,k} = 2 \int_{\Gamma_0} D(1-\nu) \left[-\frac{1}{r^3} \frac{\partial^2 B_k^{\mathcal{N}}}{\partial r \partial \theta} + \frac{1}{r^4} \frac{\partial B_k^{\mathcal{N}}}{\partial \theta} \right] \frac{\partial \overline{B_\ell^{\mathcal{N}}}}{\partial \theta} \, d\gamma \, .$$

The matrices $K_{c_D1}, \cdots, K_{c_D5}$ are defined similarly with the inclusion of the parameter c_D in the various integrals. In all definitions, the index ranges are $k, \ell = 1, \cdots, \mathcal{N}$; hence component matrices and vectors have the dimensions $\mathcal{N} \times \mathcal{N}$ and $\mathcal{N} \times 1$, respectively. Again, the reader is referred to [37, 170] for details regarding the derivation of this system and construction of component matrices.

Multiplication by the inverted system mass matrix yields the Cauchy equation

$$\dot{z}(t) = Az(t) + Bu(t) + g(t)$$
$$z(0) = z_0 \tag{8.15}$$

where $z(t) = [\vartheta(t), \dot{\vartheta}(t)]^T \in \mathbb{R}^N$, $N \equiv 2\mathcal{N}$ and $g(t) \equiv [0, M^{-1} \tilde{g}(t)]^T$. We note that the superscript N's denoting the discretization level have been repressed in accordance with standard finite dimensional control notation. In the case of periodic steady state dynamics, the initial condition is replaced by the periodicity condition $z(0) = z(\tau)$ while $g \equiv 0$ in systems with no exogenous force.

While the PDE model and system (8.15) obtained through a spatial discretization of that model are specific to the plate application, their development illustrates modeling and approximation issues which must be considered when constructing PDE-based controllers for general smart material systems. In developing a PDE model, care must be taken to incorporate both passive and active actuator contributions, accurate boundary conditions, coupling between components, and any other contributions which affect system dynamics.

The spatial discretization techniques must be sufficiently accurate so as to be efficient for implementation. More specifically, higher accuracy implies that system dynamics can be resolved with smaller approximating systems. This in turn facilitates the real-time approximation of state estimates, tracking components and feedback controls. The choice of accurate and efficient numerical methods becomes even more important when discretizing models of complex systems having two and three dimensional components. In such cases, exponentially accurate spectral methods may need to be considered to reduce system sizes.

Finally, the approximation methods must satisfy the various assumptions discussed in Chapter 7 which guarantee convergence of Riccati solutions, gains and ultimately, controls. For example, the approximation method must uniformly preserve stability and stabilizability margins and provide adjoint convergence to guarantee the desired controller convergence (see Theorem 7.10, Lemma 7.13 and Theorem 7.18). As illustrated by results in [19], these criteria may not be met by some approximation methods which are, on the other hand, quite adequate for open loop simulations (e.g., finite differences and finite elements). This further emphasizes that care be taken when choosing approximation methods for control applications.

8.3.2   Parameter Estimation

Once the model and approximation method have been developed for a specific process, physical parameters in the equations must be estimated using fit-to-data techniques to ensure an accurate model fit to the process. Details regarding the estimation of parameters in the circular plate model are given in Chapter 5, and the discussion here is intended only to give a sense as to how the estimation of parameters fits into the experimental implementation of the PDE-based control method.

Although the PDE models are often developed for general forces and responses and are typically frequency independent, it is usually prudent to estimate parameters in the environment in which control is to be implemented so as to minimize limitations in the models and unmodeled factors. In addition to minimizing the effects of model limitations, the estimation of parameters in the control environment reduces model inaccuracies due to time-dependent parameters (e.g., certain parameters may be thermally dependent). This latter issue also motivates current research on adaptive parameter estimation methods for control applications.

As detailed in [4, 43], transient plate vibrations were excited through an impact hammer strike or the input of a voltage spike to the patches, and acceleration data was measured. The parameter values summarized in Table 8.1 were obtained through a least squares minimization of the difference between the model response and the measured data. The accuracy of the reported patch constants $\mathcal{K}^B$ was also verified by repeating the experiments with sinusoidal voltages to the patches. The values in Table 8.1 were then employed when constructing the component matrices used during the experimental implementation of the controller.

A comparison of the parameter values in Table 8.1 with those reported in Table 5.4 of Chapter 5 indicates differences in the damping coefficients. This is due to differences in conditions under which the experiments were performed and highlights the necessity of estimating parameters in the environment used when controlling the structure.

		Physical Parameters
$\rho \cdot h$	Plate	3.170
(kg/m^2)	Plate + Pzt	3.216
D	Plate	11.151
$(N \cdot m)$	Plate + Pzt	11.506
c_D	Plate	1.443-4
$(N \cdot m \cdot sec)$	Plate + Pzt	2.031-4
ν	Plate	.326
	Plate + Pzt	.325
γ $(sec \cdot N/m^3)$		17.021
$\mathcal{K}^B$ (N/V) – Controlling Patch		.016
$\mathcal{K}^B$ (N/V) – Driving Patch		.017

Table 8.1: Physical parameters used in the plate control experiments.

8.3.3 Compensator Design and Matrix Construction

LQG and H^∞/MinMax compensators appropriate for the finite dimensional system (8.15), resulting from the discretization of the the model (8.12), were summarized in Section 8.1 and discrete-time approximations were discussed in Section 8.2. The difference equations used to march the compensator, tracking component and feedback control when single step discretization techniques are employed are summarized in Table 8.2.

As noted in Algorithms 8.5 and 8.6, the matrices arising from backward Euler discretizations of the equations are given by

$$R(A_c) = (I - \Delta t A_c)^{-1} \qquad R(A_{tr}) = (I - \Delta t A_{tr})^{-1}$$
$$R(A_c)B_r = (I - \Delta t A_c)^{-1} B_r \qquad R(A_c)F = (I - \Delta t A_c)^{-1} F$$

where

$$A_c = A - BK - FC \qquad K = R^{-1}B^T$$
$$A_{tr} = -[A - BK]^T \qquad F = \hat{\Pi}C^T\hat{R}^{-1}$$
$$B_r = BR^{-1}B^T .$$

We explicitly describe here the construction of the matrices and filters for the circular plate system under consideration. These matrices are then employed in Algorithms 8.5 or 8.6 to create implementation matrices and filters which are ultimately used in the experiments. While details regarding the construction of system matrices will vary according to the process being controlled, the constructions considered here illustrate some of the issues to be considered when constructing similar matrices for other PDE-based control applications.

We first note that $\tilde{N} = 16$ modified cubic splines (see (8.13)) were sufficient for resolving the plate dynamics in the frequency range under consideration.

Due to the axisymmetric excitation and response of the plate, the Fourier limit $\tilde{M} = 0$ was used in all calculations. Hence a total of $\tilde{N} = 16$ basis functions were used which led to 32 coefficients in the vector z.

The formulation and sizes of all components in the control system for the circular plate are summarized in Figures 8.2 and 8.3 below.

No Primary Input	Periodic Exogenous Force
$z_{c_{j+1}} = R(A_c)z_{c_j} + R(A_c)Fz_{ob}(t_j)$ $z_{c_0} = 0$	$z_{c_{j+1}} = R(A_c)z_{c_j} + R(A_c)Fz_{ob}(t_j)$ $\qquad\qquad + R(A_c)B_r r_j + R(A_c)g(t_j)$ $z_c(k\tau) = 0$
	$r_{j+1} = R(A_{tr})r_j + R(A_{tr})\Pi g(t_j)$ $r(\tau) = 0$
$u(t_{j+1}) = -Kz_{c_{j+1}}$	$u(t_{j+1}) = -Kz_{c_{j+1}} + R^{-1}B^T r_{j+1}$

Table 8.2: Discrete compensator $z_{c_{j+1}}$, tracking component r_{j+1} and controlling voltage $u(t_{j+1})$ for systems with no primary input or a periodic exogenous force.

Component	Size	Comments
$A = \begin{bmatrix} 0 & I \\ -M^{-1}K_D & -M^{-1}K_{c_D} \end{bmatrix}$	32×32	The elements comprising the 16×16 matrices M, K_D and K_{c_D} are summarized in (8.14)
$B = \begin{bmatrix} 0 \\ M^{-1}\tilde{B} \end{bmatrix}$	32×1	See (8.14) for the description of $\tilde{B}$
$g(t) = \begin{bmatrix} 0 \\ M^{-1}\tilde{g}(t) \end{bmatrix}$	32×1	The elements of $\tilde{g}(t)$ are detailed in (8.14)
$C = \begin{bmatrix} \mathcal{Z} & B_1^N(r_1,\theta_1), \cdots, B_N^N(r_1,\theta_1) \\ \vdots & \vdots & \vdots \\ \mathcal{Z} & B_1^N(r_p,\theta_p), \cdots, B_N^N(r_p,\theta_p) \end{bmatrix}$ with $\mathcal{Z} = [0,\cdots,0]$	$p \times 32$	In the experiments, acceleration data was integrated to obtain velocity values which are the second component of the state in the first-order formulation. Since one accelerometer was used, $p=1$.

Figure 8.2: System and control matrices used in the circular plate experiments.

Component	Size	Comments
$Q = \begin{bmatrix} d_1 I^{\mathcal{N}} & 0 \\ 0 & d_2 I^{\mathcal{N}} \end{bmatrix}$ $\bullet \begin{bmatrix} K_D & 0 \\ 0 & M \end{bmatrix}$	32×32	A weighted mass matrix was used for the penalty term Q ($I^{\mathcal{N}}$ is the $\mathcal{N} \times \mathcal{N}$ identity). As discussed in [12], this provides a means of weighting the kinetic and potential energy of the plate. Example of §8.3.5: $d_1 = d_2 = 1$ Example of §8.3.6: $d_1 = d_2 = 5$
$R = \begin{bmatrix} R_{11} & & \\ & \ddots & \\ & & R_{ss} \end{bmatrix}$	$s \times s$	In the plate experiments, one controlling patch was used so $s = 1$. Example of §8.3.5: $R_{11} = 10^{-7}$ Example of §8.3.6: $R_{11} = 10^{-10}$
$\hat{Q} = \begin{bmatrix} c_1 I^{\mathcal{N}} & 0 \\ 0 & c_2 I^{\mathcal{N}} \end{bmatrix}$	32×32	For the plate experiments, $\hat{Q}$ was treated solely as a design parameter as compared with the choice in [10] where physical arguments were used to construct the matrix. The identity weights were taken to be $c_1 = c_2 = 1$ in the experiments.
$\hat{R} = \begin{bmatrix} \hat{R}_{11} & & \\ & \ddots & \\ & & \hat{R}_{pp} \end{bmatrix}$	$p \times p$	In the experiments, $p = 1$ since one accelerometer was used for data collection. The weight was taken to be $\hat{R}_{11} = 1$.

Figure 8.3: Control and observation matrices used in the circular plate experiments. The experimental results in Section 8.3.5 demonstrate control of transient dynamics while the periodic case is illustrated in Section 8.3.6.

8.3.4   Integration of Experimental Data

Accelerometers are commonly used as sensors in structural applications since they are lightweight and compact, sufficiently sensitive and relatively inexpensive. For models in which displacement and velocity are states, however, the accelerometer data must be numerically integrated to obtain state values for control computations. An issue of paramount importance when approximately integrating experimental data concerns the robustness of the integrator with respect to inexact initial conditions and DC biases (added constants) in the data. The inexact initial conditions can be due to unknown system contributions, static shocks during system connections, et cetera. While careful calibration can alleviate some of the uncertainty in initial conditions, it cannot fully eliminate the problem. The problem of biases due to small DC voltages in the system can also be minimized but never fully eliminated. The use of a simple addition routine to approximately integrate data containing such biases leads to accumulated errors which quickly swamp the acceleration signal. Hence an integrator which is minimally affected by uncertain initial conditions and DC biases in the data is crucial for success when approximately integrating the data.

We summarize here two techniques for approximately integrating acceleration data to obtain velocity values in accordance with the relation

$$\dot{v}(t) = a(t) \ .$$

Essentially, the idea is to replace the integration by either the first-order differential equation

$$\dot{v} + \Omega v = \frac{1}{RC} a \tag{8.16}$$

or the second-order equation

$$\ddot{v} + \Omega \dot{v} + \Omega^2 v = \frac{1}{RC} \dot{a} \tag{8.17}$$

(see [102]). The design parameters Ω and RC are frequency and time constants, respectively, which are chosen so that $RC = 1$ and $\omega > 6\Omega$, where ω is the smallest observed frequency. For solution, (8.17) is written as the first-order system

$$\begin{bmatrix} \dot{v} \\ e \end{bmatrix} = \begin{bmatrix} -\Omega & \Omega \\ -\Omega & 0 \end{bmatrix} \begin{bmatrix} v \\ e \end{bmatrix} + \begin{bmatrix} \frac{a(t)}{RC} \\ 0 \end{bmatrix} \tag{8.18}$$

subject to the initial conditions

$$\begin{bmatrix} v(0) \\ e(0) \end{bmatrix} = \begin{bmatrix} v_0 \\ e_0 \end{bmatrix} \ .$$

As detailed in [40, 42], the first-order integrator (8.16) is a single pole filter which is robust with respect to inexact initial conditions in the sense that perturbations in initial conditions decay exponentially in time. DC biases in the data are propagated, however, thus limiting the utility of the filter for integrating experimental data.

In the second-order integrator (8.17) or system (8.18), both disturbances in initial conditions and DC gains in acceleration data are exponentially attenuated with the degree of attenuation increasing with increasing Ω. Moreover, the frequency response of this approximator is very close to that of the original signal for $\omega > 6\Omega$. Hence this latter method provides an accurate and robust means of approximately integrating experimental data when Ω is chosen as large as possible while satisfying $\Omega < \omega/6$.

As described in [42], a backward Euler discretization of the system (8.18) was employed in the plate control experiments described in the next section. Numerical examples in [42] demonstrate that this provides sufficiently accurate state values so as to yield effective control authority.

8.3.5 Control of Transient Vibrations with no Exogenous Force

We demonstrate first the capability of the PDE-based compensator to control transient plate vibrations. The plate was excited by a centered impact hammer strike which generated a decaying response in the first three axisymmetric modes (with frequencies of 60 Hz, 227 Hz and 512 Hz). For the results discussed here, data was collected from an accelerometer located 2″ from the plate center as depicted in Figure 8.4. Hence $P_1 = (r_1, \theta_1) = (2'', 0)$ in the construction of the observation matrix described in Figure 8.3. The experiments were also repeated with the accelerometer fixed at the plate center, and the reader is referred to [40] for a survey of those results. As noted there, nearly identical control attenuation was obtained with the accelerometer at the two locations. This illustrates the distributed nature of the model and demonstrates that collocation between the sensor and actuator is unnecessary in this control method.

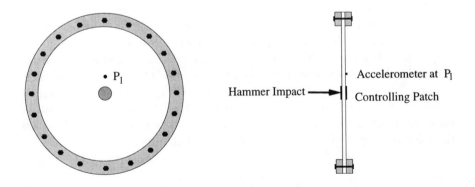

Figure 8.4: Patch, accelerometer and impact locations for the plate control experiments.

Since no exogenous force was applied to the plate, the compensator and control law summarized in Algorithm 8.5 were used to compute the controlling voltage to the patch. The component matrices as well as the cost functional and observation parameters used in these experiments are summarized in Figures 8.2 and 8.3.

A PC-based Texas Instruments TMS 320-C30 digital signal processing (DSP) board, as depicted schematically in Figure 8.5, was used for data acquisition and processing. As detailed in [40, 42], the voltage produced by the accelerometer was amplified to improve A/D conversion and low-pass filtered to minimize the effects of aliasing. The code for the conversion of data to physical units, integration to obtain $u(t_j)$ and calculation of the compensator $z_{c_{j+1}}$ and voltage $u(t_{j+1})$ was written in assembler in order to attain sufficiently fast sample rates for resolving the transient and periodic steady state frequencies excited in the experiments (in this manner, sample rates on the order of 7 kHz were obtained). The control voltage was then filtered to smooth the discontinuities introduced in D/A conversion and was amplified to physical units before input to the patches. This final amplification was required since the maximum voltage output by the DSP is 2.5 V whereas $60-70$ V were dictated at the patches by the control algorithm. The reader is again referred to [40, 42] for further details regarding the experimental setup.

The uncontrolled and controlled velocities obtained from the integrated data from the accelerometer at P_1 are plotted in Figure 8.6. The two responses were generated by hammer impacts differing by less than 2% (see [40, 42]) which corroborates the nearly identical initial energies levels noted in Figure 8.6. The reduction levels at $t = 0.5, 1.0, 1.5$ seconds are summarized in Table 8.3. As illustrated by the results in the table and figure, the velocity level in the controlled case has been reduced by 50% before 0.5 second and is essentially fully attenuated by 1.5 second.

The voltage levels to the patches must also be considered since excessive voltages can destroy the actuator capabilities of the patches. In practice, it has been observed that the patches can be used for extended periods at the frequencies of interest without damage or degradation of performance if the voltage levels are maintained below $8-10$ rms V/mil [143]. For the experiment reported here, the controlling voltage $u(t_{j+1}) = -Ky_{c_{j+1}}$ had a maximum magnitude of $70\,V$ as noted in Figure 8.6. Hence the PDE-based control method provides significant vibration attenuation with voltages that are well within the tolerance for the $7\,mil$ patches used in the experiments.

Time	Acceleration	Velocity
.5 sec	69.5	68.2
1 sec	88.9	84.7
1.5 sec	93.8	97.8

Table 8.3: The percent reductions in acceleration and velocity levels at the point $P_2 = (2'', 0)$ when feedback control is implemented.

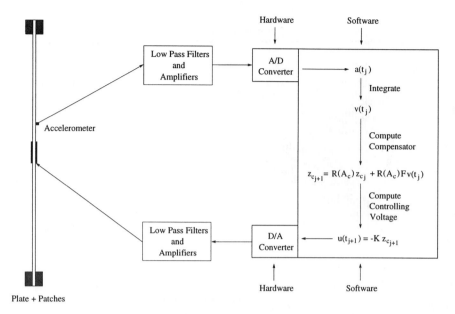

Figure 8.5: Amplifiers, DSP configuration and PC Algorithm 8.5 for controlling a plate excited by an initial impact. Component matrices are defined in Figures 8.2 and 8.3.

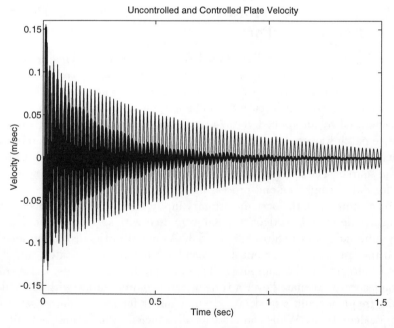

Figure 8.6: Uncontrolled and controlled plate velocity at $(2'', 0)$ in response to an impact hammer hit; —— (uncontrolled), ▬▬ (controlled).

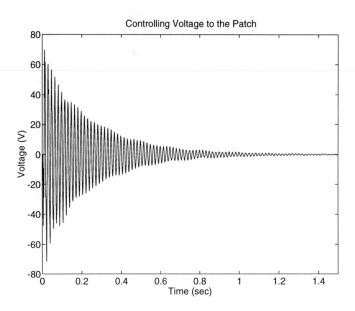

Figure 8.7: The controlling voltage for acceleration data observed at $(2'', 0)$.

8.3.6   Control of Vibrations Generated by a Periodic Exogenous Force

The implementation of the PDE-based compensator for systems driven by a periodic exogenous force is illustrated by a summary of control experiments for the circular plate driven by a periodic input. A schematic of the experimental setup is given in Figure 8.8. The exogenous disturbance to the plate was generated by an applied voltage to one of the centered patches while the opposite patch was used as the actuator. Data was collected from an off-center accelerometer located at the point $P_1 = (2'', 0)$ as well as from the primary voltage source. The PC-based DSP board discussed in Section 8.3.5 was again used for data acquisition and processing.

To accommodate the periodic primary input, the discrete compensator and output law described in Algorithm 8.6 were used with specific component matrices summarized in Figures 8.2 and 8.3. As noted in the discussion for that algorithm, tracking values r_j must be calculated before compensator values and control voltages can be computed. In these initial experiments, these tracking values were calculated over a time interval commensurate with the 350 Hz driving frequency and stored in a circular buffer for use in compensator and control calculations. While computationally efficient, this technique limits the robustness of the method with regard to variations in the driving force, and current studies are directed toward the simultaneous solution of the tracking and compensator difference equations.

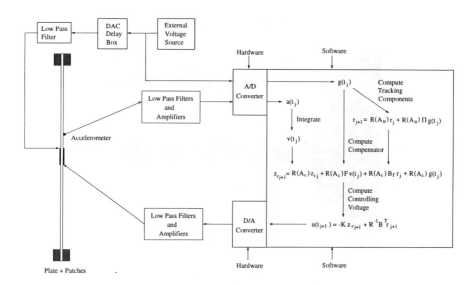

Figure 8.8: Amplifiers, DSP configuration and PC Algorithm 8.6 for controlling a plate driven by a periodic exogenous force. Component matrices are defined in Figures 8.2 and 8.3.

An unavoidable byproduct of the low-pass filters and A/D, D/A hardware is the introduction of delays and phase shifts in the filtered signals. While the phase shifts are frequency dependent, experiments indicated that at 350 Hz, $30-40^{\circ}$ phase shifts were introduced by the hardware. This was sufficient to destabilize the controller if left unrectified. In the experiments, we compensated by first conducting an offline, numerical 'identification' to determine the amount of added delay necessary for stabilizing the controller in the presence of phase shifts of the order introduced by the experimental hardware. The numerically predicted delays were then implemented via the delay box depicted in Figure 8.8. We note that the offline determination of added delays and tuning of the system in this manner is somewhat analogous to the online tuning and phase locking necessary for ensuring stability in currently implemented frequency response input/output control methods. A current topic of research in this area concerns the development of online techniques for delay compensation.

To illustrate the effectiveness of the controller, plate responses to two different experimental exogenous input signals are given. In the first case, an external oscillator was used to generate the voltage to the driving patch and control was initiated after the plate had attained a fully period steady state motion. The uncontrolled and controlled acceleration measurements at the point $P_1 = (2'', 0)$ for this case are plotted in Figure 8.9. In a second set of experiments, the PC itself was used to generate the signal and, as seen in Figure 8.10, both transient and steady state behavior is present in the plate response. In

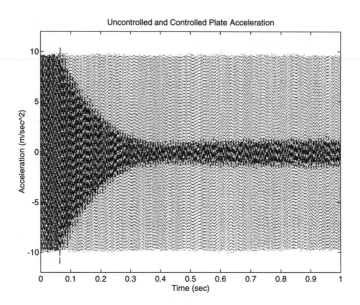

Figure 8.9: Uncontrolled and controlled plate acceleration at $P_1 = (2'', 0)$ for system driven to steady state by an external oscillator; —— (uncontrolled), ▬▬ (controlled).

both figures, it can be noted that approximately 0.06 seconds were required to calculate and store the tracking values. The calculation of the compensator and feedback control then began with $82\% - 85\%$ reduction attained by 0.4 seconds (an 85% reduction corresponds to a $20 \log(a_{con}/a_{uncon}) \approx -16.5$ dB reduction in acceleration levels). This demonstrates the feasibility and effectiveness of using a PDE-based compensator to attenuate both transient and steady state vibrations which results from a periodic exogenous disturbance. Further details and control results from these experiments can be found in [42] while related numerical studies are reported in [40].

8.4 Implementation Issues Regarding PDE-Based Compensators

The emphasis in this chapter was on the demonstration of techniques that can be employed for implementing PDE-based compensators. The approximation and implementation strategies discussed here are intended as examples and are not meant to provide an exhaustive survey of such techniques; in many cases the choice of a specific compensator design and implementation strategy will depend upon the application under consideration. For example, the modified backward Euler discretization of the tracking and compensator equations pro-

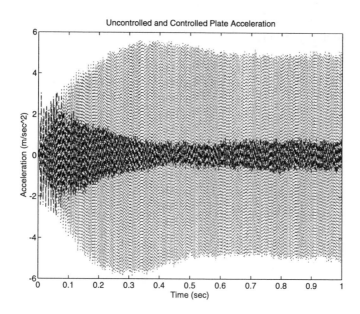

Figure 8.10: Uncontrolled and controlled plate acceleration at $P_1 = (2'', 0)$ for system with both transient and steady state components driven by a PC-generated periodic exogenous force; —— (uncontrolled), ▬▬ (controlled).

vided adequate and efficient approximations when implementing the method to reduce plate vibrations through piezoceramic patch inputs. Moreover, this discretization illustrated the form of the difference equations common to all single step discretizations of the equations. On the other hand, in applications with slower sample rates (and hence large time steps Δt) or applications requiring additional accuracy, the trapezoid or hybrid methods discussed in Section 8.2.3 may be necessary.

Similarly, various strategies can be employed for approximating the tracking solutions prior to computation of the compensator values. For applications with steady driving forces, calculation and storage of the tracking values in the manner described in Section 8.3.6 is computationally efficient and facilitates high sample rates. This 'one time' calculation of tracking values limits the robustness with regards to variable forces, however, and if such variations are expected, simultaneous solution of the tracking and compensator equations should be investigated. This latter option is facilitated by the advent of multiple processor DSP boards currently being introduced into many laboratories.

Compensation for the delays and phase shifts introduced by hardware is a crucial issue when implementing any control method. For some applications, offline delay 'identification' in the manner indicated in Section 8.3.6 and illustrated in [40] provides a viable technique for compensating for these delays. Online identification techniques and adaptive strategies are under current investigation and indicate a direction to be pursued when implementing PDE-based controllers for complex systems in which hardware delays are unavoidable and significant.

Chapter 9

Modeling and Control in Coupled Systems

In previous chapters, modeling, parameter estimation and control of structures in the context of smart material technology was considered. A basic tenet throughout the development was the careful incorporation of the physics contributed by various components including boundary conditions, actuator/structure interactions, et cetera. By incorporating the physics and utilizing the distributed nature of the PDE models, very successful damage detection and feedback control strategies were developed.

The same philosophy can be utilized in applications involving structural interactions with an adjacent fluid or acoustic field. In both cases, physical coupling between the structure and field leads to energy exchange. By incorporating this coupling in the modeling equations, strategies for controlling fluid dynamics or acoustic sound pressure levels through control of structural vibrations can be developed. In this chapter, the modeling and development of smart material controllers for systems involving a structure coupled with such adjacent fields are considered.

A typical structural acoustic application is illustrated by considering the problem of reducing interior noise levels in an aircraft fuselage. In many aircraft, a major component of the interior sound pressure field is due to structure-borne noise from the engines. The rotating blades in turboprop or turbofan engines generate low frequency, high displacement acoustic fields which couple effectively with the fuselage dynamics (this problem is exacerbated in new classes of fuel efficient engines having a large number of high velocity blades). This same structural acoustic coupling then produces interior sound pressure oscillations which, if left uncontrolled, lead to undesirable passenger environments. As discussed in [23], one means of attenuating this interior field is through the use of loudspeakers which generate an interfering secondary field. While effective in certain regimes, this strategy suffers from the disadvantage that it is inherently local in nature and requires a large array of external control hardware. Moreover, it suffers from the potential for *enhanced* noise if tuning is not extremely precise.

A second strategy for reducing sound pressure levels is through the control of structural vibrations using piezoceramic or electrostrictive elements mounted on or embedded in the fuselage. In this active structural acoustic control (ASAC) approach, structure-borne noise is reduced at its source so that global attenuation can be achieved. Moreover, by embedding the piezoceramic or electrostrictive patches in the fuselage and utilizing their actuator/sensor capabilities, the controllers can be implemented with negligible weight or passive loading contributed by the control hardware.

The experimental success of using piezoceramic actuators to control structure-borne noise is documented in [92, 168] and references therein. The emphasis in these studies was on the use of frequency input/output analysis to control temporally periodic steady state fields.

The development of a PDE-based methodology for attenuating interior noise in structural acoustic systems was initiated in [12] where a 2-D acoustic field coupled with a vibrating beam was considered. Numerical examples for this system were reported in [33] while an LQR control strategy for the model with fully nonlinear coupling conditions was developed in [36]. Suitable compensators using LQG and H^∞/MinMax theory were reported in [10], and the robustness of the method with regard to practical implementation criteria was investigated in [11]. The models and control methods were extended to a 3-D hardwalled cylinder with a vibrating end plate in [35] and a 3-D acoustic field enclosed in a vibrating shell in [39]. These latter cases model experimental setups in the Acoustics Division, NASA Langley Research Center, which are currently being used in parameter estimation and control experiments.

When developing the models for both the 2-D and 3-D systems, care was taken to incorporate the physics of the acoustic field, the structure (including actuator/sensor contributions), and dynamic coupling between the two. The careful modeling of the coupling conditions is of upmost importance in such systems since it is ultimately the transmission of energy from the structure to the acoustic field which must be reduced in order to effectively control structure-borne noise. By incorporating the coupling, a successful controller attenuates noise through a combination of reduced structural vibration levels and maintenance of structural responses which couple less effectively with the acoustic field. The incorporation of coupling dynamics in the model also facilitates control strategies which reduce control hardware by employing the distributed and predictive capabilities of the PDE models. For example, simulations reported in [10] and summarized in Section 9.3 demonstrate that through careful modeling, interior noise levels in the 2-D system can be effectively reduced using only structural measurements in the compensator with the acoustic response predicted by the coupling and acoustic model.

While the use of careful modeling to reduce or eliminate hardware, such as microphones, is of great practical concern in interior noise applications, it is of perhaps even more importance in applications involving the control of exterior noise fields. For example, arrays of exterior microphones are unwieldy for controlling noise generated by a transformer and are completely impractical for controllers designed to reduce noise transmitted from a submarine to an exterior underwater acoustic field.

Finally, by considering coupled models for structural acoustic systems, the complementary problems of noise control and fatigue reduction can be solved simultaneously. This is of paramount importance when using actuators such as piezoceramic patches since experiments have demonstrated high levels of fatigue along the edges of actuators that are programmed for optimal noise control. By incorporating both the sound pressure and fatigue levels as states to be reduced, controllers reducing both can be developed.

As will be detailed in subsequent sections, the acoustic field in many structural acoustic applications can be adequately modeled by a linear wave equation. For applications involving turbulence-induced noise, high speed flows, or fields in which viscosity is significant, the field must be modeled by Euler, Navier-Stokes or other equations appropriate for the application. While it is expected that smart material technology will profoundly affect controller design in a large number of these applications, the analysis and development of such controllers lags behind that for structural acoustic problems due to the strongly nonlinear behavior of the systems (and hence modeling equations). Modeling of the coupling between the structure and adjacent field is again crucial when considering smart material controllers, and the reader is referred to [87, 88] for issues associated with structure/fluid coupling. Initial analysis regarding the development of a PDE-based strategy for controlling recirculation flow in a channel is presented in [18]. In this latter investigation, the mechanism for control is the shear generated by in-plane excitation of piezoceramic patches bonded to the channel wall.

In this chapter, we focus on PDE-based smart material controllers for structural acoustic systems. A typical model is outlined in Section 9.1 and well-posedness issues are summarized. Numerical examples demonstrating the application of LQR, LQG and H^∞/MinMax methods from Chapters 7 and 8 to structural acoustic models are given in the next two sections. These numerical examples illustrate that by incorporating the structural acoustic coupling in the models, effective PDE-based controllers employing smart material actuators and sensors can be developed.

9.1   Modeling a Structural Acoustic System

We describe here the modeling of an experimental structural acoustic apparatus used as a prototype for fuselage vibration/interior noise investigations in the Acoustics Division, NASA Langley Research Center. The apparatus consists of a thin cylindrical shell that is supported by heavy end caps as depicted in Figure 9.1. Loudspeakers on either side of the shell generate an external acoustic field which, through structural acoustic coupling, leads to high interior sound pressure levels. Control of both structural vibrations and interior sound pressure levels is attained through the excitation of surface-mounted piezoceramic patches. For sensing, the patches are augmented by accelerometers on the shell and microphones in the interior cavity. The apparatus also provides capability for inclusion of various fuselage-type components such as ribs, stiffeners and a floor. However, prototype experiments are often performed without these structural additions, and we simplify the exposition here by considering the structure as consisting of shell, boundary conditions and actuators.

For modeling purposes, the thin cylindrical shell is assumed to have length ℓ, thickness h and radius R with the axial direction taken along the x-axis (see Figure 9.2). The unperturbed middle surface of the shell is denoted by Γ_0. To remain consistent with notation in Chapter 2, the displacements of the middle surface in the axial, circumferential and radial directions are taken to be u, v and w, respectively. Furthermore, the shell is assumed to have mass density ρ, Young's modulus E, Poisson ratio ν, and damping coefficient c_D. Bonded to the shell are s pairs of identical piezoceramic patches with the center of the ith patch pair located at the point $(\overline{x}_i, \overline{\theta}_i)$.

The heavy supporting caps at the ends of the shell are denoted by Γ in the model. Due to their mass and construction, they are assumed to provide nearly perfect clamped-edge boundary conditions for the shell and hard wall boundary conditions for the acoustic field (their vibration is negligible). The acoustic cavity enclosed by the shell and end caps is denoted by $\Omega(t)$ with the time dependence due to the volume changes which result when the shell vibrates. Throughout this analysis, the equilibrium density and speed of sound for the atmosphere in the acoustic cavity are denoted by ρ_f and c, respectively.

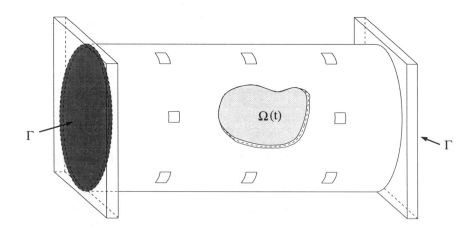

Figure 9.1: Cylindrical shell with hard end caps Γ enclosing an acoustic cavity $\Omega(t)$. Bonded to the shell are s piezoceramic patch pairs.

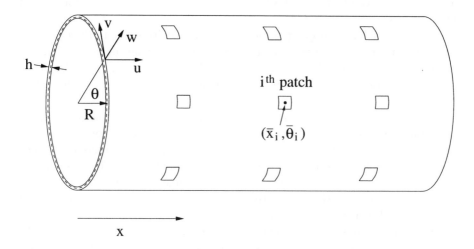

Figure 9.2: Thin cylindrical shell with s pairs of surface-mounted piezoceramic patches.

9.1.1   Strong Form of Modeling Equations

To construct a model describing the dynamics of the coupled system, it is necessary to incorporate the acoustic response, shell dynamics, piezoceramic patch/shell interactions, and the coupling acoustic/structure interactions. For the applications under consideration, it is reasonable to assume that internal sound pressure levels are below 150 dB (see Section 9.4 for a discussion of scales and units employed in acoustic applications). As detailed in [41, 153], acoustic dynamics at such levels are adequately modeled by wave equations involving either the acoustic velocity potential ϕ or pressure p. The potential satisfies $\nabla \phi = -\vec{u}$, where $\vec{u}$ denotes field velocity, and is related to the pressure through the relationship $p = \rho_f \frac{\partial \phi}{\partial t}$. The acoustic dynamics inside the cavity are then modeled by the undamped wave equation

$$\frac{\partial^2 \phi}{\partial t^2} = c^2 \Delta \phi \qquad , \quad (r, \theta, x) \in \Omega(t) \, , t > 0 \, ,$$

$$\nabla \phi \cdot \hat{n} = 0 \qquad , \quad (r, \theta, x) \in \Gamma \, , t > 0$$

where $\hat{n}$ is the outward axial unit normal to the cavity end caps, and the Laplacian in cylindrical coordinates is given by

$$\Delta \phi = \frac{\partial^2 \phi}{\partial r^2} + \frac{1}{r} \frac{\partial \phi}{\partial r} + \frac{1}{r^2} \frac{\partial^2 \phi}{\partial \theta^2} + \frac{\partial^2 \phi}{\partial x^2} \, .$$

We point out that the formulation in terms of the potential provides a natural means of incorporating natural boundary and coupling conditions. It also facilitates the formulation of the problem in first-order form for control applications.

In this model, air damping inside the cavity is omitted due to the relatively small dimensions (on the order of 1-3 meters) of the experimental cavity. For significantly larger cavities or cases in which the cavity contains absorbent objects, interior cavity damping can be readily incorporated as detailed in [41].

The choice of modeling shell equations depends on the relative shell dimensions and the required accuracy. To simplify the discussion, we assume that the shell's length is relatively short in relation to its radius and hence the Donnell-Mushtari equations can be used to approximate its motion. As detailed in Chapter 2, higher-order equations such as the Byrne-Flügge-Lur'ye equations can be substituted for those of Donnell and Mushtari if warranted by the shell dimensions or accuracy requirements.

To simplify notation when extending the models of Section 3.1 for shells with one patch pair to the case of s patch pairs, it is advantageous to consolidate patch effects through the characteristic and general indicator functions

$$\chi_{pe_i}(x,\theta) \equiv \chi_{[x_{1i},x_{2i}]}(x)\chi_{[\theta_{1i},\theta_{2i}]}(\theta)$$

$$S_{pe_i}(x,\theta) \equiv S_{1,2}(x)\hat{S}_{1,2}(\theta)$$

which are assumed to be centered over the ith patch. The individual indicator functions $S_{1,2}(x)$ and $\hat{S}_{1,2}(\theta)$ are defined in (3.9). The Donnell-Mushtari equations can then be written as

$$R\rho h \frac{\partial^2 u}{\partial t^2} - R\frac{\partial N_x}{\partial x} - \frac{\partial N_{\theta x}}{\partial \theta} = -R\sum_{i=1}^{s} \frac{\partial (N_x)_{pe_i}}{\partial x} S_{pe_i}(x,\theta)$$

$$R\rho h \frac{\partial^2 v}{\partial t^2} - \frac{\partial N_\theta}{\partial \theta} - R\frac{N_{x\theta}}{\partial x} = -\sum_{i=1}^{s} \frac{\partial (N_\theta)_{pe_i}}{\partial \theta} S_{pe_i}(x,\theta)$$

$$R\rho h \frac{\partial^2 w}{\partial t^2} - R\frac{\partial^2 M_x}{\partial x^2} - \frac{1}{R}\frac{\partial^2 M_\theta}{\partial \theta^2} - 2\frac{\partial^2 M_{x\theta}}{\partial x \partial \theta} + N_\theta$$ (9.1)

$$= R\hat{q}_n - \sum_{i=1}^{s}\left[R\frac{\partial^2 (M_x)_{pe_i}}{\partial x^2} + \frac{1}{R}\frac{\partial^2 (M_\theta)_{pe_i}}{\partial \theta^2}\right]$$

where $\hat{q}_n$ is the external, normal load on the shell. The internal force resultants $N_x, N_\theta, N_{x\theta},\ N_{\theta x}$ and moment resultants $M_x, M_\theta, M_{x\theta}, M_{\theta x}$ are defined in a manner analogous to that used when defining the expressions (3.3) for an undamped shell with a single patch pair. Similarly, the external moments and forces

$$(M_x)_{pe_i} = [(M_x)_{pe_{i1}} + (M_x)_{pe_{i2}}] \chi_{pe_i}(x, \theta)$$

$$(M_\theta)_{pe_i} = [(M_\theta)_{pe_{i1}} + (M_\theta)_{pe_{i2}}] \chi_{pe_i}(x, \theta)$$

$$(N_x)_{pe_i} = [(N_x)_{pe_{i1}} + (N_x)_{pe_{i2}}] \chi_{pe_i}(x, \theta) S_{pe_i}(x, \theta)$$

$$(N_\theta)_{pe_i} = [(N_\theta)_{pe_{i1}} + (N_\theta)_{pe_{i2}}] \chi_{pe_i}(x, \theta) S_{pe_i}(x, \theta)$$

(see (3.8)) reflect the contributions of each patch in the manner described in (3.7). For example,

$$(M_x)_{pe_{i1}} = \frac{-E_{pe_{i1}}}{1 - \nu_{pe_{i1}}} \left[\frac{1}{8} \left(4 \left(\frac{h}{2} + h_{pe_i} \right)^2 - h^2 \right) \right.$$

$$\left. + \frac{1}{R} \frac{1}{24} \left(8 \left(\frac{h}{2} + h_{pe_i} \right)^3 - h^3 \right) \right] \frac{d_{31_i}}{h_{pe_i}} u_{i1}$$

$$(N_x)_{pe_{i1}} = \frac{-E_{pe_{i1}}}{1 - \nu_{pe_{i1}}} \left[h_{pe_i} + \frac{1}{R} \frac{1}{8} \left(4 \left(\frac{h}{2} + h_{pe_i} \right)^2 - h^2 \right) \right] \frac{d_{31_i}}{h_{pe_i}} u_{i1}$$

with definitions similar to those of (3.7) for the remaining external resultants. In these expressions, $E_{pe_{i1}}, \nu_{pe_{i1}}, h_{pe_i}$ and d_{31_i} denote the Young's modulus, Poisson ratio, thickness and strain constant for the outer patch in the ith pair while u_{i1} denotes the input voltage to that patch.

Under the assumption of clamped-edge conditions due to the heavy end caps, the boundary conditions for the shell are taken to be

$$u(t, \theta, x) = v(t, \theta, x) = w(t, \theta, x) = \frac{\partial w}{\partial x}(t, \theta, x) = 0$$

at $x = 0, \ell$. These conditions may need to be altered depending upon the specific experimental setup. For example, if slight energy loss is suspected at the shell edge, almost fixed boundary conditions analogous to those discussed in [45] may be employed.

For the experimental shell subjected to an external acoustic field $\hat{g}$ generated by loudspeakers, the normal load $\hat{q}_n$ on the shell is given by

$$\hat{q}_n(t, \theta, x) = \hat{g}(t, \theta, x) - \rho_f \frac{\partial \phi}{\partial t}(t, w(t, \theta, x), v(t, \theta, x), u(t, \theta, x)) \ .$$

The second contribution in this expression is the backpressure from the internal acoustic field which is in general nonlinear since it occurs at the shell's surface. A second coupling condition is the velocity constraint

$$\frac{\partial \phi}{\partial r}(t, w(t, \theta, x), v(t, \theta, x), u(t, \theta, x)) = \frac{\partial w}{\partial t}(t, \theta, x)$$

which simply states that the shell is impermeable to air. Note that this second constraint also provides a boundary condition for the acoustic response.

Under the assumption of small displacements, which is inherent in the linear shell theories, the variable domain $\Omega(t)$ is approximated by the fixed domain $\Omega = \{(r, \theta, x) \mid 0 \le r < R, 0 \le \theta < 2\pi, 0 < x < \ell\}$, and the general nonlinear coupling conditions are approximated by their linear approximations. The strong form of the linear model for the coupled structural acoustic system is then

$$\frac{\partial^2 \phi}{\partial t^2} = c^2 \Delta \phi \quad , \quad (r, \theta, x) \in \Omega \,, t > 0 \,,$$

$$\nabla \phi \cdot \hat{n} = 0 \quad , \quad (r, \theta, x) \in \Gamma \,, t > 0$$

$$\frac{\partial \phi}{\partial r}(t, R, \theta, x) = \frac{\partial w}{\partial t}(t, \theta, x) \quad , \quad (\theta, x) \in \Gamma_0 \,, t > 0$$

$$R\rho h \frac{\partial^2 u}{\partial t^2} - R\frac{\partial N_x}{\partial x} - \frac{\partial N_{\theta x}}{\partial \theta} = -R \sum_{i=1}^{s} \frac{\partial (N_x)_{pe_i}}{\partial x} S_{pe_i}(x, \theta)$$

$$R\rho h \frac{\partial^2 v}{\partial t^2} - \frac{\partial N_\theta}{\partial \theta} - R\frac{N_{x\theta}}{\partial x} = -\sum_{i=1}^{s} \frac{\partial (N_\theta)_{pe_i}}{\partial \theta} S_{pe_i}(x, \theta)$$

$$R\rho h \frac{\partial^2 w}{\partial t^2} - R\frac{\partial^2 M_x}{\partial x^2} - \frac{1}{R}\frac{\partial^2 M_\theta}{\partial \theta^2} - 2\frac{\partial^2 M_{x\theta}}{\partial x \partial \theta} + N_\theta \qquad \left.\begin{array}{c} (\theta, x) \in \Gamma_0 \\ t > 0 \end{array}\right.$$

$$= R\left[\hat{g}(t, \theta, x) - \rho_f \frac{\partial \phi}{\partial t}(t, R, \theta, x)\right] \qquad (9.2)$$

$$-R\sum_{i=1}^{s} \left\{\frac{\partial^2 (M_x)_{pe_i}}{\partial x^2} + \frac{1}{R^2}\frac{\partial^2 (M_\theta)_{pe_i}}{\partial \theta^2}\right\}$$

$$u = v = w = \frac{\partial w}{\partial x} = 0 \quad , \quad \text{for all} \ (t, \theta, 0) \ \text{and} \ (t, \theta, \ell)$$

$$\phi(0, r, \theta, x) = \phi_0(r, \theta, x) \quad , \quad \frac{\partial \phi}{\partial t}(0, r, \theta, x) = \phi_1(r, \theta, x) \quad , \quad (r, \theta, x) \in \Omega$$

$$u(0, \theta, x) = u_0(\theta, x) \quad , \quad \frac{\partial u}{\partial t}(0, \theta, x) = u_1(\theta, x)$$

$$v(0, \theta, x) = v_0(\theta, x) \quad , \quad \frac{\partial v}{\partial t}(0, \theta, x) = v_1(\theta, x) \qquad \left.\begin{array}{c} \\ \\ \end{array}\right\} \quad , \quad (\theta, x) \in \Gamma_0 \,.$$

$$w(0, \theta, x) = w_0(\theta, x) \quad , \quad \frac{\partial w}{\partial t}(0, \theta, x) = w_1(\theta, x)$$

As discussed in Chapter 3, the strong form of the modeling equations leads to first and second derivatives of both internal and external moment and force resultants in the shell component of the system. Due to the piecewise constant nature of parameters in these resultants, this yields derivatives of the spatial Heaviside function and Dirac delta and consequently unbounded (spatially discontinuous) control inputs. Both difficulties are alleviated in the weak formulation. Moreover, smoothness requirements on approximating elements are reduced in the weak or variational form of the modeling equations.

9.1.2 Weak Form of Modeling Equations

The state for the problem with dynamics in second-order form is taken to be $y = (\phi, u, v, w)$ in the Hilbert space $H = \bar{L}^2(\Omega) \times L^2(\Gamma_0) \times L^2(\Gamma_0) \times L^2(\Gamma_0)$. The choice of the space $\bar{L}^2(\Omega)$, defined as the quotient of $L^2(\Omega)$ over the constant functions, results from the fact that the potentials are determined only to within a constant. For $\Phi = (\phi, u, v, w)$ and $\Psi = (\xi, \eta_1, \eta_2, \eta_3)$, the H inner product is taken to be

$$\langle \Phi, \Psi \rangle_H = \int_\Omega \frac{\rho_f}{c^2} \phi \bar{\xi} d\omega + \int_{\Gamma_0} \rho h u \overline{\eta_1} d\gamma + \int_{\Gamma_0} \rho h v \overline{\eta_2} d\gamma + \int_{\Gamma_0} \rho h w \overline{\eta_3} d\gamma$$

where $d\gamma = R d\theta dx$, $d\omega = r dr d\theta dx$ and the overbars here denote complex conjugation.

To provide a class of test functions which are considered when defining a variational form of the problem, we also define the Hilbert space $V = \bar{H}^1(\Omega) \times H_0^1(\Gamma_0) \times H_0^1(\Gamma_0) \times H_0^2(\Gamma_0)$ where $\bar{H}^1(\Omega)$ is the quotient space of $H^1(\Omega)$ over the constant functions. The subscript 0 in the remaining components of the product space denotes the subset of functions in the traditional Sobolev spaces which satisfy the essential boundary conditions $u = v = w = \frac{\partial w}{\partial x} = 0$ at $x = 0, \ell$. Motivated by the form of the strain energy for the shell and wave operator for the acoustic field, the V inner product for the system in which the shell is *devoid* of patches is

$$\langle \Phi, \Psi \rangle_V = \int_\Omega \rho_f \nabla \phi \cdot \overline{\nabla \xi} d\omega$$

$$+ \int_{\Gamma_0} \frac{Eh}{(1 - \nu^2)} \left\{ \left[\frac{\partial u}{\partial x} + \frac{\nu}{R} \left(\frac{\partial v}{\partial \theta} + w \right) \right] \frac{\overline{\partial \eta_1}}{\partial x} + \frac{1}{2R}(1 - \nu) \left[\frac{\partial v}{\partial x} + \frac{1}{R} \frac{\partial u}{\partial \theta} \right] \frac{\overline{\partial \eta_1}}{\partial \theta} \right\} d\gamma$$

$$+ \int_{\Gamma_0} \frac{Eh}{(1 - \nu^2)} \left\{ \frac{1}{R} \left[\frac{1}{R} \frac{\partial v}{\partial \theta} + \frac{w}{R} + \nu \frac{\partial u}{\partial x} \right] \frac{\overline{\partial \eta_2}}{\partial \theta} + \frac{1}{2}(1 - \nu) \left[\frac{\partial v}{\partial x} + \frac{1}{R} \frac{\partial u}{\partial \theta} \right] \frac{\overline{\partial \eta_2}}{\partial x} \right\} d\gamma$$

$$+ \int_{\Gamma_0} \frac{Eh}{(1 - \nu^2)} \left\{ \frac{1}{R} \left[\frac{1}{R} \frac{\partial v}{\partial \theta} + \frac{w}{R} + \nu \frac{\partial u}{\partial x} \right] \overline{\eta_3} + \frac{h^2}{12} \left[\frac{\partial^2 w}{\partial x^2} + \frac{\nu}{R^2} \frac{\partial^2 w}{\partial \theta^2} \right] \frac{\overline{\partial^2 \eta_3}}{\partial x^2} \right.$$

$$\left. + \frac{h^2}{12R^2} \left[\frac{1}{R^2} \frac{\partial^2 w}{\partial \theta^2} + \nu \frac{\partial^2 w}{\partial x^2} \right] \frac{\overline{\partial^2 \eta_3}}{\partial \theta^2} + \frac{h^2}{6R^2}(1 - \nu) \frac{\partial^2 w}{\partial x \partial \theta} \frac{\overline{\partial^2 \eta_3}}{\partial x \partial \theta} \right\} d\gamma .$$

It should be noted that when modeling shells with surface-mounted piezo-ceramic patches, the thickness, Young's modulus and Poisson ratio vary in regions covered by patches as indicated by (3.3). The V inner product must be modified accordingly when considering such cases.

The energy principles discussed in Chapter 2, or integration in combination with the use of Green's theorem then yields the second-order variational system

$$\int_\Omega \frac{\rho_f}{c^2}\frac{\partial^2\phi}{\partial t^2}\overline{\xi}d\omega + \int_\Omega \rho_f \nabla\phi \cdot \overline{\nabla\xi}d\omega$$

$$+ \int_{\Gamma_0}\left\{ \rho h \frac{\partial^2 u}{\partial t^2}\overline{\eta_1} + N_x \overline{\frac{\partial\eta_1}{\partial x}} + \frac{1}{R}N_{\theta x}\overline{\frac{\partial\eta_1}{\partial\theta}} \right\}d\gamma$$

$$+ \int_{\Gamma_0}\left\{ \rho h \frac{\partial^2 v}{\partial t^2}\overline{\eta_2} + \frac{1}{R}N_\theta \overline{\frac{\partial\eta_2}{\partial\theta}} + N_{x\theta}\overline{\frac{\partial\eta_2}{\partial x}} \right\}d\gamma$$

$$+ \int_{\Gamma_0}\left\{ \rho h \frac{\partial^2 w}{\partial t^2}\overline{\eta_3} + \frac{1}{R}N_\theta\overline{\eta_3} - M_x \overline{\frac{\partial^2\eta_3}{\partial x^2}} - \frac{1}{R^2}M_\theta \overline{\frac{\partial^2\eta_3}{\partial\theta^2}} \right.$$

$$\left. - \frac{2}{R}M_{x\theta}\overline{\frac{\partial^2\eta_3}{\partial x\partial\theta}} \right\}d\gamma \tag{9.3}$$

$$+ \int_{\Gamma_0}\rho_f \left\{ \frac{\partial\phi}{\partial t}\overline{\eta_3} - \frac{\partial w}{\partial t}\overline{\xi} \right\}d\gamma$$

$$= \int_{\Gamma_0}\sum_{i=1}^s \left\{ (N_x)_{pe_i}\overline{\frac{\partial\eta_1}{\partial x}} + \frac{1}{R}(N_\theta)_{pe_i}\overline{\frac{\partial\eta_2}{\partial\theta}} \right.$$

$$\left. - (M_x)_{pe_i}\overline{\frac{\partial^2\eta_3}{\partial x^2}} - \frac{1}{R^2}(M_\theta)_{pe_i}\overline{\frac{\partial^2\eta_3}{\partial\theta^2}} \right\}d\gamma$$

$$+ \int_{\Gamma_0}\hat{g}\overline{\eta_3}d\gamma$$

for all $(\xi,\eta_1,\eta_2,\eta_3) \in V$.

It is noted that in this form, the derivatives have been transferred from the shell's moment and force resultants onto the test functions. This eliminates the problem of having to approximate derivatives of the Heaviside function and Dirac delta as well as reduces smoothness requirements when approximating solutions.

The system dynamics can be approximated by formally replacing the state variables in (9.3) by their finite dimensional approximations and constructing the resulting matrix system. In order to determine convergence criteria as well as establish well-posedness for the model, however, it is advantageous to pose the problem in terms of sesquilinear forms and bounded linear operators which they generate.

9.1.3   Abstract Problem Formulation and Model Well-Posedness

To define appropriate sesquilinear forms $\sigma_i : V \times V \to \mathbb{C}$, $i = 1, 2$, we group the stiffness and wave contributions separately from damping and coupling terms, thus leading to the definitions

$$\sigma_1(\Phi, \Psi) = \langle \Phi, \Psi \rangle_V$$

and

$$\sigma_2(\Phi, \Psi) =$$

$$\int_{\Gamma_0} \frac{c_D h}{(1-\nu^2)} \left\{ \left[\frac{\partial u}{\partial x} + \frac{\nu}{R} \left(\frac{\partial v}{\partial \theta} + w \right) \right] \overline{\frac{\partial \eta_1}{\partial x}} + \frac{1}{2R}(1-\nu) \left[\frac{\partial v}{\partial x} + \frac{1}{R} \frac{\partial u}{\partial \theta} \right] \overline{\frac{\partial \eta_1}{\partial \theta}} \right\} d\gamma$$

$$+ \int_{\Gamma_0} \frac{c_D h}{(1-\nu^2)} \left\{ \frac{1}{R} \left[\frac{1}{R} \frac{\partial v}{\partial \theta} + \frac{w}{R} + \nu \frac{\partial u}{\partial x} \right] \overline{\frac{\partial \eta_2}{\partial \theta}} + \frac{1}{2}(1-\nu) \left[\frac{\partial v}{\partial x} + \frac{1}{R} \frac{\partial u}{\partial \theta} \right] \overline{\frac{\partial \eta_2}{\partial x}} \right\} d\gamma$$

$$+ \int_{\Gamma_0} \frac{c_D h}{(1-\nu^2)} \left\{ \frac{1}{R} \left[\frac{1}{R} \frac{\partial v}{\partial \theta} + \frac{w}{R} + \nu \frac{\partial u}{\partial x} \right] \overline{\eta_3} + \frac{h^2}{12} \left[\frac{\partial^2 w}{\partial x^2} + \frac{\nu}{R^2} \frac{\partial^2 w}{\partial \theta^2} \right] \overline{\frac{\partial^2 \eta_3}{\partial x^2}} \right.$$

$$\left. + \frac{h^2}{12 R^2} \left[\frac{1}{R^2} \frac{\partial^2 w}{\partial \theta^2} + \nu \frac{\partial^2 w}{\partial x^2} \right] \overline{\frac{\partial^2 \eta_3}{\partial \theta^2}} + \frac{h^2}{6 R^2}(1-\nu) \frac{\partial^2 w}{\partial x \partial \theta} \overline{\frac{\partial^2 \eta_3}{\partial x \partial \theta}} \right\} d\gamma$$

$$+ \int_{\Gamma_0} \rho_f \left(\phi \overline{\eta_3} - w \overline{\xi} \right) d\gamma .$$

As detailed in [39], σ_1 satisfies

$$|\sigma_1(\Phi, \Psi)| \le c_1 |\Phi|_V |\Psi|_V \ , \ \text{for some} \ \ c_1 \in \mathbb{R} \qquad \text{(bounded)}$$

$$\text{Re} \ \sigma_1(\Phi, \Phi) \ge c_2 |\Phi|_V^2 \ , \ \text{for some} \ \ c_2 > 0 \qquad (V\text{-elliptic}) \qquad (9.4)$$

$$\sigma_1(\Phi, \Psi) = \overline{\sigma_1(\Psi, \Phi)} \qquad \qquad \text{(symmetric)}$$

for all $\Phi, \Psi \in V$. The damping term σ_2, on the other hand, satisfies

$$|\sigma_2(\Phi, \Psi)| \le c_3 |\Phi|_V |\Psi|_V \ , \ \text{for some} \ \ c_3 \in \mathbb{R} \qquad \text{(bounded)}$$

$$\qquad \qquad \qquad \qquad \qquad \qquad \qquad \qquad \qquad \qquad (9.5)$$

$$\text{Re} \ \sigma_2(\Phi, \Phi) \ge c_4 |\Phi|_H^2 \ , \ \text{for some} \ \ c_4 \ge 0 \qquad (H\text{-semielliptic})$$

with the weaker semiellipticity due to the absence of damping in the acoustic cavity. As discussed in [41], this latter condition can be strengthened to obtain a V-elliptic bound if cavity dimensions or atmospheric properties warrant the inclusion of strong damping in the model.

To account for the control contributions, we let U denote the Hilbert space containing the control inputs, and we define the control operator $B \in \mathcal{L}(U, V^*)$ by

$$\langle Bu(t), \Psi \rangle_{V^*, V} = \int_{\Gamma_0} \sum_{i=1}^{s} \left\{ (N_x)_{pe_i} \overline{\frac{\partial \eta_1}{\partial x}} + \frac{1}{R}(N_\theta)_{pe_i} \overline{\frac{\partial \eta_2}{\partial \theta}} \right.$$

$$\left. - (M_x)_{pe_i} \overline{\frac{\partial^2 \eta_3}{\partial x^2}} - \frac{1}{R^2}(M_\theta)_{pe_i} \overline{\frac{\partial^2 \eta_3}{\partial \theta^2}} \right\} d\gamma$$

for $\Psi \in V$, where $\langle \cdot, \cdot \rangle_{V^*,V}$ is the usual duality pairing. We again caution the reader that the control $u(t) \in U$ should not be confused with the axial or longitudinal displacement $u(t, \theta, x)$ appearing in the shell model. The use of this variable for both quantities is well established in the literature and its representation of displacement or control should be clear from the context.

Finally, by letting $\tilde{g} = \left(0, 0, 0, \frac{\hat{g}}{\rho h}\right)$, we can write the system (9.3) in the abstract variational form

$$\langle \ddot{y}(t), \Psi \rangle_{V^*,V} + \sigma_2(\dot{y}(t), \Psi) + \sigma_1(y(t), \Psi) = \langle Bu(t) + \tilde{g}(t), \Psi \rangle_{V^*,V} , \qquad (9.6)$$

which is of the general abstract variational form (7.1) for a damped second-order system.

Following in the framework of Chapter 4 and Chapter 7, we define the product spaces $\mathcal{H} = V \times H$ and $\mathcal{V} = V \times V$ and consider the state $z(t) = [y(t), \dot{y}(t)]^T \in \mathcal{H}$. The product space forcing terms are formulated as

$$g(t) = \begin{bmatrix} 0 \\ \tilde{g}(t) \end{bmatrix} \quad , \quad \mathcal{B}u(t) = \begin{bmatrix} 0 \\ Bu(t) \end{bmatrix}$$

and the system operator is taken to be

$$\mathrm{dom}\mathcal{A} = \{(\phi_1, \phi_2) \in \mathcal{H} | \phi_2 \in V, A_1\phi_1 + A_2\phi_2 \in H\}$$

$$\mathcal{A} = \begin{bmatrix} 0 & I \\ -A_1 & -A_2 \end{bmatrix} \qquad (9.7)$$

with $A_1, A_2 \in \mathcal{L}(V, V^*)$ defined as usual by

$$\langle A_i\phi, \eta \rangle_{V^*,V} = \sigma_i(\phi, \eta) \quad , \quad i = 1, 2 .$$

As discussed in Chapter 4 and Chapter 7, the weak form (9.6) is then formally equivalent to the strong form of the equation

$$\dot{z}(t) = \mathcal{A}z(t) + \mathcal{B}u(t) + g(t) \quad \text{in } \mathcal{V}^* = V \times V^*$$

$$z(0) = z_0 = \begin{bmatrix} y(0) \\ \dot{y}(0) \end{bmatrix} . \qquad (9.8)$$

In Chapter 4, it was established that for systems in which σ_1 and σ_2 satisfy the boundedness, ellipticity and symmetry conditions (9.4) and (9.5), the operator $\mathcal{A}$ defined in (9.7) generates a semigroup $\mathcal{T}(t)$ on $\mathcal{H}$. While that semigroup is strongly continuous, it is *not* analytic since σ_2 is not V-elliptic for models with undamped acoustic fields.

As a first step toward establishing the well-posedness of the system model, the semigroup $\mathcal{T}$ on $\mathcal{H}$ can be extended through extrapolation space techniques to the semigroup $\hat{\mathcal{T}}$ on $\mathcal{W}^* = [\mathrm{dom}\mathcal{A}^*]^* \supset \{0\} \times V^*$ so as to be compatible with the control input and external forces (see Chapter 4 for details). This guarantees that the mild solution

$$z(t) = \hat{\mathcal{T}}(t)z_0 + \int_0^t \hat{\mathcal{T}}(t-s)\left[\mathcal{B}u(s) + g(s)\right] ds \qquad (9.9)$$

is well-defined for $\mathcal{B}u + g \in L^2(0, T; \mathcal{V}^*)$. Furthermore, the results from Chapter 4 along with standard semigroup results [152, page 109] guarantee that if the mappings $t \mapsto u(t)$ from $[0, T]$ to $\mathbb{R}^s$ and $t \mapsto g(t)$ from $[0, T]$ to $\mathcal{V}^*$ are Lipschitz continuous, then (9.8) has a unique strong solution on $[0, T]$ given by (9.9). This establishes the well-posedness of the fully coupled structural acoustic system model.

9.1.4 Control Applications

By modeling the structural acoustic system in the manner described in this section, passive and active actuator contributions as well as coupling between the structural and acoustic field can be incorporated in a coupled set of PDE. Because the model is well-posed in the sense that a unique solution exists, it can be used as a basis for approximating the system dynamics for simulation (prediction) or control purposes. For physical systems that do not warrant the inclusion of strong internal damping in the acoustic field, the infinite dimensional control theory and convergence analysis of Chapter 7 does not directly apply since the correct modeling system is only weakly damped. While the extension of the infinite dimensional theory to this regime is under investigation, a strategy that has numerically been proven effective is to discretize the system and apply appropriate finite dimensional control results from Chapter 8.

The examples in the following two sections illustrate the construction of controllers in this manner. A 3-D system (modeling another experimental test device at NASA) consisting of a hardwalled cylinder with a vibrating plate at one end is considered in the next section. The system is discretized using a Fourier/Galerkin method to yield an N-dimensional matrix system of the form (8.5). The performance of an LQR full state feedback control law is then illustrated through a set of numerical examples.

The development of LQG and H^∞/MinMax compensators for structural acoustic systems is illustrated in Section 9.3. The system considered there consists of a 2-D slice from the 3-D system of Section 9.2. A Galerkin discretization of the model again yields a matrix system of the form (8.5), at which point, the LQG and H^∞/MinMax compensator theory of Sections 8.1.2 and 8.1.3 can be applied. Numerical examples provide comparisons between attenuation levels attained with the H^∞ compensator and levels reached with an LQR full state feedback control law. Taken in combination, Section 9.2 and Section 9.3 illustrate various facets that must be considered when designing smart material, PDE-based controllers for structural acoustic systems.

9.2   Noise Attenuation in a 3-D Cavity: LQR Control

To illustrate the construction of an LQR controller for a typical structural acoustic system, we consider the model for a second experimental setup in the Acoustics Division, NASA Langley Research Center. That apparatus consists of a concrete cylinder with a thin, circular aluminum plate mounted at one end. A loudspeaker adjacent to the plate excites the plate which in turn transmits energy to the interior acoustic field. Negligible energy is transmitted from the acoustic field to the enclosing cylinder due to the thickness of the concrete. Control is implemented through the excitation of piezoceramic patches mounted in pairs to the plate.

9.2.1   System Model

The experimental apparatus is modeled by a cylindrical domain Ω having length ℓ and radius a as pictured in Figure 9.3a. At one end of the cylinder is a clamped flexible plate of thickness h which is assumed to have Kelvin-Voigt damping. Bonded to the plate are sectorial and circular piezoceramic patches (see Figure 9.3b) which are placed in pairs so that a bending moment is produced when voltage is appled. The patches and glue layer are assumed to have thicknesses h_{pe} and $h_{b\ell}$, respectively.

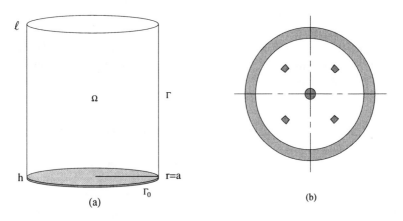

Figure 9.3: (a) The cylindrical acoustic cavity; (b) The circular plate with patches.

State variables for the plate and cavity contributions are taken to be w and ϕ, respectively, where w denotes the transverse plate displacement and ϕ is the acoustic velocity potential (the choice of w rather than u to denote axial displacements is made to preserve consistency with plate notation in Chapter 8 and avoid confusion with the control term $u(t)$). The state space and space of test functions are taken to be $H = \bar{L}^2(\Omega) \times L^2(\Gamma_0)$ and $V = \bar{H}^1(\Omega) \times H_0^2(\Gamma_0)$ where $H_0^2(\Gamma_0) = \{\psi \in H^2(\Gamma_0) \mid \psi = \psi' = 0 \text{ at } r = a\}$. As discussed in Section 9.1, the quotient spaces $\bar{L}^2(\Omega)$ and $\bar{H}^1(\Omega)$ provide a natural setting when considering the acoustic potential since it is determined only to within a constant.

Following the strategy of Section 9.1, the structural and acoustic wave equations are coupled through back pressure and continuity of velocity conditions. The system is then posed in weak form to reduce smoothness requirements on approximating elements and eliminate the problems associated with differentiating discontinuous moments. As detailed in [34, 170], the linearized variational form of the coupled system of equations modeling this setup is

$$\int_\Omega \frac{\rho_f}{c^2} \frac{\partial^2 \phi}{\partial t^2} \overline{\xi} d\omega + \int_\Omega \rho_f \nabla \phi \cdot \overline{\nabla \xi} d\omega \tag{9.10a}$$

$$+ \int_{\Gamma_0} \rho h \frac{\partial^2 w}{\partial t^2} \overline{\eta} d\gamma - \int_{\Gamma_0} M_r \overline{\frac{\partial^2 \eta}{\partial r^2}} d\gamma - \int_{\Gamma_0} \frac{1}{r^2} M_\theta \left[r \overline{\frac{\partial \eta}{\partial r}} + \overline{\frac{\partial^2 \eta}{\partial \theta^2}} \right] d\gamma \tag{9.10b}$$

$$-2 \int_{\Gamma_0} \frac{1}{r^2} M_{r\theta} \left[r \overline{\frac{\partial^2 \eta}{\partial r \partial \theta}} - \overline{\frac{\partial \eta}{\partial \theta}} \right] d\gamma \tag{9.10b}$$

$$+ \int_{\Gamma_0} \rho_f \left(\frac{\partial \phi}{\partial t} \overline{\eta} - \frac{\partial w}{\partial t} \overline{\xi} \right) d\gamma \tag{9.10c}$$

$$= \int_{\Gamma_0} \sum_{i=1}^{s} \mathcal{K}_i^B u_i(t) \chi_{pe_i}(r, \theta) \overline{\nabla^2 \eta} d\gamma + \int_{\Gamma_0} \hat{g} \overline{\eta} d\gamma \tag{9.10d}$$

for all test functions $(\xi, \eta) \in V$. Note that the differentials are $d\omega = r \, dr \, d\theta \, dx$ and $d\gamma = r \, dr \, d\theta$ whereas the overbars in (9.10) denote complex conjugation.

In this weak formulation of the model, (9.10a) contains the wave components in the weak form used in (9.3) when modeling the previous structural acoustic system. Component (9.10b) is the weak form of the circular plate equation discussed in Section 3.2.2 and summarized in (8.12). Expressions for the internal moments M_r, M_θ and $M_{r\theta}$ in terms of material properties are given in (3.14). The terms in (9.10c) model the mechanisms (backpressure and velocity) which couple the plate and acoustic dynamics. Finally, the contributions due to the activation of the piezoceramic patches through applied voltages $u_i(t)$ are found in the first component of (9.10d) while the forces $\hat{g}$ due to the external acoustic field are contained in the second component. We reiterate that while the patch constants $\mathcal{K}_i^B$ can be analytically calculated using (3.16), in applications, their values must be estimated using fit-to-data techniques due wide variations in materials and manufacturer specifications.

To approximate the system dynamics, Fourier-Galerkin schemes are used to discretize the infinite dimensional system (9.10). The plate displacement is approximated in the manner described in Section 8.3.1. Specifically, the basis is taken to be $B_k^N(r, \theta) = r^{|\hat{m}|} B_n^m(r) e^{im\theta}$ where $B_n^m(r)$ denotes the nth modified cubic spline satisfying the boundary conditions, and $\hat{m}$ is the mth Fourier index which is truncated to control conditioning. The plate displacement is then approximated by

$$w^N(t, r, \theta) = \sum_{k=1}^{N} w_k^N(t) B_k^N(r, \theta) \tag{9.11}$$

which is the same approximation (8.13) used when discretizing the system to construct a plate controller.

As detailed in [34, 170], a suitable Fourier-Galerkin expansion for the potential is

$$\phi^{\mathcal{M}}(t,r,\theta,x) = \sum_{\ell=1}^{\mathcal{M}} \phi_\ell^{\mathcal{M}}(t) \mathcal{P}_\ell^{\mathcal{M}}(r,\theta,x)$$

$$= \sum_{p=0}^{P_w} \sum_{\substack{m=-M_w}}^{M_w} \sum_{\substack{n=0 \\ p+|m|+n\neq 0}}^{N_w^{p,m}} \phi_{pmn}(t) e^{im\theta} r^{|\hat{m}|} P_n^{p,m}(r) P_p(x) \tag{9.12}$$

where

$$\hat{m} = \begin{cases} m & , \ |m| = 0, \cdots, 5 \\ 5 & , \ |m| = 6, \cdots, M_w \end{cases}$$

and

$$P_n^{p,m}(r) = \begin{cases} P_1(r) - 1/3 & , \ p = m = 0, n = 1 \\ P_n(r) & , \ \text{otherwise} . \end{cases}$$

Here $P_n(r)$ and $P_p(x)$ are the nth and p^{th} Legendre polynomials which have been mapped to the intervals $(0, a)$ and $(0, \ell)$, respectively. The term $P_1(r) - 1/3$ when $p = m = 0, n = 1$ results from the orthogonality properties of the Legendre polynomials and arises when enforcing the condition $\int_\Omega \phi^{\mathcal{M}}(t,r,\theta,x) d\omega = 0$ so as to guarantee that the functions are suitable as a basis for the quotient space. The inclusion of the weight $r^{|\hat{m}|}$ again incorporates the decay of the analytic solution near the origin while ensuring its uniqueness at that point. Finally, we note that the limit $N_w^{p,m}$ is given by $N_w^{p,m} = N_w + 1$ when $p + |m| \neq 0$ and $N_w^{p,m} = N_w$ when $p = m = 0$, which implies that $\mathcal{M} = (2M_w + 1)(N_w + 1)(P_w + 1) - 1$ basis functions are used in the wave expansion.

The substitution of the expansions (9.11) and (9.12) into (9.10) with orthogonalization against basis elements then yields the matrix system

$$M^N \dot{z}^N(t) = \overline{A}^N z^N(t) + \overline{B}^N u(t) + \overline{g}^N(t)$$
$$M^N z^N(0) = \overline{z}_0^N$$

where $N = 2(\mathcal{M}+\mathcal{N})$. The column N vector z^N has the form $z^N = [\varphi(t), \vartheta(t), \dot\varphi(t), \dot\vartheta(t)]^T$ where $\varphi(t) = [\phi_1^{\mathcal{M}}(t), \cdots, \phi_{\mathcal{M}}^{\mathcal{M}}(t)]$ and $\vartheta(t) = [w_1^{\mathcal{N}}(t), \cdots, w_{\mathcal{N}}^{\mathcal{N}}(t)]$ contain the generalized Fourier coefficients for the approximate acoustic potential and plate displacement, respectively. As in previous discussions, $u(t) = [u_1(t), \cdots, u_s(t)]^T$ contains the s patch input variables. The system matrices and vectors have the form

$$M^N = \left[\begin{array}{cc|cc} K_A & & & \\ & K_D & & \\ \hline & & M_A & \\ & & & M \end{array}\right] \quad , \quad \overline{A}^N = \left[\begin{array}{cc|cc} & & K_A & \\ & & & K_D \\ \hline -K_A & & -A_{c1} \\ & -K_D & -A_{c2} & -K_{c_D} \end{array}\right]$$

and

$$\overline{B}^N = \left[0, 0, 0, \tilde{B}\right]^T \quad , \quad \overline{g}^N(t) = [0, 0, 0, \tilde{g}(t)]^T .$$

The vector $\overline{z}_0^N = [f_1, g_1, f_2, g_2]^T$ contains the projections of the initial values into the approximating finite dimensional subspaces.

The $\mathcal{N} \times \mathcal{N}$ matrices M, K_D, K_{c_D} are the mass, stiffness and damping matrices which arise when approximating the dynamics of a damped plate with fixed-end boundary conditions using the expansion (9.11). Moreover, $\tilde{B} \in \mathbb{R}^{\mathcal{N} \times s}$ and $\tilde{g}(t) \in \mathbb{R}^{\mathcal{N}}$ are the control and forcing terms for the isolated plate while g_1, g_2 incorporate the initial plate conditions. The components of these matrices and vectors are detailed in (8.14).

The $\mathcal{M} \times \mathcal{M}$ matrices M_A and K_A have the components

$$[M_A]_{i,j} = \int_{\Omega} \frac{\rho_f}{c^2} \mathcal{P}_i^{\mathcal{M}} \overline{\mathcal{P}_j^{\mathcal{M}}} \, d\omega$$

$$[K_A]_{i,j} = \int_{\Omega} \nabla \mathcal{P}_i^{\mathcal{M}} \cdot \overline{\nabla \mathcal{P}_j^{\mathcal{M}}} \, d\omega ,$$

$i, j = 1, \cdots, \mathcal{M}$, while f_1, f_2 denote the projections of the initial acoustic components into the subspace spanned by the Legendre polynomials. It can be noted that the M_A and K_A are the mass and stiffness matrices which arise when approximating the *uncoupled* wave equation with Neumann boundary conditions on a cylindrical domain.

Finally, the contributions from the coupling terms are contained in the matrices

$$[A_{c1}]_{i,\ell} = -\int_{\Gamma_0} \rho_f B_\ell^N \overline{\mathcal{P}_i^{\mathcal{M}}} d\gamma \quad , \quad [A_{c2}]_{\ell,i} = \int_{\Gamma_0} \rho_f \mathcal{P}_\ell^{\mathcal{M}} \overline{B_i^N} d\gamma$$

where the index ranges are $i = 1, \cdots, \mathcal{M}$ and $\ell = 1, \cdots, \mathcal{N}$.

Multiplication by the inverted system mass matrix yields the equivalent Cauchy system

$$\dot{z}^N(t) = A^N z^N(t) + B^N u(t) + g^N(t)$$

$$z^N(0) = z_0^N .$$

(9.13)

In this form, the finite dimensional control problem can be discussed.

9.2.2 LQR Control Problem

To demonstrate the capabilities of PDE-based controllers for attenuating interior sound pressure levels when full state measurements are available, we consider here an LQR feedback law subject to a periodic exogenous force. Such a law can be obtained form the finite dimensional theory in Section 8.1.2 by considering observations in $\mathbb{R}^N$ and choosing C to be the identity. In this case, the compensator z_c actually yields the state z (we again use the control convention of dropping the superscript N), and one must solve only the feedback Riccati equation. As detailed in Section 8.1.2 and summarized in Algorithm 8.2, the optimal controlling voltage for the discretized model is

$$u(t) = -R^{-1}B^T\Pi z(t) + R^{-1}B^T r(t)$$

where R is an $s \times s$ diagonal matrix whose components r_{ii}, $i = 1, \cdots, s$, weight the controlling voltage to the ith patch. The Riccati solution Π is obtained by solving the algebraic Riccati equation

$$\Pi A + A^T\Pi - \Pi B R^{-1}B^T\Pi + Q = 0$$

while $r(t)$, solving the differential equation

$$\dot{r}(t) = -\left[A - BR^{-1}B^T\Pi\right]^T r(t) + \Pi g(t)$$
$$r(0) = r(\tau),$$

incorporates information regarding the external force. Finally, the optimal trajectory solves the differential equation

$$\dot{z}(t) = \left[A - BR^{-1}B^T\Pi\right] z(t) + BR^{-1}B^T r(t) + g(t)$$
$$z(0) = z(\tau).$$

As discussed in Chapter 8, the matrix Q can be chosen to emphasize the minimization of particular state variables. It follows from energy considerations (see [12]) that an appropriate choice for Q in this application is

$$Q = M\mathcal{D} = M\mathrm{diag}\left[d_1 I^{\mathcal{M}}, d_2 I^{\mathcal{N}}, d_3 I^{\mathcal{M}}, d_4 I^{\mathcal{N}}\right] \qquad (9.14)$$

where M is the mass matrix, I^k, $k = \mathcal{N}, \mathcal{M}$, denote $k \times k$ identity matrices, and d_i are parameters which are chosen to enhance stability and performance of the feedback.

9.2.3   Numerical Example – LQR Control Law

The attenuation that can be obtained using the LQR control law is illustrated here through a numerical example. So that physical dimensions in the model are consistent with those for the experimental setup, the length and radius of the cavity were taken to be $1.067\,m$ ($42''$) and $a = 0.229\,m$ ($9''$), respectively with a plate having thickness $h = 0.00127\,m$ ($0.05''$) mounted at one end of the enclosing cylinder. A pair of circular piezoceramic patches having thickness $h_{pe} = 0.00018\,m$ ($0.007''$) and radius $rad = 0.019\,m$ ($0.75''$) were located at the center of the plate as depicted in Figure 9.4.

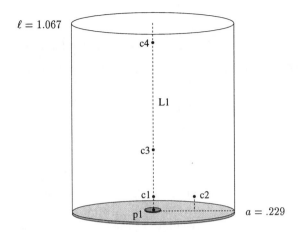

Figure 9.4: The acoustic cavity with a pair of centered circular patches and the observation points $p1 = (0,0)$, $c1 = (0,0,.05)$, $c2 = (a/2,\pi/2,.05)$, $c3 = (0,0,.35)$ and $c4 = (0,0,1.0)$.

The physical parameters that were chosen for the structure and acoustic cavity are summarized in Table 9.1. The flexural rigidity D for the plate was obtained using the "handbook" value $E = 7.1 \times 10^{10}\, N/m^2$ for the Young's modulus of aluminum. The remaining choices are comparable to values found when estimating parameters for the isolated plate with a similar patch configuration (see Table 8.1). We emphasize once again that a first step when determining gains to be used during experimental implementation of the scheme is the estimation of these parameters for the system in the form in which it is going to be controlled. While robustness of a control scheme might allow for some leeway in these values, the results will be degraded and potentially destabilized by the use of inaccurate system parameters.

The acoustic source driving the plate was taken to be

$$\hat{g}(t,r,\theta) = 28.8\sin(500\pi t)$$

which models a periodic plane wave having an rms sound pressure level of 120 dB (again, see Section 9.4 for a summary of acoustical units). As illustrated in examples of [32], the excitation frequency of 250 Hz is close to the natural frequency, 238 Hz, of the third system mode (it should be emphasized that when dealing with coupled systems, it is necessary to consider system modes as compared with modes for the isolated components; while the two sets are often similar, contributions due to coupling will lead to differences in frequency and mode shape).

Structure			Acoustic Cavity	
Parameter	Plate	Plate + Pzt	Parameter	Cavity
$\rho_p \cdot Thickness \; (kg/m^2)$	3.429	3.489	$\rho_f \; (kg/m^3)$	1.21
$D \; (N \cdot m)$	13.601	13.901		
$c_D \; (N \cdot m \cdot sec)$	1.150-4	2.250-4		
ν	.33	.32		
$\mathcal{K}^B \; (N/V)$		.027		
			c (m/sec)	343

Table 9.1: Physical parameters for the structure and acoustic cavity.

When approximating the dynamics of the system, it was determined that the choices $\mathcal{N} = 12$ and $\mathcal{M} = 99$ (see (9.11) and (9.12)) were sufficient for resolving the range of frequencies under consideration and the following results were obtained with those limits. For both the uncontrolled and controlled cases, time histories of the system response over the time interval $[0, .16]$ were calculated at the plate point $p1 = (0,0)$ and cavity points $c1 = (0,0,.05)$, $c2 = (a/2, \pi/2, .05)$, $c3 = (0,0,.35)$ and $c4 = (0,0,1.0)$ (see Figure 9.4). The rms values of the acoustic pressure were also calculated along the axial line $L1 = \{(r, \theta, x) \,|\, r = 0, 0 \leq x \leq 1.067\}$.

For the uncontrolled and controlled responses, the rms pressure levels at the cavity points $c1, c2, c3, c4$ and plate displacement at the point $p1$ are summarized in Table 9.2. These results indicate reductions ranging from 8.5 dB at the point $c1$ near the plate to 18.2 dB at $c3$ in the middle of the cavity. To illustrate the spatial behavior of the uncontrolled and controlled interior acoustic fields, the rms sound pressure levels along the line $L1$ are plotted in Figure 9.5 (both linear and decibel scales are included). These plots show that near the plate, the sound pressure level is reduced by approximately 13 dB. The reduction in pressure is then fairly uniform as one moves toward the back of the cavity with the amount of reduction actually increasing near the back wall.

The temporal behavior of the plate displacement at $p1$ and acoustic pressure at $c3$ is indicated by the time history and frequency response plots in Figures 9.6 and 9.7. It can be seen that following a slight transient stage, the controlled displacement and acoustic pressure are maintained at a nearly uniform level throughout the time interval. The frequency plots illustrate the coupling between the plate and acoustic components as well as the attenuation due to the controller. As noted in the frequency response in Figure 9.7, the transient acoustic dynamics are nearly eliminated in the controlled case, with the remaining response due solely to the 250 Hz driving force.

These results numerically illustrate that very effective sound pressure attenuation can be realized using a PDE-based full state feedback controller. They also illustrate the global acoustic reductions that can be attained by using piezoceramic patches to alter structural dynamics in a manner which significantly reduces the coupling and transmission of energy to the interior acoustic field. From an implementation perspective, however, this approach is not practical since full state measurements are typically not available using current sensor technology. The construction of a compensator for structural acoustic systems is illustrated in the next section.

	$p1$	$c1$	$c2$	$c3$	$c4$
Uncontrolled System	4.1e-05 m	98.2 dB	94.8 dB	104.4 dB	107.9 dB
Controlled System	9.2e-06 m	89.7 dB	84.2 dB	86.2 dB	90.0 dB

Table 9.2: Rms displacement and decibel levels in the uncontrolled and controlled cases as calculated at the plate point $p1$ and cavity points $c1, c2, c3, c4$.

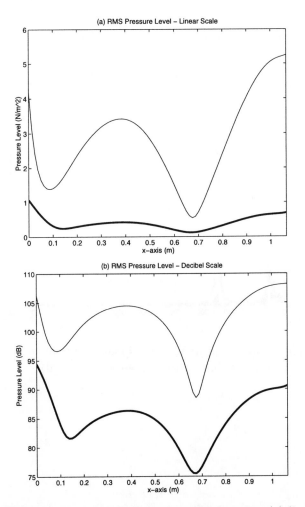

Figure 9.5: RMS sound pressure levels along the line $L1$: (a) linear, and (b) decibel pressure scales; —— (uncontrolled), ▬▬ (controlled).

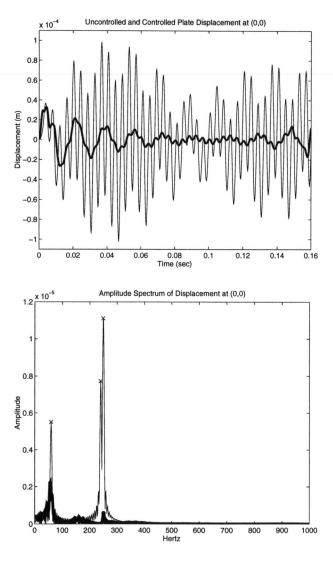

Figure 9.6: Time history and frequency response of the plate displacement at the point $p_1 = (0, 0)$; —— (uncontrolled), ▬▬ (controlled).

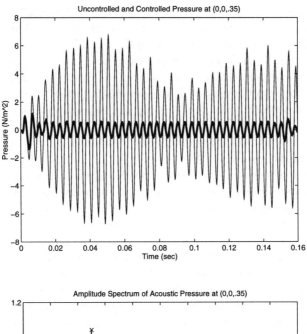

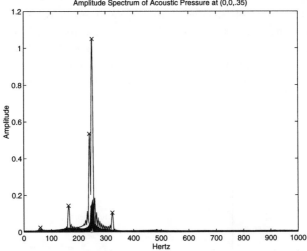

Figure 9.7: Time history and frequency response of the acoustic pressure at the point $c3 = (0, 0, .35)$; ——— (uncontrolled), ━━━ (controlled).

9.3 Noise Attenuation in a 2-D Model: LQG and MinMax Controllers

A 2-D slice from the 3-D experimental apparatus and model of Section 9.2 is used as a model for comparing noise attenuation obtained with LQG and H^∞/MinMax compensators and an LQR full state feedback law. Specifically, the structural acoustic system consists of a 2-D cavity $\Omega = [0, a] \times [0, \ell]$ enclosed on three sides by hard walls Γ and on the fourth by a flexible beam whose unperturbed middle surface is denoted by Γ_0 (see Figure 9.8a). A Cartesian coordinate system is employed so as to remain consistent with models in cited references; the correspondence (x, y) to (r, x) indicates the relationship of the model to a slice from the previously discussed 3-D system. An external field $\hat{g}$ drives the beam which, through structural acoustic coupling, leads to interior acoustic waves. Control is implemented through the excitation of piezoceramic patches on the beam which are driven out-of-phase in order to generate pure bending moments (see Figure 9.8b).

The system is modeled by a linear acoustic wave equation coupled through backpressure and velocity conditions with an Euler-Bernoulli beam equation. As detailed in the discussion of Section 9.2, the acoustic potential ϕ and transverse displacement w are convenient choices for the states. Fixed-end boundary conditions, modeling the conditions in the corresponding 3-D experimental setup, are assumed for the beam while zero velocity conditions, $\nabla\phi \cdot \hat{n} = 0$, are used to model the hard cavity walls Γ. Throughout this discussion, the acoustic field is assumed to have density ρ_f and wave speed c. The density, flexural rigidity and Kelvin-Voigt damping parameters for the beam are denoted by ρ, EI and $c_D I$, respectively. For the simulation examples reported on here, we ignored mass, stiffness and damping loading of the beam due to the patches. This was simply for expediency and, as noted in Chapters 5, 6 and 8, we have not made such a simplification when dealing with experimental data or implementation.

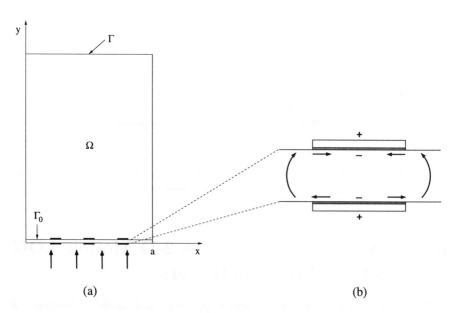

(a)   (b)

Figure 9.8: (a) Structural acoustic system; (b) piezoceramic patches driven out-of-phase to create pure bending moments.

9.3.1   Modeling Equations and Discrete System

The strong form of the coupled system equations can be derived in a manner analogous to that detailed in Section 9.2 for the 3-D model. For approximation and control purposes, it is advantageous to consider the weak form of the model with derivatives transferred from discontinuous moments to sufficiently smooth test functions. The state space for the weak formulation is taken to be $H = \bar{L}^2(\Omega) \times L^2(\Gamma_0)$ while test functions are chosen in $V = \bar{H}^1(\Omega) \times H_0^2(\Gamma_0)$. Again, $\bar{L}^2(\Omega)$ and $\bar{H}^1(\Omega)$ denote quotient spaces while $H_0^2(\Gamma_0)$ is the subset $H^2(\Gamma_0)$ whose elements satisfy the fixed boundary conditions.

As detailed in [12, 36], the resulting weak form of the system equations is

$$\int_\Omega \frac{\rho_f}{c^2} \frac{\partial^2 \phi}{\partial t^2} \xi d\omega + \int_\Omega \rho_f \nabla \phi \cdot \nabla \xi d\omega \tag{9.15a}$$

$$+ \int_{\Gamma_0} \rho \frac{\partial^2 w}{\partial t^2} \eta d\gamma + \int_{\Gamma_0} EI \frac{\partial^2 w}{\partial x^2} \frac{\partial^2 \eta}{\partial x^2} d\gamma + \int_{\Gamma_0} c_D I \frac{\partial^3 w}{\partial x^2 \partial t} \frac{\partial^2 \eta}{\partial x^2} d\gamma \tag{9.15b}$$

$$+ \int_{\Gamma_0} \rho_f \left(\frac{\partial \phi}{\partial t} \eta - \frac{\partial w}{\partial t} \xi \right) d\gamma \tag{9.15c}$$

$$= \int_{\Gamma_0} \sum_{i=1}^s \mathcal{K}_i^B u_i(t) \chi_{pe_i}(x) \frac{\partial^2 \eta}{\partial x^2} d\gamma + \int_{\Gamma_0} \hat{g} \eta d\gamma \tag{9.15d}$$

for all $(\xi, \eta) \in V$. The components in (9.15a) and (9.15c) contain the variational or energy form of the wave equation and coupling conditions discussed in Section 9.1 and Section 9.2. The internal beam dynamics are expressed in (9.15b) while the external contributions due to the piezoceramic patches and driving force are contained in (9.15d). Details regarding the internal contributions due to the patches (which we have ignored here) as well as the moments generated by an input voltage can be found in Section 3.3.2.

A corresponding discrete system is obtained using a Galerkin method analogous to that described for the 3-D system in Section 9.2, with cubic splines and Legendre polynomials again used as basis elements for the displacement and potential. Specifically, the beam displacement is approximated by the expansion

$$w^\mathcal{N}(t, x) = \sum_{k=1}^\mathcal{N} w_k^\mathcal{N}(t) B_k^\mathcal{N}(x) \quad , \quad \mathcal{N} = N_b - 1$$

where the basis $\{B_k^\mathcal{N}\}_{k=1}^{N_b-1}$ consists of standard cubic splines with $B_1^\mathcal{N}, B_{N_b-1}^\mathcal{N}$ modified to satisfy the fixed-end boundary conditions [12]. The modification of the cubic splines to satisfy the four boundary essential conditions reduces the number of basis functions from the standard number of $N_b + 3$ to $N_b - 1$.

A suitable Galerkin expansion for the potential is

$$\phi^{\mathcal{M}}(t,x,y) \;=\; \sum_{m=0}^{M_w}\sum_{\substack{n=0 \\ m+n\neq 0}}^{N_w} \phi_{mn}(t)P_n(x)P_m(y)$$

$$=\; \sum_{\ell=1}^{\mathcal{M}} \phi_\ell(t)\mathcal{P}_\ell^{\mathcal{M}}(x,y) \;\;,\quad \mathcal{M}=M_w\cdot N_w-1$$

where M_w, N_w again denote wave indices and $P_n(x)$ and $P_m(y)$ denote the standard Legendre polynomials that have been scaled by transformation to the intervals $[0,a]$ and $[0,\ell]$, respectively. In addition to yielding exponential accuracy for the acoustic approximations, the Legendre polynomials efficiently provide a basis for the quotient spaces $\bar{L}^2(\Omega)$ and $\bar{H}^1(\Omega)$ through enforcement of the condition $m+n\neq 0$ which eliminates the constant function.

By restricting the infinite dimensional system (9.15) to $H^{\mathcal{M}+\mathcal{N}}\times H^{\mathcal{M}+\mathcal{N}}$, where $H^{\mathcal{M}+\mathcal{N}}=H_c^{\mathcal{M}}\times H_b^{\mathcal{N}}$, $H_c^{\mathcal{M}}=\operatorname{span}\{\mathcal{P}_\ell^{\mathcal{M}}\}_{\ell=1}^{\mathcal{M}}$, $H_b^{\mathcal{N}}=\operatorname{span}\{B_k^{\mathcal{N}}\}_{k=1}^{\mathcal{N}}$, and choosing $(\xi,\eta)\in H^{\mathcal{M}+\mathcal{N}}$ to be basis functions, one obtains the finite dimensional system

$$\int_\Omega \frac{\rho_f}{c^2}\frac{\partial^2\phi^{\mathcal{M}}}{\partial t^2}\mathcal{P}_\ell^{\mathcal{M}}d\omega + \int_\Omega \rho_f\nabla\phi^{\mathcal{M}}\cdot\nabla\mathcal{P}_\ell^{\mathcal{M}}d\omega$$

$$+\int_{\Gamma_0}\rho\frac{\partial^2 w^{\mathcal{N}}}{\partial t^2}B_k^{\mathcal{N}}d\gamma + \int_{\Gamma_0}EI\frac{\partial^2 w^{\mathcal{N}}}{\partial x^2}\frac{\partial^2 B_k^{\mathcal{N}}}{\partial x^2}d\gamma + \int_{\Gamma_0}c_DI\frac{\partial^3 w^{\mathcal{N}}}{\partial x^2\partial t}\frac{\partial^2 B_k^{\mathcal{N}}}{\partial x^2}d\gamma$$

$$+\int_{\Gamma_0}\rho_f\left(\frac{\partial\phi^{\mathcal{M}}}{\partial t}B_k^{\mathcal{N}} - \frac{\partial w^{\mathcal{N}}}{\partial t}\mathcal{P}_\ell^{\mathcal{M}}\right)d\gamma$$

$$=\int_{\Gamma_0}\sum_{i=1}^{s}\mathcal{K}_i^B u_i(t)\chi_{pe_i}(x)\frac{\partial^2 B_k^{\mathcal{N}}}{\partial x^2}d\gamma + \int_{\Gamma_0}\hat{g}B_k^{\mathcal{N}}d\gamma$$

(9.16)

where $\ell=1,\cdots,\mathcal{M}$ and $k=1,\cdots,\mathcal{N}$. Reformulating (9.16) in first-order form yields the matrix system

$$\dot{z}^N(t)=A^N z^N(t)+B^N u(t)+g^N(t)$$

$$z^N(0)=z_0^N$$

(9.17)

where $N=2(\mathcal{M}+\mathcal{N})$ and $z^N(t)=[\varphi(t),\vartheta(t),\dot\varphi(t),\dot\vartheta(t)]^T$ contains the generalized Fourier coefficients for the first-order state variables. The system matrices for this system are constructed in a manner analogous to that of (9.13); further details can be found in [12].

9.3.2   LQG and H^∞/MinMax Compensators

In typical applications, one has available observations z_{ob} in $\mathbb{R}^p, p<N$, and from these values, state estimates and controls must be constructed. Under

the assumption of no input or output disturbances, the observed system corresponding to the discrete system (9.17) is

$$\dot{z}(t) = Az(t) + Bu(t) + g(t) \quad , \quad z(0) = z(\tau)$$
$$z_{ob}(t) = Cz(t) \tag{9.18}$$
$$z_{out}(t) = Dz(t) + Eu(t)$$

which is of the form (8.5) considered in Section 8.1.2. To obtain the control u, the quadratic functional

$$J_\tau(u, z_0) = \frac{1}{2} \int_0^\tau \{\langle Qz(t), z(t)\rangle + \langle Ru(t), u(t)\rangle\} \, dt \,, \qquad Q = D^T D \,, \ R = E^T E$$

is minimized subject to (9.18). The exogenous force g in (9.18) is again assumed to be temporally periodic. Specific choices of the observation matrix C and the performance matrices Q and R for the structural acoustic system will be discussed in the next section.

The incorporation of input and output disturbances or uncertainties in the model leads to an observed system of the form

$$\dot{z}(t) = Az(t) + Bu(t) + \hat{D}v(t) + g(t) \quad , \quad z(0) = z(\tau)$$
$$z_{ob}(t) = Cz(t) + \hat{E}v(t) \tag{9.19}$$
$$z_{out}(t) = Dz(t) + Eu(t)$$

with the performance index

$$J_\tau(u, z_0) = \frac{1}{2} \int_0^\tau \left\{ \langle Qz(t), z(t)\rangle + \langle Ru(t), u(t)\rangle - \gamma^2 \langle v(t), v(t)\rangle \right\} dt \,.$$

The structure of the vectors $\hat{D}v(t)$ and $\hat{E}v(t)$ is motivated by the expected state or measurement uncertainties for the application under consideration. To attain robustness with respect to these uncertainties, H^∞/MinMax theory is employed when constructing a compensator. In the discussion of Section 8.1.3, it was noted that the construction of an H^∞ compensator is simplified by the assumption that input and output disturbances are independent and hence $\hat{D}\hat{E}^T = 0$. For ease of implementation, we consider here the special case when $\hat{D}v(t)$ and $\hat{E}v(t)$ are of the form $\hat{D}v_1(t)$ and $\hat{E}v_2(t)$ with $v_1(t)$ and $v_2(t)$ independent. We note that this is a matter of convenience, and the dependent case can be analyzed similarly with only slight modifications in the theory (see [117]).

As discussed in Sections 8.1.2 and 8.1.3 and summarized in Algorithms 8.2 and 8.4, compensators in both the LQG and H^∞/MinMax theories are determined by the general differential equation

$$\dot{z}_c(t) = A_c z_c(t) + F z_{ob}(t) + B_r r(t) + g(t)$$
$$z_c(0) = z_c(\tau) \,.$$

Once the compensator has been obtained, the controlling voltage is given by the dynamic feedback law

$$u(t) = -Kz_c(t) + R^{-1}B^T r(t)$$

where $r(t)$ satisfies the equation

$$\dot{r}(t) = A_{tr}r(t) + \Pi g(t)$$
$$r(0) = r(\tau) \ .$$

The component matrices for the two theories are summarized in Table 9.3.

LQG Theory: (see Algorithm 8.2)

    Matrices

$$A_c = A - FC - BK \ , \qquad K = R^{-1}B^T\Pi$$
$$B_r = BR^{-1}B^T \ , \qquad\qquad F = \hat{\Pi}C^T\hat{R}^{-1}$$
$$A_{tr} = -[A - BK]^T$$

  Riccati Equations

$$\Pi A + A^T\Pi - \Pi BR^{-1}B^T\Pi + Q = 0$$
$$\hat{\Pi}A^T + A\hat{\Pi} - \hat{\Pi}C^T\hat{R}^{-1}C\hat{\Pi} + \hat{Q} = 0$$

H^∞/**MinMax Theory:** (see Algorithm 8.4)

    Matrices

$$A_c = A - FC - BK + \gamma^{-2}\hat{Q}\Pi \ , \qquad K = R^{-1}B^T\Pi$$
$$B_r = BR^{-1}B^T - \gamma^{-2}\hat{Q} \ , \qquad\qquad F = \left[I - \gamma^{-2}\hat{\Pi}\,\Pi\right]^{-1}\hat{\Pi}C^T\hat{R}^{-1}$$
$$A_{tr} = -[A - BK - \gamma^{-2}\hat{Q}\Pi]^T$$

  Riccati Equations

$$\Pi A + A^T\Pi - \Pi\left[BR^{-1}B^T - \gamma^{-2}\hat{Q}\right]\Pi + Q = 0$$
$$\hat{\Pi}A^T + A\hat{\Pi} - \hat{\Pi}\left[C^T\hat{R}^{-1}C - \gamma^{-2}Q\right]\hat{\Pi} + \hat{Q} = 0$$

Table 9.3: Matrices for the LQG and H^∞/MinMax compensators.

9.3.3   Design Operators for the Structural Acoustic System

In the observed systems (9.18) or (9.19), the structure of the matrices/vectors A, B and g are determined by the model for the structural acoustic system and the approximation method used to discretize that model. The $p \times N$ matrix C is dictated by the manner in which state observations are measured. The matrix operators $Q, R, \hat{Q}, \hat{R}, \hat{D}, \hat{E}$ are determined according to design criteria specified for the application under consideration. We illustrate here choices for the observation matrix and design operators that are appropriate for the structural acoustic system. We point out that while these choices prove adequate in the simulations for the considered 2-D model, they are by no means unique, and further tuning of parameters and operator design according to the application can optimize performance when implementing the method.

Observation Operators

Since the potential is not a state which is readily measured, the observations are assumed to consist of N_d beam displacement measurements, N_p cavity pressure values and N_v beam velocity values at the points

$$
\left.
\begin{aligned}
&\text{Displacement}: \; x_{i_d}, & i_d &= 1, \cdots, N_d \\
&\text{Pressure}: \; \left(x_{i_p}, y_{i_p}\right), & i_p &= 1, \cdots, N_p \\
&\text{Velocity}: \; x_{i_v}, & i_v &= 1, \cdots, N_v
\end{aligned}
\right\} \Rightarrow p = N_d + N_p + N_v
$$

as depicted in Figure 9.9. In experiments, such displacement, pressure and velocity measurements could be obtained from proximity sensors, microphones and laser vibrometers, respectively or the displacement and velocity could be computed from accelerometer data digitally integrated in the manner described in Section 8.3.4. It is also assumed that sensor errors proportional to the open loop signal are present in the observed state. For example, the observed displacement at x_{i_d} is assumed to have the form

$$
w(t, x_{i_d}) + \alpha_2(t) \max_{t \in [0, \tau]} w(t, x_{i_d})_{OL}
$$

where $w(t, x_{i_d})_{OL}$ denotes the open loop signal and $\alpha_2(t)$ is a uniformly distributed scalar that determines the noise or disturbance level.

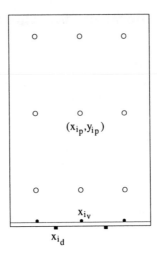

Figure 9.9: Structural acoustic cavity with pressure sensors at (x_{i_p}, y_{i_p}), displacement sensors at x_{i_d}, and velocity sensors at x_{i_v}.

The observation matrix incorporating point evaluations, in accordance with the discrete sensors, is given by

$$C = \begin{bmatrix} 0 & C_{disp} & 0 & 0 \\ 0 & 0 & C_{pres} & 0 \\ 0 & 0 & 0 & C_{vel} \end{bmatrix} \quad , \quad \begin{matrix} C_{disp} \in \mathbb{R}^{N_d \times \mathcal{N}} \\ C_{pres} \in \mathbb{R}^{N_p \times \mathcal{M}} \\ C_{vel} \in \mathbb{R}^{N_v \times \mathcal{N}} \end{matrix}$$

where

$$[C_{disp}]_{i_d,k} = B_k^{\mathcal{N}}(x_{i_d}) = \int_{\Gamma_0} \delta(x - x_{i_d}) B_k^{\mathcal{N}}(x) d\gamma$$

$$[C_{pres}]_{i_p,\ell} = \mathcal{P}_\ell^{\mathcal{M}}(x_{i_p}, y_{i_p}) = \int_{\Omega} \delta(x - x_{i_p}, y - y_{i_p}) \mathcal{P}_\ell^{\mathcal{M}}(x, y) d\omega$$

$$[C_{vel}]_{i_v,k} = B_k^{\mathcal{N}}(x_{i_v}) = \int_{\Gamma_0} \delta(x - x_{i_v}) B_k^{\mathcal{N}}(x) d\gamma \ .$$

The disturbance contributes the component

$$\hat{E}v_2(t) = \alpha_2(t) \begin{bmatrix} \hat{E}_{disp} \\ \hat{E}_{pres} \\ \hat{E}_{vel} \end{bmatrix}_{p \times 1} \quad , \quad \begin{matrix} \hat{E}_{disp} \in \mathbb{R}^{N_d} \\ \hat{E}_{pres} \in \mathbb{R}^{N_p} \\ \hat{E}_{vel} \in \mathbb{R}^{N_v} \end{matrix}$$

where

$$\left[\hat{E}_{disp}\right]_{i_d} = \max_{t \in [0,\tau]} w^{\mathcal{N}}(t, x_{i_d})_{OL}$$

$$\left[\hat{E}_{pres}\right]_{i_p} = \max_{t \in [0,\tau]} \phi_t^{\mathcal{M}}(t, x_{i_p}, y_{i_p})_{OL}$$

$$\left[\hat{E}_{vel}\right]_{i_v} = \max_{t \in [0,\tau]} w_t^{\mathcal{N}}(t, x_{i_v})_{OL} \ .$$

We reiterate that the form of $\hat{E}v_2(t)$ depends upon assumptions specific to the application, and while robustness permits latitude when constructing the operators, the performance of the method is enhanced by physically motivated choices.

With C and $\hat{E}v_2(t)$ thus defined, the observations of the finite dimensional system (9.19) have the form

$$z_{ob}(t) = Cz = \begin{bmatrix} C_{disp}\vec{w}(t) \\ C_{pres}\vec{\phi}_t(t) \\ C_{vel}\vec{w}_t(t) \end{bmatrix} + \alpha_2(t) \begin{bmatrix} \hat{E}_{disp} \\ \hat{E}_{pres} \\ \hat{E}_{vel} \end{bmatrix} .$$

Hence, the i_d^{th} observation of the approximate displacement is given by

$$[z_{ob}(t)]_{i_d} = [C_{disp}\vec{w}(t)]_{i_d} + \alpha_2(t)\left[\hat{E}_{disp}\right]_{i_d}$$

$$= w^{\mathcal{N}}(t, x_{i_d}) + \alpha_2(t) \max_{t\in[0,\tau]} w^{\mathcal{N}}(t, x_{i_d})_{OL}$$

with analogous expressions for the observed pressure and velocity. These approximations can be used when numerically investigating the controllers. When experimentally implementing the method, the same matrices C and $\hat{E}$ are used in the control routines, but the observations $z_{ob}(t)$ consist of experimental measurements rather than numerical approximations.

Modeling Uncertainties

The operator $\hat{D}v_1(t)$, which incorporates modeling uncertainties, should ideally reflect the behavior of the modeling system $\dot{z}(t) = Az(t)$ or $\dot{z}(t) = Az(t) + Bu(t) + g(t)$. For the simulations of the 2-D system, the disturbance was taken of the form $\hat{D}v_1(t) = M^{-1}\bar{D}v_1(t)$ where M denotes the mass matrix and

$$\bar{D}v_1(t) = \begin{bmatrix} \bar{D}_{pot} & & & \\ & \bar{D}_{disp} & & \\ & & \bar{D}_{pres} & \\ & & & \bar{D}_{vel} \end{bmatrix}_{N\times4} \begin{bmatrix} v_{pot}(t) \\ v_{disp}(t) \\ v_{pres}(t) \\ v_{vel}(t) \end{bmatrix}_{4\times1} . \tag{9.20}$$

The components are given by

$$\left[\bar{D}_{pot}\right]_\ell = \int_\Omega 1.0 \cdot \nabla \mathcal{P}_\ell^{\mathcal{M}} d\omega , \qquad \ell = 1, \cdots, \mathcal{M}$$

$$\left[\bar{D}_{disp}\right]_k = \int_{\Gamma_0} 1.0 \cdot \frac{d^2 B_k^{\mathcal{N}}}{dx^2} d\gamma , \qquad k = 1, \cdots, \mathcal{N}$$

$$\left[\bar{D}_{pres}\right]_\ell = \int_\Omega 1.0 \cdot \mathcal{P}_\ell^{\mathcal{M}} d\omega , \qquad \ell = 1, \cdots, \mathcal{M}$$

$$\left[\bar{D}_{vel}\right]_k = \int_{\Gamma_0} 1.0 \cdot B_k^{\mathcal{N}} d\gamma , \qquad \ell = 1, \cdots, \mathcal{N}$$

and

$$v_{pot}(t) = 0$$
$$v_{disp}(t) = 0$$
$$v_{pres}(t) = wt_{pres} \cdot \alpha_1(t)$$
$$v_{vel}(t) = wt_{vel} \cdot \alpha_1(t)$$

where wt_{pres} and wt_{vel} weight the pressure and velocity components, and $\alpha_1(t)$ is a uniformly distributed random number which indicates the modeling uncertainty (see [10] for further details).

While incorporating information about the system, the choice (9.20) does not follow directly from the constitutive laws or moment and force principles used to derive the modeling equations. A second choice for $\hat{D}$ which incorporates uncertainties at the constitutive level is a weighted stiffness matrix [9]. Perturbations, errors or uncertainties in the exogenous force $g(t)$ can also be incorporated in the input term $\hat{D}v_1(t)$ as demonstrated through examples in [11]. These choices illustrate some of the criteria that can be used when designing terms $\hat{D}v_1(t)$ for specific applications.

Quadratic Control and Observation Criteria

The final operators to be specified are the quadratic control functional matrices Q, R and observation matrices $\hat{Q}, \hat{R}$. The control components are specified in the same manner described in Figure 8.3 for the isolated plate and Section 9.2.2 for the 3-D structural acoustic system. Specifically, R is an $s \times s$ diagonal matrix whose components r_{ii}, $i = 1, \cdots, s$ weight the voltage to the ith patch while energy considerations were used to obtain the form of Q specified in (9.14).

As detailed in [10], the $N \times N$ observation matrix $\hat{Q}$ is given by $\hat{Q} = \hat{D}\hat{D}^T$ for $\hat{D}$ defined in terms of $\bar{D}$ of (9.20). Finally, the $p \times p$ matrix $\hat{R}$, which weights the various sensors, was specified as the identity. If the application dictates the weighting of specific sensors, a diagonal matrix analogous to R can be specified for $\hat{R}$.

9.3.4   Numerical Examples

To illustrate the capabilities of the H^∞/MinMax compensator, simulation results comparing an open loop response with LQR full state and H^∞/MinMax output feedback responses are given here. The dimensions and physical parameters for the system are summarized in Table 9.4. The force driving the beam was taken to be

$$g(t, x) = 2.04 \sin(470\pi t)$$

which models a plane wave with an rms sound pressure level of 117 dB. As detailed in [33], the 235 Hz driving frequency strongly couples with the 181.3 Hz second system mode. This mode exhibits primarily acoustic wave behavior since it corresponds to the 171.5 Hz natural frequency for the *isolated* wave (recall from the discussion of Section 9.2.3 that the frequencies and shapes of the system modes differ slightly from the modes for the isolated structure and wave due to the structural acoustic coupling). This provides a significantly more severe test for the controller than the 3-D case considered in Section 9.2.3 in which the driving force coupled less strongly with *acoustic-like* modes.

For this example, a centered pair of patches covering 1/2 of the beam were used as actuators (hence $s=1$). The quadratic cost functional parameters were taken to be $R = 10^{-6}$ and $d_1 = d_2 = d_4 = 1$ and $d_3 = 1 \times 10^4$, with d_3 much larger than d_1, d_2 or d_4 in order to penalize large pressure variations. Also, the parameter γ was taken to have the value $\gamma = 1$.

The basis limits were taken to be $N_b = 8$ and $M_w = N_w = 6$. Hence $\mathcal{N} = 7$ cubic splines were used to approximate beam dynamics while $\mathcal{M} = 48$ Legendre polynomials were used in the potential expansion. This yielded 110 generalized Fourier coefficients or equivalently, implied an order of 110 for the control system in (9.17).

The open and closed loop dynamics of the structural acoustic system were then calculated over the time interval $[0, \tau] = [0, 10/75]$ with various disturbance levels considered in the model and observations. When constructing the H^∞/MinMax compensator, pressure values at the points $p_1, \cdots, p_5$ depicted in Figure 9.10 and displacement and velocity values at $x = .1, .3, .5$ were used to reconstruct the state. For the time interval $[3/75, 10/75]$, the rms sound pressure levels at the point $c_1 = (.3, .1)$ with $0\%, 1\%$ and 5% disturbance levels are compiled in Table 9.5. It is noted that for the previously described configuration of sensors and choices for design parameters, the H^∞/MinMax compensator performed nearly as well as the LQR full state controller with a 12 dB (75% on a linear scale – see (9.21)) sound pressure level reduction attained in the presence of 5% noise. Moreover, it should be noted that observations at the point c_1 were *not* used when estimating the state for the H^∞/MinMax compensator, and the reductions at that point are typical of those observed throughout the cavity.

The reduction attained by the H^∞/MinMax compensator is further illustrated in Figure 9.11 where the time history at the point c_1 is compared with the open loop response and LQR full state feedback trajectory. These results further demonstrate that the H^∞/MinMax compensator provides significant attenuation in spite of the strong coupling between the driving force and the acoustic-like system modes.

Beam		Acoustic Cavity	
Dimensions	Parameters	Dimensions	Parameters
$a = .6\ m$	$\rho = 1.35\ kg/m$	$a = .6\ m$	$\rho_f = 1.21\ kg/m^3$
$b = .1\ m$	$EI = 73.96\ N \cdot m^2$	$\ell = 1\ m$	$c = 343\ m/sec$
$h = .005\ m$	$c_D I = .001\ Kg \cdot m^3/sec$		
	$\mathcal{K}^B = .0023\ N \cdot m/V$		

Table 9.4: Dimensions and physical parameters used in the numerical simulations of the 2-D coupled structural acoustic system.

(rms) Sound Pressure Levels at $c_1 = (.3, .1)$ – (dB)			
% Disturbance	Open Loop	LQR Full State	H^∞ Compensator
0	82.6	70.1	70.3
1	82.6	70.1	70.3
5	82.6	70.2	70.6

Table 9.5: Open and closed loop sound pressure levels at the point $c_1 = (.3, .1)$ over the time interval $[3/75, 10/75]$.

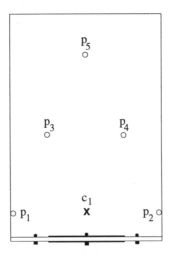

○ Pressure Observations

• Velocity Observations

■ Displacement Observations

X Reported Pressure

Figure 9.10: Sensor orientation on the beam and in the acoustic cavity. The velocity and displacement observations were obtained at $x = .1, .3, .5$. The observation points for pressure are at $p_1 = (0, .1), p_2 = (.6, .1), p_3 = (.15, .45)$, $p_4 = (.45, .45)$ and $p_5 = (.3, .8)$. Pressures calculated at $c_1 = (.3, .1)$ indicate performance of the controllers.

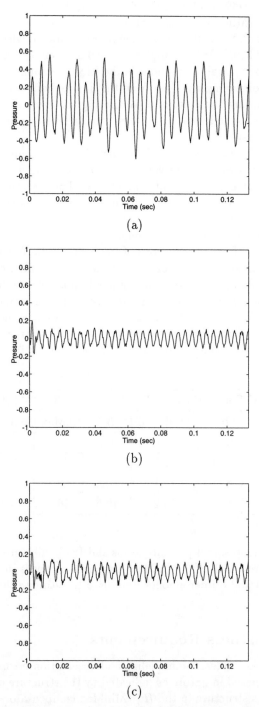

Figure 9.11: Pressure at $c_1 = (.3, .1)$ with a 5% disturbance level; (a) open loop, (b) LQR full state feedback and (c) H^∞/MinMax compensator.

To illustrate the effects of microphone placement and number on the performance of the H^∞/MinMax compensator, sound pressure levels obtained with observations from $0, 1, 3$ and 5 microphones are compared with open loop and LQR levels in Table 9.6. When simulating the 0 microphone case, it was necessary to include one virtual microphone in the Riccati computations to alleviate numerical ill-conditioning; the data from this microphone was not used, however, when estimating the state and calculating the controlling voltage. The one and three microphone responses were calculated using observations from p_1 and p_1, p_2, p_5, respectively.

As expected the performance of the H^∞/MinMax compensator declines as the number of pressure observations is reduced since less information is available for reconstructing the state. In all cases, however, the sound pressure levels are always at least 3 dB lower than the open loop response, even when no pressure information is used when estimating the state. In this latter case, the inclusion of coupling effects in the model permits a partial estimation of the acoustic state solely from observations of the beam displacement and velocity.

While the elimination of pressure sensors through careful modeling of structural acoustic or fluid/structure coupling has important ramifications for applications involving interior fields, it is perhaps even more important in applications involving the control of an external field. As we have noted previously, in such cases (e.g., control of noise generated by a transformer or submarine), arrays of external microphones are often unwieldy and/or infeasible.

(rms) Sound Pressure Levels at $c_1 = (.3, .1)$ – (dB)						
Disturbance	Open	LQR Full	H^∞/MinMax Compensator			
%	Loop	State	5 mics	3 mics	1 mic	0 mic
0	82.6	70.1	70.3	70.3	76.1	79.3
1	82.6	70.1	70.3	70.3	76.1	79.4
5	82.6	70.2	70.6	70.6	76.3	79.7

Table 9.6: Open loop sound pressure levels and LQR full state responses from Table 9.5 compared with H^∞/MinMax compensator levels in which observations from 0, 1, 3 and 5 microphones were used to estimate the state.

9.3.5   Robustness Requirements

In the preceding examples, the modeling and observation errors were assumed to be primarily spatial in nature as indicated by the structure of $\hat{D}$ and $\hat{E}$. By incorporating this structure in the H^∞/MinMax compensator, a certain measure of robustness with regards to these types of errors or perturbations was attained. In experimental applications, the controller must also contend with frequency uncertainties, phase shifts and time delays due to filters and hardware. As noted in Section 8.3.6, these "temporal" disturbances or uncertainties can destabilize a controller if it is not sufficiently robust or compensation is not provided for the disturbances.

Some of these perturbations can be incorporated as uncertainty signals in the H^∞/MinMax cost functional to yield a compensator that is more robust than the LQG or Kalman filter. Examples in [11] illustrate initial efforts to incorporate the temporal uncertainties for the structural acoustic system into the perturbed model and cost functional. These examples also illustrate, however, the difficulties associated with the design of a feedback controller that is stable and effective in the presence of frequency uncertainties, phase shifts and delays, and a significant aspect of current research is directed toward that goal.

9.4 Acoustic Units and Scales

When quantifying sound levels, commonly used measures are the power, intensity or average power transmitted per unit area in the direction of wave propagation, and pressure. While the consideration of acoustic power or intensity is advantageous in certain applications, the pressure is more easily measured and this makes it a natural choice for a state variable.

For modeling purposes, the instantaneous acoustic pressure p with units of force per unit area is often considered (this is defined as the difference between the total pressure and a static pressure p_0). This provides a means of quantifying time-dependent dynamics of an acoustic field. A second measure of pressure which is commonly employed is the root mean square (rms) or effective pressure p_{rms}. For the continuous-time pressure function on a time interval $[t_0, t_f]$, this is defined by

$$p_{rms} = \sqrt{\frac{1}{t_f - t_0} \int_{t_0}^{t_f} [p(t)]^2 dt} \ .$$

From this expression, it can be verified that for a single-frequency, harmonic, plane pressure wave, the instantaneous and rms pressures are related through the expression

$$(p_{rms})^2 = \frac{1}{2}|p|^2 \ .$$

For N_t uniformly sampled discrete pressure measurements at times t_i, the rms pressure is calculated through the expression

$$p_{rms} = \sqrt{\frac{1}{N_t} \sum_{i=1}^{N_t} [p(t_i)]^2}$$

which is simply the quadratic mean of the data. For steady state applications involving multiple frequencies, the time averaging rms pressure provides a natural means of quantifying the level of the acoustic field. It is also consistent with many meters and data acquisition devices which automatically time averages the data. Due to this averaging, however, it does not provide adequate quantification of transient dynamics and the instantaneous pressure p should be used to describe field dynamics in such cases.

Since the pressure has units of force per unit area, it can be expressed in terms of *pascals* (N/m^2) or other consistent units. Due to the wide range of levels, however it is more common to employ a logarithmic scale for pressure or intensity measurements (the response of the human ear is also more closely attuned to a logarithmic scale than a linear intensity scale). This motivates the consideration of sound pressure levels defined by

$$SPL = 20 \log_{10} \left(\frac{p_{rms}}{p_{ref}} \right) \tag{9.21}$$

with nondimensional units termed decibels (dB). Here p_{ref} denotes a reference pressure which is consistent with the units and applications. For the examples in Section 9.2 and Section 9.3, sound pressure levels were calculated using the reference level

$$p_{ref} = 2.04 \times 10^{-5} \ N/m^2$$

which corresponds to the pressure amplitude of a plane wave having an intensity level of zero (see [173]). Other acousticians simply use the value $p_{ref} = 2 \times 10^{-5} \ N/m^2$ as the refence level for sound pressure transmissions in air. For underwater sound, a common choice for the reference pressure is $p_{ref} = 1 \times 10^{-6} \ N/m^2$ [153]. While these choices are often considered conventional, possible ambiguity is avoided by reporting the units and level for p_{ref} (e.g., dB *re* $p_{ref} \ N/m^2$ where *re* implies "refer to"). For a more complete discussion of the units used in acoustic applications, the reader is referred to [53]. Further details regarding the development of acoustic models can be found in [41, 114, 146, 153, 173].

Bibliography

[1] P. Akella, X. Chen, W. Cheng, D. Hughes and J.T. Wen, Modeling and control of smart structures with bonded piezoelectric sensors and actuators, *Smart Materials and Structures,* 3, 1994, pp. 344-353.

[2] D. Armon, Y. Ben-Hain, and S. Braun, Crack detection in beams by rank-ordering of eigenfrequency shifts, *Mechanical Systems and Signal Processing,* to appear.

[3] B. Azvine, G.R. Tomlinson and R. J. Wynne, Initial studies in the use of active constrained layer damping for controlling resonant vibration, Proceedings of the SPIE Conference on Smart Structures and Materials, Orlando, 1994; *Smart Materials and Structures,* submitted.

[4] H.T. Banks, D.E. Brown, V. Metcalf, R.J. Silcox, R.C. Smith and Y. Wang, A PDE-based methodology for modeling, parameter estimation and feedback control in structural and structural acoustic systems, Proceedings of the North American Conference on Smart Structures and Materials, Orlando, FL, 1994, pp. 311-320.

[5] H.T. Banks, D.E. Brown, R.J. Silcox, R.C. Smith, and Y. Wang, Modeling and estimation of boundary parameters for imperfectly clamped structures, to appear.

[6] H.T. Banks and J. Burns, Hereditary control problems: numerical methods based on averaging approximations, *SIAM J. Control and Optimization,* 16(2), 1978, pp. 169-208.

[7] H.T. Banks and J. Burns, *Introduction to Control of Distributed Parameter Systems,* CRSC Lecture Notes, 1996, to appear.

[8] H.T. Banks, D. Cioranescu and D.A. Rebnord, Homogenization models for 2-D grid structures, CRSC Technical Report CRSC-TR92-4, June, 1992; *J. Asymptotic Analysis,* to appear.

[9] H.T. Banks, M.A. Demetriou and R.C. Smith, H^∞ control of noise in a 3-D structural acoustic system, *Proc. 34th IEEE Conf. on Decision and Control,* New Orleans, LA, December 1995, pp. 3719-3724.

[10] H.T. Banks, M.A. Demetriou and R.C. Smith, An H^∞/MinMax periodic control in a $2-D$ structural acoustic model with piezoceramic actuators, CRSC Technical Report, CRSC-TR94-9, June 1994; *IEEE Trans. Auto Control*, to appear.

[11] H.T. Banks, M.A. Demetriou and R.C. Smith, Robustness studies for H^∞ feedback control in a structural acoustic model with periodic excitation, CRSC Technical Report CRSC-TR95-12, March 1995; *International Journal of Robust and Nonlinear Control*, to appear.

[12] H.T. Banks, W. Fang, R.J. Silcox and R.C. Smith, Approximation methods for control of acoustic/structure models with piezoceramic actuators, *Journal of Intelligent Material Systems and Structures*, 4(1), 1993, pp. 98-116.

[13] H.T. Banks, D.S. Gilliam, V.I. Shubov, Global solvability for damped abstract nonlinear hyperbolic systems, CRSC Technical Report CRSC-TR95-26, August, 1995; *Differential and Integral Equations*, to appear.

[14] H.T. Banks and D.W. Iles, On compactness of admissible parameter sets: convergence and stability in inverse problems for distributed parameter systems, ICASE Report 86-38, 1986; *Proc. Conf. on Control Systems Governed by PDE's*, Feb., 1986, Gainesville, FL, Springer Lecture Notes in Control and Inf. Science, 97, 1987, pp. 120-142.

[15] H.T. Banks, D.J. Inman, D.J. Leo and Y. Wang, An experimentally validated damage detection theory in smart structures, CRSC Technical Report CRSC-TR95-7, January, 1995; *Journal of Sound and Vibration*, to appear.

[16] H.T. Banks and K. Ito, A unified framework for approximation in inverse problems for distributed parameter systems, *Control-Theory and Advanced Technology*, 4, 1988, pp. 73-90.

[17] H.T. Banks and K. Ito, Approximation in LQR problems for infinite dimensional systems with unbounded input operators, CRSC Technical Report CRSC-TR94-22, November 1994; *Journal of Mathematical Systems, Estimation and Control*, to appear.

[18] H.T. Banks and K. Ito, Structural actuator control of fluid/structure interactions, *Proc. 33rd IEEE Conf. on Decision and Control*, Lake Buena Vista, FL, Dec. 14-16, 1994, pp. 283-288.

[19] H.T. Banks, K. Ito and C. Wang, Exponentially stable approximations of weakly damped wave equations, in *Distributed Parameter Systems Control and Applications*, (F. Kappel et al, eds.), ISNM Vol. 100, Birkhäuser, 1991, pp. 1-33.

[20] H.T. Banks, K. Ito, and Y. Wang, Well-posedness for damped second order systems with unbounded input operators, CRSC Technical Report CRSC-TR93-10, June, 1993; *Differential and Integral Equations*, 8(3), 1995, pp. 587-606.

[21] H.T. Banks and F. Kappel, Spline approximation for functional differential equations, *Journal of Differential Equations*, 34, 1979, pp. 496-522.

[22] H.T. Banks, F. Kappel and C. Wang, Weak solutions and differentiability for size structured population models, in *Distributed Parameter Systems Control and Applications*, (F. Kappel et al, eds.), ISNM Vol. 100, Birkhäuser, 1991, pp. 35-50.

[23] H.T. Banks, S.L. Keeling and R.J. Silcox, Optimal control techniques for active noise suppression, *Proc. 27th IEEE Conf. on Decision and Control*, Austin, TX, December 7-9, 1988, pp. 2006-2011.

[24] H.T. Banks, S.L. Keeling, R.J. Silcox and C. Wang, Linear quadratic tracking problems in Hilbert Space: applications to optimal active noise control, Proceedings of the 5th IFAC Symposium on Control of DPS, Perpignan, France, June 1989, pp. 17-22.

[25] H.T. Banks and K. Kunisch, The linear regulator problem for parabolic systems, *SIAM J. Control and Optimization*, 22, 1984, pp. 684-698.

[26] H.T. Banks and K. Kunisch, *Estimation Techniques for Distributed Parameter Systems*, Birkhäuser, Boston, 1989.

[27] H.T. Banks, A.J. Kurdila, and G. Webb, Identification of hysteretic control influence operators representing smart actuators, Part I, Formulations; Part II, Approximation, to appear.

[28] H.T. Banks, N.J. Lybeck, B. Munõz, and L. Yanyo, Nonlinear elastomers: modeling and estimation, CRSC Technical Report CRSC-TR95-19, May, 1995; Proc. 3rd IEEE Mediterranean Symp. on New Directions in Control and Automation, July, 1995, 1, pp. 1-7.

[29] H.T. Banks and D.A. Rebnord, Estimation of material parameters for grid structures, *Journal of Mathematical Systems, Estimation and Control*, 1, 1991, pp. 107-130.

[30] H.T. Banks and D.A. Rebnord, Analytic semigroups: applications to inverse problems for flexible structures, CAMS Technical Report 90-3, January, 1990, University of Southern California; in *Differential Equations with Applications*, (*Intl. Conf. Proc.*, F. Kappel et al, eds., Retzhof, Austria), Marcel Dekker,Inc., New York, 1991, pp. 21-35.

[31] H.T. Banks, I.G. Rosen and K. Ito, A spline-based technique for computing Riccati operators and feedback controls in regulator problems for
delay systems, *SIAM J. Sci. Stat. Comp.*, 5, 1984, pp. 830-855.

[32] H.T. Banks, R.J. Silcox and R.C. Smith, Numerical simulations of a
coupled 3-D structural acoustics system, Proceedings of the Second Conference on Recent Advances in Active Control of Sound and Vibration,
Blacksburg, VA, 1993, pp. 85-97.

[33] H.T. Banks, R.J. Silcox and R.C. Smith, The modeling and control of
acoustic/structure interaction problems via piezoceramic actuators: 2-D
numerical examples, *ASME Journal of Vibration and Acoustics*, 116(3),
1994, pp. 386-396.

[34] H.T. Banks and R.C. Smith, Modeling and approximation of a coupled 3-D structural acoustics problem, *Computation and Control III*,
(K.L. Bowers and J. Lund, eds.), Birkhäuser, Boston, 1993, pp. 29-48.

[35] H.T. Banks and R.C. Smith, Noise control in a 3-D structural acoustic system: numerical simulations, Proceedings of the Second International Conference on Intelligent Materials, Williamsburg, VA, June 1994,
pp. 128-139.

[36] H.T. Banks and R.C. Smith, Feedback control of noise in a 2-D nonlinear
structural acoustics model, *Discrete and Continuous Dynamical Systems*,
1(1), 1995, pp. 119-149.

[37] H.T. Banks and R.C. Smith, The modeling and approximation of a structural acoustics problem in a hard-walled cylindrical domain, CRSC Technical Report CRSC-TR94-26, December 1994; in *Dynamics and Control
of Distributed Systems*, (H.S. Tzou and L.A. Bergman, eds.), Cambridge
University Press, to appear.

[38] H.T. Banks and R.C. Smith, Active control of acoustic pressure fields
using smart material technologies, in *Flow Control*, (M. Gunzburger,
ed.), Institute for Mathematics and Its Applications (IMA) Volume 68,
Springer-Verlag, 1995, pp. 1-33.

[39] H.T. Banks and R.C. Smith, Well-posedness of a model for structural
acoustic coupling in a cavity enclosed by a thin cylindrical shell, *Journal
of Mathematical Analysis and Applications*, 191, 1995, pp. 1-25.

[40] H.T. Banks and R.C. Smith, Implementation issues regarding PDE-based
controllers – control of transient and periodic plate vibrations, CRSC
Technical Report, CRSC-TR95-16, April 1995.

[41] H.T. Banks and R.C. Smith, Parameter estimation in a structural acoustic system with fully nonlinear coupling conditions, *Mathematical and
Computer Modeling*, 23(4), 1996, pp. 17-50.

[42] H.T. Banks, R.C. Smith, D.E. Brown, R.J. Silcox, and V.L. Metcalf, Experimental confirmation of a PDE-based approach to design of feedback controls, ICASE Report 95-42; *SIAM J. Control and Optimization*, submitted.

[43] H.T. Banks, R.C. Smith, D.E. Brown, V.L. Metcalf and R.J. Silcox, The estimation of material and patch parameters in a PDE-based circular plate model, CRSC Technical Report CRSC-TR95-24, July 1995; *Journal of Sound and Vibration*, submitted.

[44] H.T. Banks, R.C. Smith and Y. Wang, The modeling of piezoceramic patch interactions with shells, plates and beams, *Quart. Appl. Math*, 53, 1995, pp. 353-387.

[45] H.T. Banks, R.C. Smith and Y. Wang, Modeling and parameter estimation for an imperfectly clamped plate, CRSC Technical Report CRSC-TR95-2, January 1995; *Computation and Control IV,* (K.L. Bowers and J. Lund, eds.), Birkhäuser, Boston, 1995, pp. 23-42.

[46] H. T. Banks and Y. Wang, Damage detection and characterization in smart material structures, in *Control and Estimation of Distributed Parameter Systems: Nonlinear Phenomena* (W. Desch, F. Kappel and K. Kunisch, eds.), ISNM Vol. 118, Birkhäuser, 1994, pp. 21-44.

[47] H.T. Banks, Y. Wang and D.J. Inman, Bending and shear damping in beams: frequency domain estimation techniques, *ASME Journal of Vibration and Acoustics*, 116(2), 1994, pp. 188-197.

[48] H.T. Banks, Y. Wang, D.J. Inman and H. Cudney, Parameter identification techniques for the estimation of damping in flexible structure experiments, *Proc. 26th IEEE Conf. on Decision and Control*, Los Angeles, CA, December 1987, pp. 1392-1395.

[49] H.T. Banks, Y. Wang, D.J. Inman and J.C. Slater, Variable coefficient distributed parameter system models for systems with piezoceramic actuators and sensors, CRSC Technical Report CRSC-TR92-9, September, 1992; *Proc. 31st IEEE Conf. on Decision and Control*, Tucson, December 1992, pp. 1803-1808.

[50] H.T. Banks, Y. Wang, D.J. Inman and J.C. Slater, Approximation and parameter identification for damped second order systems with unbounded input operators, CRSC Technical Report CRSC-TR93-9, May, 1993; *Control: Theory and Advanced Technology*, 10(4), 1994, pp. 873-892.

[51] T. Başar and P. Bernhard, H^∞-*Optimal Control and Related Minimax Design Problems*, Birkhäuser, Boston, 1991.

[52] D.S. Bayard, Statistical additive uncertainty bounds using Schroeder-phased input designs, Internal Document JPL D-8145; JPL EM 343-1214, December 19, 1990.

[53] L.L. Beranek, Letter symbols and conversion factors for acoustical quantities, in *American Institute of Physics Handbook*, (D.E. Gray, ed.), McGraw-Hill, New York, 1957, pp. 3.18ff.

[54] D.S. Bernstein and D.C. Hyland, The optimal projection equations for finite-dimensional fixed-order dynamic compensation of infinite-dimensional systems, *SIAM J. Control and Optimization*, 24(1), 1986, pp. 122-151.

[55] H. Block and J.P. Kelly, Electro-rheology, *J. Phys. D., Appl. Phys.*, 21, 1988, pp. 1661-1677.

[56] L.C. Brinson, One dimensional constitutive behavior of shape memory alloys: thermomechanical derivation with nonconstant material functions, *Journal of Intelligent Material Systems and Structures*, 4, 1993, pp. 229-242.

[57] L.C. Brinson and R. Lammering, Finite element analysis of the behavior of shape memory alloys and their applications, *Int. J. Solids Structures*, 30, 1993, pp. 3261-3280.

[58] J.A. Burns and H. Marrekchi, Optimal fixed-finite-dimensional compensator for Burgers' equation with unbounded input/output operators, ICASE Report 93-19; *Computation and Control III*, (K.L. Bowers and J. Lund, eds.), Birkhäuser, Boston, 1993, pp. 83-104.

[59] J.A. Burns and R.D. Spies, A numerical study of parameter sensitivities in Landau-Ginzburg models of phase transitions in shape memory alloys, *Journal of Intelligent Material Systems and Structures*, 5, 1994, pp. 321-332.

[60] C.D. Butter and G.B. Hocker, Fiber optics strain gauge, *Applied Optics*, 17, 1978, pp. 2867-2869.

[61] W.G. Cady, *Piezoelectricity*, Volume 1, Dover Publications, New York, 1964.

[62] P. Cawley and R. Ray, A comparison of the natural frequency changes produced by cracks and slots, *ASME Journal Vibration, Acoustics, Stress and Reliability in Design*, 110, 1988, pp. 366-370.

[63] G. Chavent, About the stability of the optimal control solution of inverse problems, *Inverse and Improperly Posed Problems in Differential Equations*, (G. Anger, Ed.), Akademic-Verlag, Berlin, 1979, pp. 45-58.

[64] P.G. Ciarlet, *Mathematical Elasticity Vol. 1: Three Dimensional Elasticity*, North Holland, Amsterdam, 1988.

[65] R.L. Clark and C.R. Fuller, Control of sound radiation with adaptive structures, *Journal of Intelligent Material Systems and Structures*, 2, 1991, pp. 431-452.

[66] R.L. Clark and C.R. Fuller, Modal sensing of efficient acoustic radiators with polyvinylidene fluoride distributed sensors in active structural acoustic control approaches, *J. Acoust. Soc. Am.*, 91(6), 1992, pp. 3321-3329.

[67] R.L. Clark and C.R. Fuller, Optimal placement of piezoelectric actuators and polyvinylidene fluoride error sensors in active structural acoustic control approaches, *J. Acoust. Soc. Am.*, 92(3), 1992, pp. 1521-1533.

[68] R.L. Clark, Jr., C.R. Fuller and A. Wicks, Characterization of multiple piezoelectric actuators for structural excitation, *J. Acoust. Soc. Am.*, 90(1), 1991, pp. 346-357.

[69] F. Colonius and K. Kunisch, Stability for parameter estimation in two point boundary value problems, Inst. für Mathematik, Universität Graz und Technische Universität Graz, Bericht No. 50-1984, October 1984.

[70] E.F. Crawley and E.H. Anderson, Detailed models of piezoceramic actuation of beams, AIAA Paper 89-1388-CP, 1989.

[71] E.F. Crawley and J. de Luis, Use of piezoelectric actuators as elements of intelligent structures, *AIAA Journal*, 25(10), October, 1987, pp. 1373-1385.

[72] E.F. Crawley, J. de Luis, N.W. Hagood and E.H. Anderson, Development of piezoelectric technology for applications in control of intelligent structures, Applications in Control of Intelligent Structures, American Controls Conference, Atlanta, June 1988, pp. 1890-1896.

[73] R.F. Curtain, Finite dimensional compensators for parabolic distributed systems with unbounded control and observation, *SIAM J. Control and Optimization*, 22(2) 1984, pp. 255-276.

[74] R.F. Curtain and D. Salamon, Finite dimensional compensators for infinite dimensional systems with unbounded input operators, *SIAM J. Control and Optimization*, 24(4), 1986, pp. 797-816.

[75] G. Da Prato, Synthesis of optimal control for an infinite dimensional periodic problem, *SIAM J. Control and Optimization*, 25(3), 1987, pp. 706-714.

[76] G. DaPrato and P. Grisvard, Maximal regularity for evolution equations by interpolation and extrapolation, *J. Func. Anal.*, 58, 1984, pp. 107-124.

[77] J.E. Dennis, Jr. and R.B. Schnabel, *Numerical Methods for Unconstrained Optimization and Nonlinear Equations*, Prentice-Hall, Inc., Englewood Cliffs, NJ, 1983.

[78] A.F. Devonshire, Advances in physics, *Philosophical Magazine*, 3(10), April 1954, pp. 86-130.

[79] E.K. Dimitriadis, C.R. Fuller and C.A. Rogers, Piezoelectric actuators for distributed noise and vibration excitation of thin plates, *ASME Journal of Vibration and Acoustics*, 13, 1991, pp. 100-107.

[80] J. Dosch, D.J. Inman, and E. Garcia, A self-sensing piezoelectric actuator for collocated control, *Journal of Intelligent Material Systems and Structures*, 3, 1992, pp. 166-185.

[81] T.W. Duerig, K.N. Melton, D. Stöckel and C.M. Wayman (eds), *Engineering Aspects of Shape Memory Alloys*, Butterworth-Heinemann, London, 1990.

[82] C.L. Dym, *Introduction to the Theory of Shells,* Pergamon Press, New York, 1974.

[83] EDO Acoustics Corporation Division, 2645 South 300 West, Salt Lake City, UT 84115, (801) 486-2115, personal communications.

[84] A.B. Flatau, D.L. Hall and J.M. Schlesselman, Magnetostrictive vibration control systems, *Journal of Intelligent Material Systems and Structures*, 4(4), 1993, pp. 560-565.

[85] W. Flügge, *Stresses in Shells,* Second Edition, Springer-Verlag, New York, 1973.

[86] R.W. Fox and A.T. McDonald, *Introduction to Fluid Mechanics,* John Wiley and Sons, New York, 1992.

[87] A. Frendi, L. Maestrello and A. Bayliss, On the coupling between a supersonic laminar boundary layer and a flexible surface, *AIAA Journal*, 31(4), April 1993, pp. 703-713.

[88] A. Frendi, L. Maestrello and A. Bayliss, Coupling between plate vibration and acoustic radiation, *Journal of Sound and Vibration*, 177(2), 1994, pp. 207-226.

[89] M. Fripp and N. Hagood, Comparison of electrostrictive and piezoceramic actuators for vibration suppression, Proceedings of the 1995 North American Conference on Smart Structures and Materials, February 26 - March 3, San Diego, CA, 1995.

[90] M. Fripp, N. Hagood and L. Luoma, Distributed structural actuation with electrostrictors, Proceedings of the 1994 Smart Structures and Materials Conference, Orlando, FL, February 14-16, (N. Hagood, ed.), SPIE 2190, 1994, pp. 571-585.

[91] C.R. Fuller, G.P. Gibbs and R.J. Silcox, Simultaneous active control of flexural and extensional waves in beams, *Journal of Intelligent Material Systems and Structures*, 1(2), 1990, pp. 235-247.

[92] C.R. Fuller, S.D. Snyder, C.H. Hansen and R.J. Silcox, Active control of interior noise in model aircraft fuselages using piezoceramic actuators, Paper 90-3922, AIAA 13th Aeroacoustics Conference, Tallahassee, FL, October 1990.

[93] M.V. Gandhi and B.S. Thompson, *Smart Materials and Structures*, Chapman and Hall, New York, 1992.

[94] A.P. Gast and C.F. Zukoski, Electrorheological fluids as colloidal suspensions, *Advances in Colloid and Interface Science*, 30, 1989, pp. 153-202.

[95] J.S. Gibson and A. Adamian, Approximation theory for linear-quadratic-Gaussian optimal control of flexible structures, *SIAM J. Control and Optimization*, 29(1), 1991, pp. 1-37.

[96] P.E. Gill, W. Murray and M.H. Wright, *Practical Optimization*, Academic Press, New York, 1981.

[97] A.E. Green and J.E. Adkins, *Large Elastic Deformations and Nonlinear Continuum Mechanics*, Clarendon Press, Oxford, 1960.

[98] A.E. Green and W. Zerna, *Theoretical Elasticity*, Clarendon Press, Oxford, 1968.

[99] N.W. Hagood and A. von Flotow, Damping of structural vibrations with piezoelectric materials and passive electrical networks, *Journal of Sound and Vibration*, 146(2), 1991, pp. 243-268.

[100] D.L. Hall and A.B. Flatau, Nonlinearities, harmonics and trends in dynamic applications of Terfenol-D, Proceedings of the SPIE Conference on Smart Structures and Intelligent Materials, Vol. 1917, Part 2, 1993, pp. 929-939.

[101] D.L. Hall and A.B. Flatau, Broadband performance of a magnetostrictive shaker, *Active Control of Noise and Vibration 1992*, DSC Volume 38, Am. Soc. of Mech. Eng., 1992, pp. 95-104.

[102] W.L. Hallauer and S.E. Lamberson, Experimental active vibration damping of a plane truss using hybrid actuation, *Proc. 30th AIAA/ASME/ASCE/AHS/ASC Structural Dynamics and Materials Conference*, Paper 89-1169-CP, 1989, pp. 80-90.

[103] A. Haraux, Linear semigroups in Banach spaces, in *Semigroups, Theory and Applications, II*, (H. Brezis, et al., eds.), Pitman Res. Notes in Math, Vol. 152, Longman, London, 1986, pp. 93-135.

[104] N.J. Hoff, The accuracy of Donnell's equations, *J. Appl. Mech.*, 22, 1955, pp. 329-334.

[105] K.H. Hoffman and M. Niezgodka, Mathematical models of dynamical martensitic transformations in shape memory alloys, *Journal of Intelligent Material Systems and Structures*, 1, 1990, pp. 355-373.

[106] C.L. Hom and N. Shankar, A fully coupled constitutive model for electrostrictive ceramic materials, *Journal of Intelligent Material Systems and Structures*, 5, November 1994, pp. 795-801.

[107] C.L. Hom and N. Shankar, Finite element modeling of multilayered electrostrictive actuators, *Smart Materials and Structures*, submitted.

[108] C.L. Hom, S.M. Pilgrim, N. Shankar, K. Bridger, M. Massuda and S.R. Winzer, Calculation of quasi-static electromechanical coupling coefficients for electrostrictive ceramic materials, *IEEE Transactions on Ultrasonics, Ferroelectrics, and Frequency Control*, 41(4), July 1994, pp. 542-551.

[109] K. Ito, Finite-dimensional compensators for infinite-dimensional systems via Galerkin-type approximations, *SIAM J. Control and Optimization*, 28(6), 1990, pp. 1251-1269.

[110] K. Ito and H.T. Tran, Linear quadratic optimal control problem for linear systems with unbounded input and output operators: numerical approximations, *Proceedings of the 4th International Conference on Control of Distributed Systems*, (F. Kappel et al, eds.), ISNM Vol. 91, Birkhäuser, 1989, pp. 171-195.

[111] D.W. Jensen, J. Pascual, and J.M. Cory, Jr., Dynamic strain sensing of a composite lattice with an integrated optical fiber, *Journal of Intelligent Material Systems and Structures*, 2, 1991, pp. 198-214.

[112] J. Jia and C.A. Rogers, Formulation of a laminated shell theory incorporating embedded distributed actuators, The American Society of Mechanical Engineers, Reprinted from *AD-Vol. 15, Adaptive Structures*, Editor: B.K. Wada, Book No. H00533, 1989.

[113] D. Jiles, *Introduction to Magnetism and Magnetic Materials*, Chapman and Hall, New York, 1991.

[114] M.C. Junger and D. Feit, *Sound, Structures, and Their Interaction*, Published by the Acoustical Society of America through the American Institute of Physics, 1993.

[115] F. Kappel and D. Salamon, Spline approximation for retarded systems and the Riccati equation, *SIAM J. Control and Optimization*, 25(4), 1987, pp. 1082-1117.

[116] F. Kappel and D. Salamon, On the stability properties of spline approximations for retarded systems, *SIAM J. Control and Optimization*, 27(2), 1989, pp. 407-431.

[117] B. van Keulen, H_∞-*Control for Distributed Parameter Systems: A State-Space Approach*, Birkhäuser, Boston, 1993.

[118] S.J. Kim and J.D. Jones, Optimal design of piezo-actuators for active noise and vibration control, AIAA 13th Aeroacoustics Conference, Tallahassee, FL, October 1990.

[119] M. Krasnoselskii and A. Pokrovskii, *Systems with Hysteresis*, Springer, Berlin, 1989; Russian Edition, Nauka, Moscow, 1983.

[120] H. Kraus, *Thin Elastic Shells: An Introduction to the Theoretical Foundations and the Analysis of Their Static and Dynamic Behavior*, John Wiley and Sons, Inc., New York, 1967.

[121] C. Kravaris and J.H. Seinfeld, Identification of parameters in distributed parameter systems by regularization, *SIAM J. Control and Optimization*, 23, 1985, pp. 217-241.

[122] E. Kreyszig, *Introductory Functional Analysis with Applications*, Wiley Classics Edition, John Wiley and Sons, New York, 1989.

[123] J.E. Lagnese and J.-L. Lions, *Modelling Analysis and Control of Thin Plates*, Collection *Recherches en Mathématiques Appliquées*, Masson, Paris, 1989.

[124] J.D. Lambert, *Numerical Methods for Ordinary Differential Systems: The Initial Value Problem*, John Wiley and Sons, Inc., New York, 1991.

[125] L.D. Landau and E.M. Lifshitz, *Theory of Elasticity*, Volume 7 of *Course of Theoretical Physics*, Translated from the Russian by J.B. Sykes and W.H. Reid, Pergamon Press, London, 1959.

[126] C.M. LaPeter, Application of distributed measurements for finite elements model verification, Master's Thesis, Virginia Polytechnic Institute and State University, May 1992.

[127] I. Lasiecka, Galerkin approximations of infinite-dimensional compensators for flexible structures with unbounded control action, *Acta Applicandae Mathematicae*, 28, 1992, pp. 101-133.

[128] I. Lasiecka, Finite element approximations of compensator design for analytic generators with fully unbounded controls/observations, *SIAM J. Control and Optimization*, 33(1), 1995, pp. 67-88.

[129] I. Lasiecka, D. Lukes and L. Pandolfi, Input dynamics and nonstandard Riccati equations with applications to boundary control of damped wave and plate equations, *Journal of Optimization Theory and Applications*, 84(3), 1995, pp. 549-574.

[130] I. Lasiecka and R. Triggiani, Numerical approximations of algebraic Riccati equations for abstract systems modelled by analytic semigroups, and applications, *Mathematics of Computation*, 57(196), 1991, pp. 639-662 and S13-S37.

[131] I. Lasiecka and R. Triggiani, Algebraic Riccati equations arising from systems with unbounded input-solution operator: applications to boundary control problems for wave and plate equations, *Nonlinear Analysis, Theory, Methods and Applications*, 20(6), 1993, pp. 659-695.

[132] C.-K. Lee and F.C. Moon, Modal sensors/actuators, *Journal of Applied Mechanics*, Transaction of the ASME, 57 June 1990, pp. 434-441.

[133] A.W. Leissa, *Vibration of Plates*, NASA SP-160, 1969, Reprinted by the Acoustical Society of America through the American Institute of Physics, 1993.

[134] A.W. Leissa, *Vibration of Shells*, NASA SP-288, 1973, Reprinted by the Acoustical Society of America through the American Institute of Physics, 1993.

[135] H.C. Lester and S. Lefebvre, Piezoelectric actuator models for active sound and vibration control of cylinders, Proceedings of the Conference on Recent Advance in Active Control of Sound and Vibration, Blacksburg, VA, 1991, pp. 3-26.

[136] C. Liang and C.A. Rogers, One dimensional thermomechanical constitutive relations for shape memory materials, *Journal of Intelligent Material Systems and Structures*, 1, 1990, pp. 207-234.

[137] J.L. Lions, *Optimal Control of Systems Governed by Partial Differential Equations*, Springer-Verlag, New York, 1971.

[138] J.L. Lions and E. Magenes, *Non-Homogeneous Boundary Value Problems and Applications*, Springer-Verlag, New York, 1972.

[139] A.E.H. Love, *A Treatise on the Mathematical Theory of Elasticity*, Cambridge University Press, Fourth Edition, 1927.

Bibliography   293

[140] D.G. Luenberger, An introduction to observers, *IEEE Trans. Auto. Control*, AC-16, 1971, pp. 596-602.

[141] S. Markuš, *The Mechanics of Vibrations of Cylindrical Shells,* Elsevier, New York, 1988.

[142] J.E. Marsden and T.J.R. Hughes, *Mathematical Foundations of Elasticity*, Prentice Hall, Englewood Cliffs, 1983.

[143] V.L. Metcalf, U.S. Army Research Laboratory, NASA Langley Research Center, Personal Communication.

[144] D.W. Miller, S.A. Collins and S.P. Peltzman, Development of spatially convolving sensors for structural control applications, AIAA Paper AIAA-90-1127-CP, 1990, pp. 2283-2297.

[145] L.S.D. Morley, An improvement of Donnell's approximation for thin-walled circular cylinders, *Quarterly Journal of Mechanics and Applied Mathematics*, 12, 1959, pp. 89-99.

[146] P.M. Morse, *Vibration and Sound,* Published by the Acoustical Society of America through the American Institute of Physics, 1991.

[147] K.D. Murphy, L.N. Virgin and S.A. Rizzi, Free vibration of thermally loaded panels including initial imperfections and post-buckling effects, NASA Technical Memorandum 109097, April 1994.

[148] C.G. Namboodri, Jr. and C.A. Rogers, Tunable vibration/strain sensing with electrostrictive materials, Proceedings of the Conference on Recent Advances in Adaptive and Sensory Materials and Their Applications, Blacksburg, VA, April 27-29, 1992, C.A. Rogers and R.C. Rogers, eds., Technomic Publishing Company, pp. 285-297.

[149] V.V. Novozhilov, *Thin Shell Theory,* Second Augmented and Revised Edition, Translated from the Second Russian Edition by P.G. Lowe, Edited by J.R.M Radok, P. Noordhoff Ltd., Groningen, The Netherlands, 1964.

[150] R.W. Ogden, *Nonlinear Elastic Deformations,* Ellis Horwood, Ltd, Chichester, 1984.

[151] A.B. Palazzollo, S. Jagannathan, A.F. Kascak, T. Griffin, J. Giriunas, G.T. Montague, Piezoelectric actuator-active vibration control of the shaft line for a gas turbine engine test stand, Proceedings of the International Gas Turbine and Aeroengine Congress and Exposition, Cincinnati, OH, May 24-27, 1993, ASME Paper 93-GT-262.

[152] A. Pazy, *Semigroups of Linear Operators and Applications to Partial Differential Equations,* Springer-Verlag, New York, 1983.

[153] A.D. Pierce, *Acoustics: An Introduction to Its Physical Principles and Applications*, McGraw-Hill, New York, 1981.

[154] Piezo Kinetics Corporation, (814) 355-1593, personal communications.

[155] J. Pratt and A.B. Flatau, Development and analysis of a self-sensing magnetostrictive actuator design, *Journal of Intelligent Material Systems and Structures*, 6(5), 1995, pp. 639-648.

[156] P.M. Prenter, *Splines and Variational Methods*, Wiley, New York, 1975.

[157] A.J. Pritchard and D. Salamon, The linear quadratic control problem for infinite dimensional systems with unbounded input and output operators, *SIAM J. Control and Optimization*, 25(1), 1987, pp. 121-144.

[158] D.A. Rebnord, Parameter estimation for two-dimensional grid structures, Ph.D. Thesis, Brown University, May, 1989.

[159] J.H. Robinson, Acoustics Division, NASA Langley Research Center, personal communications.

[160] J.H. Robinson, S.A. Rizzi, S.A. Clevenson and E.F. Daniels, Large deflection random response of flat and blade stiffened carbon panels, Proceedings of the 33rd AIAA/ASME/ASCE/AHS/ASC Structures, Structural Dynamics and Materials Conference, Dallas, TX, 1992.

[161] J.A. Rongong, J.R. Wright, G.R. Tomlinson and R.J. Wynne, Modeling of hybrid constrained layer/piezoceramic approach to active damping, 1995, preprint.

[162] D.L. Russell, On mathematical models for the elastic beam with frequency proportional damping, in *Control and Estimation of Distributed Parameter Systems,* (H.T. Banks, ed.), SIAM, Philadelphia, 1992, pp. 125-169.

[163] A.S. Saada, *Elasticity Theory and Applications,* Robert E. Krieger Publishing Company, Malabar, FL, 1987.

[164] H. Sato, Free vibration of beams with abrupt changes of cross-section, *Journal Sound and Vibration*, 89, 1993, pp. 59-64.

[165] J.M Schumacher, A direct approach to compensator design for distributed parameter systems, *SIAM J. Control and Optimization*, 21(6), 1983, pp. 823-836.

[166] L. Schumaker, *Spline Functions: Basic Theory,* John Wiley and Sons, New York, 1987.

[167] R.E. Showalter, *Hilbert Space Methods for Partial Differential Equations,* Pitman Publishing Ltd., London, 1977.

[168] R.J. Silcox, S. Lefebvre, V.L. Metcalf, T.B. Beyer and C.R. Fuller, Evaluation of piezoceramic actuators for control of aircraft interior noise, Proceedings of the DGLR/AIAA 14th Aeroacoustics Conference, Aachen, Germany, May 11-14, 1992.

[169] J.S. Sirkis and H.W. Haslach, Jr., Complete phase-strain model for structurally embedded interferometric optical fiber sensors, *Journal of Intelligent Material Systems and Structures*, 2, 1991, pp. 3-24.

[170] R.C. Smith, A Galerkin method for linear PDE systems in circular geometries with structural acoustic applications, ICASE Report No. 94-40; *SIAM Journal on Scientific Computing*, to appear.

[171] W. Soedel, *Vibrations of Shells and Plates,* Second Edition, Marcel Dekker, Inc., New York, 1993.

[172] H. Tanabe, *Equations of Evolution,* Pitman Publishing Ltd., London, 1979.

[173] S. Temkin, *Elements of Acoustics*, Wiley, New York, 1981.

[174] S. Timoshenko and S. Woinowsky-Krieger, *Theory of Plates and Shells,* Second Edition, McGraw-Hill Book Company, Inc., New York, 1987.

[175] E. du Trémolet de Lacheisserie, *Magnetostriction: Theory and Applications of Magnetoelasticity*, CRS Press, Ann Arbor, 1993.

[176] H.S. Tzou and M. Gadre, Theoretical analysis of a multi-layered thin shell coupled with piezoelectric actuators for distributed vibration controls, *Journal of Sound and Vibration*, 132(3), 1989, 433-450.

[177] H.S. Tzou and R.V. Howard, A piezothermoelastic thin shell theory applied to active structures, *ASME Journal of Vibration and Acoustics*, 116(3), July 1994, pp. 295-302.

[178] R.D. Turner, T. Valis, W.D. Hogg, and R.M. Measures, Fiber-optic strain sensors for smart structures, *Journal of Intelligent Material Systems and Structures*, 1, 1990, pp. 26-49.

[179] A. Visintin, *Differential Models of Hysteresis*, Springer-Verlag, New York, 1994.

[180] Y. Wang, Damping modeling and parameter estimation in Timoshenko beams, Ph.D. Thesis, Brown University, May, 1990.

[181] I.M. Ward, *Mechanical Properties of Solid Polymers*, John Wiley, New York, 1983.

[182] J.T. Weissenburger, Effect of local modifications on the vibration char-
      acteristics of linear systems, *ASME Journal of Applied Mechanics*, 35(1),
      1968, pp. 327-332.

[183] F.B. Weissler, Semilinear evolution equations in Banach spaces, *J. Func.
      Anal.*, 32, 1979, pp. 277-296.

[184] W.M. Winslow, Induced vibration of suspensions, *J. Appl. Phys.*, 20,
      1949, pp. 1137-1140.

[185] J. Wloka, *Partial Differential Equations,* Cambridge University Press,
      Cambridge, 1987.

[186] Y.S. Yoon and W.W.-G. Yeh, Parameter identification in an inhomoge-
      neous medium with the finite element method, *Soc. Petr. Engr. J.*, 16,
      1976, pp. 217-226.

Notation

Some symbols are used in this monograph for two distinct quantities. This always occurs in different contexts and only when no confusion should result. For symbols with multiple definitions, page numbers are separated by a semicolon. Commas are used to indicate multiple references to single definitions.

Symbol	Meaning	Page
$\hookrightarrow$	continuous and dense embedding	97
$\cong$	Riesz equivalence	97
$*$	H^* is the conjugate dual of the Hilbert space H	96
$\langle \cdot, \cdot \rangle_H$	inner product in Hilbert space H	96
$\lvert \cdot \rvert_H$	norm in Hilbert space H	97
$\langle \cdot, \cdot \rangle_{V^*, V}$	duality product, extension by continuity of $\langle \cdot, \cdot \rangle_H$ from $V \times H$ to $V^* \times H$	97
A	Lamé constant	29
A_1, A_2	bounded, linear operators generated by σ_1, σ_2	97
$\mathcal{A}$	system operator	107
$\mathcal{A}^*$	adjoint operator of $\mathcal{A}$	110
$\hat{\mathcal{A}}$	extension of $\mathcal{A}$ from dom$\mathcal{A}$ to $\mathcal{H}_1$	112
a	radius of a circular plate	63
a_2	patch constant, $a_2 \equiv (h/2 + h_{pe})^2 - h^2/8$	76
a_3	patch constant, $a_3 \equiv (h/2 + h_{pe})^3 - h^3/8$	76
B	magnetic flux density; Lamé constant	12; 29
$\mathcal{B}$	control input operator	179
b	beam width	66
$C^1[0, T]$	class of continuously differentiable functions on $[0, T]$	100
$C((0, T), H)$	class of H-valued continuous functions on $(0, T)$	100
$\mathcal{C}$	observation operator	181
$\mathbb{C}$	set of complex numbers	97
c	speed of sound in air	246
c_D	Kelvin-Voigt damping coefficient	34
D	electric flux density; elastic coefficient	4; 61
d, d_{31}	piezoelectric strain constant	4

Index

MASSON Éditeur
120, boulevard Saint-Germain
75280 Paris Cedex 06
Dépôt légal : juillet 1996

SNEL S.A.
Rue Saint-Vincent 12 – 4020 Liège
tél. 32(0)41 43 76 91 - fax 32(0)41 43 77 50
juin 1996